# THE MEASUREMENT OF APPEARANCE

# THE MEASUREMENT OF APPEARANCE

RICHARD S. HUNTER

President, Hunter Associates Laboratory
Fairfax, Virginia

A WILEY-INTERSCIENCE PUBLICATION

JOHN WILEY & SONS New York / London / Sydney / Toronto

***Library of Congress Cataloging in Publication Data***

Hunter, Richard Sewall, 1909–
The measurement of appearance.

"A Wiley-Interscience publication."
Bibliography: p.
Includes indexes.
1. Materials—Appearance. 2. Commercial products.
3. Optical instruments. 4. Colorimetry. I. Title.

TA418.H86 620.1′1295 75-20429
ISBN 0-471-42141-3

Printed in the United States of America

10 9 8 7 6 5 4 3 2 1

# PREFACE

This book has been prepared to identify appearance attributes of objects and the methods available for measuring them. It is designed to serve as both a readable textbook and a reference work. As such it makes available much material not previously organized for ready reference.

The primary premise of the book is that "object appearance" involves not only color, but also such attributes as gloss, luster, and translucency. Color is undeniably the most noticed of the appearance attributes, but the contributions of the others are vital to the identification of object appearance. Object size and shape are also obviously important factors in overall appearance, but they are outside the scope of this book. Nonc of these factors can be ignored when making a judgment about the appearance of an object, whether the judgment is visual or instrumental.

In addition to being economically important, the way things look is inevitably a fascinating subject to everyone, since it is a major factor is one's daily experience. Because it has been such a constant part of our lives since birth, our recognition of the appearance of objects and our judgments concerning them are astute, fast, and highly discriminating. To simulate these appearance discriminations by measurements, analysis of the technical basis of appearance must be known. An instrument to measure an appearance attribute cannot be built without knowing the physical sources of that attribute and the predictable human response to it.

With knowledge comes the realization that a complete physical specification of all the factors that contribute to an object's appearance is too complicated and cumbersome to be either attainable or useful. However, it is usually feasible to make measurements of the specific attributes important for any given problem. Specific methods for the measurement of color, gloss, opacity, lightness, and the like are in widespread use in science and industry, and have proved to be valid and extremely useful in identifying and controlling product appearance.

Appearance specification is interdisciplinary in approach. The established scientific fields of physics, physiology, psychology, psychophysics, and materials technology must all be brought into any complete discussion of appearance description and measurement. Consequently the first part of the book, which concerns the nature of appearance, draws from the fields of physiology and psychology, as it considers the eye-brain combination and the way it receives and interprets light signals. Then the field of physics enters by way of the optical properties of objects responsible for the way things look and the effects of the viewing situation upon appearance.

The second part of the book deals with the numerical scales used to measure object appearance. It draws upon psychophysics in describing the uses of physical techniques to give numbers having psychological significance. The third part of the book deals with instruments for the measurement of the attributes of object appearance, their principles of design, and a survey of the major ones in use. The final chapter discusses specific applications of appearance measurement.

The appendices of the book are intended to supply, in quick reference tabular form, many of the important dimensions, relationships, terms, and formulas needed by a person engaged in the analysis of object appearance. There is a glossary, which serves as a list of terms and a subject index, and a bibliography identifying references that supply more detail than this elementary book can include.

In the writing of this book I am indebted to many friends and associates who over the years have helped me appreciate the challenges offered by the study of visual appearance. I am indebted to others who have helped point the way to satisfactory solutions to these challenges. In particular I owe my education in color to the late Dr. Deane B. Judd and Dr. K. S. Gibson. They guided me in my formative years at the National Bureau of Standards. The late M. Rae Paul of the former National Lead Company, a member of ASTM Committee D-1 on Paint and Protective Coatings, was one of the first to call my attention to the needs of science and industry for practical, yet accurate, techniques for color and appearance measurement. At the Gardner Laboratory Dr. H. A. Gardner, who died several years ago, helped me greatly with his willingness to encourage an enthusiastic young worker in a new technology. More recently, at the Hunter Associates Laboratory I have had many earnest helpers. I will name only Margaret Burns because she not only evaluated the textbook material for logic and understandability but also did most of the hard detailed work necessary to put the manuscript together and prepare it for publication.

RICHARD S. HUNTER

*Fairfax, Virginia*
*May 1975*

# CONTENTS

# THE MEASUREMENT OF APPEARANCE

# PART ONE

# APPEARANCE ATTRIBUTES AND THEIR BASES IN PHYSICS, PSYCHOLOGY, AND PHYSIOLOGY

# CHAPTER ONE

# ATTRIBUTES OF THE APPEARANCE OF OBJECTS

It is interesting to stop and analyze our daily evaluations of all the things we see. When we consider the complexity of the light patterns that strike the eye and the amount of signal sorting necessary to convert the nerve impulses to identifications of objects, space, location, and movement, we begin to realize how sophisticated the visual sense is. Experience plays a large part in our evaluation of what is seen, and no doubt at an early age we tend to develop our abilities of visual discrimination by using other senses to confirm visual analysis. However, this is probably done quite unconsciously, so that as adults we are certain that an object is shiny or dull, dirty or clean, transparent or opaque, without being aware of how we arrive at these decisions.

Usually we examine an object with our eyes to determine its desirability or usefulness to us. Is it old or new, fresh or stale, beautiful or ugly, unused or worn? Almost always the judgment is made without recognizing the actual optical processes that give us the clues.

The driver of an automobile on a highway appraises the surface of the highway ahead, but not for its darkness of color or for its gloss. Instead, he is looking for the presence of water or ice on the highway revealed by dark color or gloss. A farmer looks at the soil in his field to determine its readiness for plowing, planting, or cultivating. He is not trying to determine color and texture, although these are the specific optical attributes that reveal to him the soil quality. A man examines a shirt in his closet for cleanness as an all-inclusive quality. He does not look consciously for patterns and degrees of yellowness and grayness that, nevertheless, are the appearance attributes used to judge lack of cleanness. It is obvious that very little conscious analysis is made by a person of his actual visual sensations of the appearance of things about him. He simply comes to a conclusion based on an untold number of previous experiences. He enjoys the glossiness of his new car without associating it with any particular phenomenon involving the interaction of light and object, just as he picks the "freshest" meat without consciously realizing that the redness of the meat influenced his choice.

To understand appearance specification and measurement, however, we must appreciate the actual optical and visual events that occur—in light sources, in objects we see, in the eye, and in the human nervous system. Therefore, we begin our discussion of the nature of appearance by considering the different ways in which we see things and how this affects the judgments we make about how things look.

## EVALUATING LIGHT

When we "see" things, we see by means of light. Select any object such as a book cover, table top, or a flat sheet of pink paper. Look at any single point or uniform area of the object. Now let your eye wander across the whole object. Notice how the aspects of lights you see vary over the surface from one place to another. Note also how your brain interprets that variation. Through such evaluations of patterns of light distribution made over and over again during the course of each day we make most of our judgments concerning the things we see. Patterns of light entering the eye are the stimuli on which appearance judgments are based.

Until we have more completely established our terminology, let us use the term "color" to identify the aspect of object appearance by which we recognize it as red, orange, yellow, green, blue, violet, black, white, gray, or intermediate. Let us use the term "geometric attributes of appearance" for those aspects such as gloss and texture that, with selected conditions of illumination and view, will cause perceived light to vary from point to point over a surface of uniform color. This division of what we see into color and spatial (or geometric) aspects is the basic and fundamental classification that underlies the entire science of appearance technology. It provides the basis for our division later on of appearance attributes into two main categories—those associated with color, and those, such as gloss, that result from the spatial distribution of light from the object, and which are called the geometric attributes.

There is an even greater sophistication in how we make these color and spatial evaluations, resulting from the years of experience that the eye and brain have had in working together to report our surroundings to us. Consider some of the astute deductions resulting from the simple, everyday experience of looking at a white piece of paper. Place the paper in front of you and hold your hand several inches above it. Some parts of the paper will look darker than others. We say the darker part is "in the shadow." We still see the sheet of paper as being uniform in its properties. We would never conclude that some unseen hand had painted part of the sheet gray. Similarly, if a white paper is illuminated by a red Christmas-tree light, we would not say that the paper is red, but rather that it is white paper under red light.

## MODES OF APPEARANCE

Our ability to see red light and still describe the paper as being white, or to see a light and dark area of the same surface and still recognize the surface as being uniform, has produced a need for a means of characterizing appearance situations. We do this by classifying the apparent visual condition or "mode" of each thing seen. Consider three such visual conditions:

1. The illuminant mode—we see the stimulus as a source of light, for example, an incandescent lamp.
2. The object mode—we see the stimulus as being illuminated by some light source, as when lamp light falls upon a wall.
3. The aperture mode—we see the stimulus simply as light. This would be the case when a door is opened just a crack and we see a glimmer of light through it.

There is often ambiguity in our minds regarding the mode of the stimulus. The same stimulus may look two different ways, depending on our interpretation. As Ralph Evans puts it, it "looks different the second time [Evans, 1948]." Imagine yourself in a field on a dark night, with a light visible at a distance. At that point the stimulus appears to be a lighted window (aperture mode). But as we move closer, we see that it is really a lighted wall inside a house (object mode), seen through the window. Then, as we look through the window, we see a lamp, and we call that the light source, only to realize that what we just saw was the lamp shade, which is an object and not the light source.

The eye always tries to pin down what it sees as either object or light source. From a distance the light from the window appears yellowish in color but, when we get closer and look through the window, we find that the wall is in fact white, and that it is a yellow light shining on it that made it appear yellow. Thus the same thing (the wall) looks either white or yellow, depending on the "mode" we see it in. When we see it as a light source (that is, in the illuminant mode), it appears yellow, but when seen as an object in reference to a light source or other objects (in the object mode) it looks white.

The third situation (aperture mode) obtains when the eye looks through a small hole or aperture, usually at a uniformly colored area, so that the observer has no way of deciding what is the color of the illuminant and what is the color of the screen being viewed. Because it eliminates the influence of spatial distribution of light and involves observation of only one elemental area at a time, the aperture mode is normally used in visual color-matching experiments to obtain simplified and standardized data on the performance of the eye.

The question of mode has greatest significance when it comes to relating light measured instrumentally to what the eye sees. To interpret measurements one must be aware of the mode in which the eye is operating since it so much affects the visual evaluation. Our interest focuses on the object mode because most of the practical problems of appearance analysis involve objects. We are operating in this mode when we look around a room and instantaneously make thousands of judgments based on the patterns of light that reach our eyes. In the normal object viewing situation we are aware of the illuminant, and we see not only the object but also many of its surroundings. The light is reflected or transmitted by the object to the eye, and from this light we make our appraisal of the object. Evaluations of color, haze, polish, clarity, gloss, whiteness, and opacity are all made by observing what the object does to the light shining on it. Further, we always strive to place the visual response "out there" in the object. For instance, we do not talk about seeing red light in our eye when we look at a apple. Because we place color in the object and consider it to be part of the object, we tend to see the same object as being the same color even though the actual light received by the eye varies considerably. This phenomenon is called color constancy, and occurs only in the object mode. A white piece of paper is said to be white whether it is illuminated by a bluish daylight, or by a yellowish lamplight. The color constancy phenomenon tends to compensate for the effect of the color of the illuminant on the perceived color of the object. Color constancy is so much a part of everyday experience that we have a special term "metamerism" for the special condition of a spectral difference between two similarly colored specimens that causes them to change color differently when the illuminant is changed.

## EVALUATING THE APPEARANCE OF OBJECTS

Since our visual process works to put appearance "in" the object, we evaluate object appearance in terms of attributes or specific visual qualities of the object. Our simple division of light into color and spatial properties can now be expanded into a list of object-appearance attributes. The color properties encompass three color attributes, and the spatial properties relate to a number of geometric attributes, such as gloss, turbidity, and texture.

## Color Attributes

We consider color to be a composite, three-dimensional characteristic consisting of a lightness attribute, and two chromatic attributes, which in this discussion we shall call "hue" and "saturation." Since color is three-dimensional, it takes a solid-coordinate system to represent color graphically. The term "color solid" is widely used to identify dimensional arrangements that are usefully related to visual attributes of color.

Let us analyze a little more closely the manner in which we perceive color by considering the color of a yellow school bus. The most obvious thing about it is that it is a shade of yellow rather than blue or purple. This characteristic is generally called hue. Many people are content to just call this color and let it go at that. A color specification, however, consists of more than just a designation of hue. Colors of the same hue can differ in how much color there appears to be. This concept, which we will refer to as saturation, is conveyed by words such as depth, vividness, and purity. Saturation is subtler and less obvious than hue, but we are very much aware of it, whether consciously or not, as we describe to ourselves the color of the school bus. Chances are that we call it a vivid yellow, noting that much of the conspicuousness of the color lies not so much in the yellow hue as in the vibrancy and purity of the color. Saturation can be thought of as a measure of how different the color is from gray.

The colors of the rainbow arranged in a circle would make a hue circle such as that shown in Figure 1.1. The concept, or dimension, of saturation can be incorporated into the hue circle if the center point of the circle is considered to be neutral gray and the most saturated color is to be located at the greatest

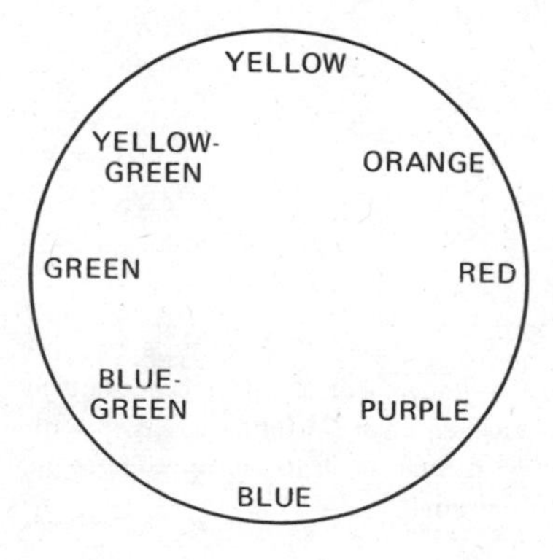


**Figure 1.1.** Arrangement of colors in a hue circle.

distance from the center. Our hue circle, including the less saturated colors, now looks like Figure 1.2. A change in hue is indicated by a move around the circle, while an increase in saturation is indicated by a greater distance outward from the gray point at the center. Hue and saturation are the two dimensions of color called the chromatic attributes.

The school-bus-yellow color would seem to belong on the orange side of yellow and about as far out from the center of the hue circle as possible. But how about the color of a fresh lemon, or that of mustard? The school bus is darker (and slightly redder) than the first but lighter than the second. Thus all of the three colors may be of about the same saturation and not much different in hue either. We need a third dimension—the lightness of the color, to finish our specification. The two-dimensional hue circle cannot fully accomodate the colors of the bus, the lemon and the mustard. We cannot place them all properly without introducing a third dimension. This is the luminous attribute, which we call "lightness" when dealing with objects.

Now we begin to realize how really subtle is our appreciation of color. We are aware, not only of the hue and saturation but also that there are many lightness levels. We can think of the lightness dimension as being entirely separate from the chromaticity attributes (hue and saturation). In fact, it is easier to grasp the concept if we consider first the colors white, gray, and black. Let us arrange these "achromatic" colors (those without hue or saturation) vertically with white at the top through light gray and dark gray to black at the bottom. Our hue circle can now be moved up and down, using the achromatic colors as a vertical axis. In this way we generate the hue-saturation-lightness color solid (Figure 1.3), which is frequently used in color identification and analysis.

There are many measurement systems for the specification of colors in three-

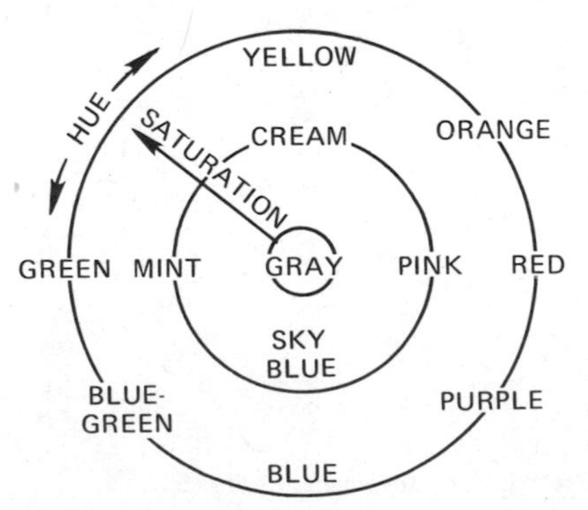


**Figure 1.2.** Arrangement of colors in a hue and saturation surface. Saturation increases with distance outward from the gray neutral point at the center of the hue circle.

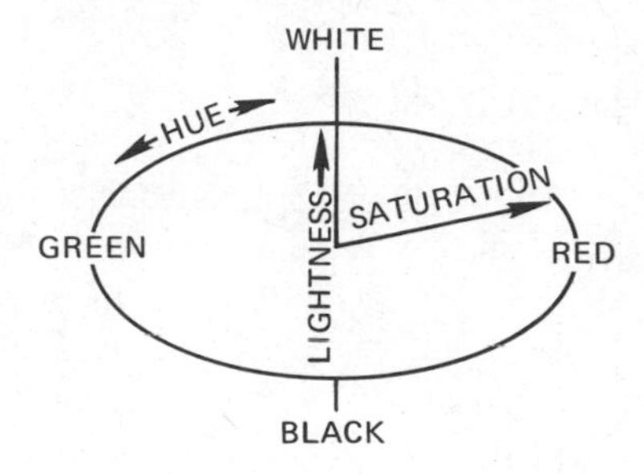


**Figure 1.3.** Three-dimensional-color space. The lightness dimension provides an achromatic center axis on which the hue circle can be positioned at varying lightness levels.

dimensional space. They are all covered by the general terms "color space" and "color coordinates." The way a person is involved in the use of color often determines the type of three-dimensional-specification system he prefers to use. For example, the Munsell Color System, devised by the artist A. H. Munsell around 1905, has been used extensively by artists and designers through the years. This system closely resembles the one just described, except that the saturation dimension is called "chroma" and lightness is termed "value," while hue remains "hue." Figure 1.4 gives a cylindrical representation of the three-dimensional Munsell system for color notation, with the hues arranged around the vertical axis of achromatic colors, with the more saturated colors to the outside of the solid.

The product technologist, such as a dye or color formulator, whose daily work involves formulating or changing colors of paints, textiles, papers, or other products, tends to use different dimensions in his color space. He thinks of hue in the same manner as everyone else. However, saturation and lightness do not correspond to the attributes with which he works. He starts with a white base material of pigment or fiber that he colors by addition of either dye or pigment. When colored dyes or pigments are added to the white base the product becomes simultaneously more saturated and darker. The color formulator calls this change "depth" and measures it in color space as distance from white at the top of the color solid.

The product technologist is often concerned with getting the brightest and most saturated colors available. Where there is a choice between colorants, he normally chooses those that give the highest saturation. He calls this attribute cleanness, vividness, or brightness. In his concept of color space this attribute is measured upward and outward from black or dark gray at the bottom of the color solid, not horizontally out from the center as in the Munsell system.

Because it possesses three dimensions familiar to the color formulator, the Ostwald Color System and Color Solid are widely used. Figure 1.5 shows a

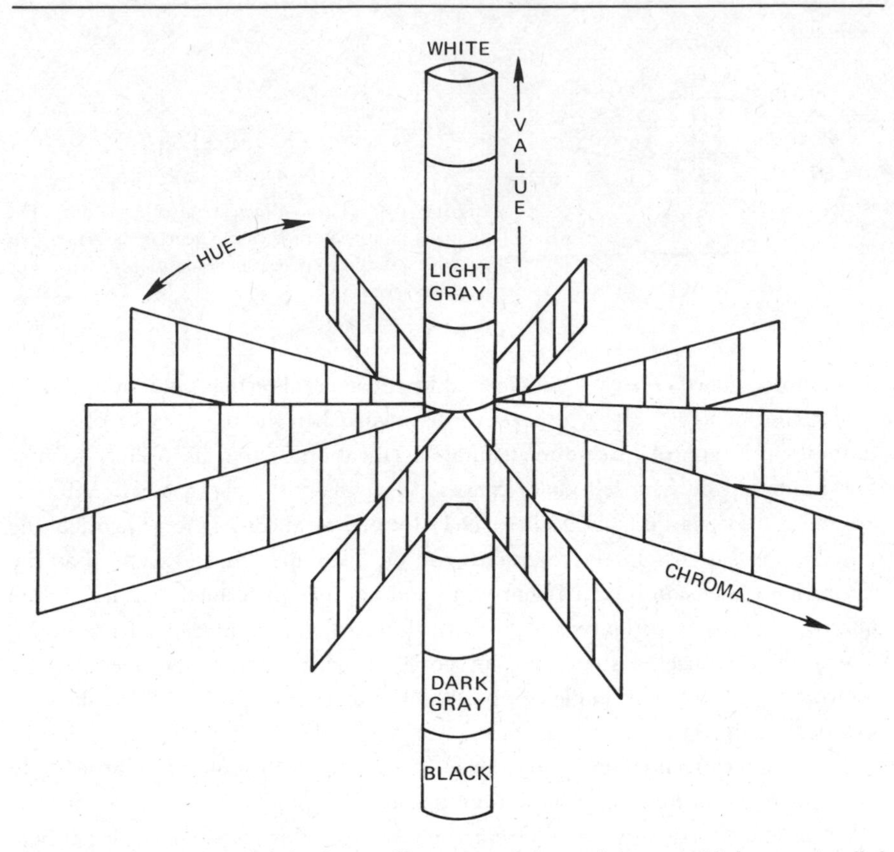


**Figure 1.4.** The Munsell Color System. Hues are spaced around the vertical lightness axis called "value." There is an increase in saturation (chroma) with horizontal distance from the central axis to the outside of the color solid.

"leaf" from the Ostwald color tree (similar to the Munsell tree in that each hue constitutes a leaf radiating out from the achromatic axis). Notice that this color solid does not divide easily into two chromatic attributes and one luminous attribute. Hue is still a chromatic attribute, but depth and vividness, because of their diagonal nature, are neither strictly chromatic nor luminous but combinations of the two. Depth and the vividness are less accurately located in the coordinate systems of color measurement because they relate directly to the technologists' experience with actual ingredients. Ingredient content is not uniformly convertible to any three-dimensional system.

In addition to artists and color formulators, there is another group of people who, because they are concerned with the measurement of color, use yet

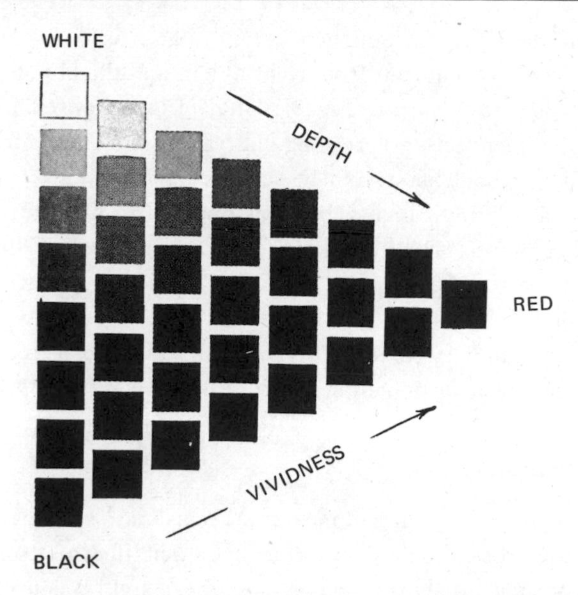


**Figure 1.5.** Ostwald Triangle, showing modification of a full-color red with white and black. The color formulator thinks in these diagonal dimensions of depth and vividness as he mixes materials together.

another arrangement of attributes in color space. A foreman for a textile bleach plant or for a paper machine thinks in terms of the common visual experience of seeing red, green, yellow, blue, white, and black as paired opposites; red-green, yellow-blue, and white-black. For example, he sees his green product as being too much on the yellow side, needing to be bluer. The dimensions of his system, called the opponent-colors system, are shown in Figure 1.6. The black-

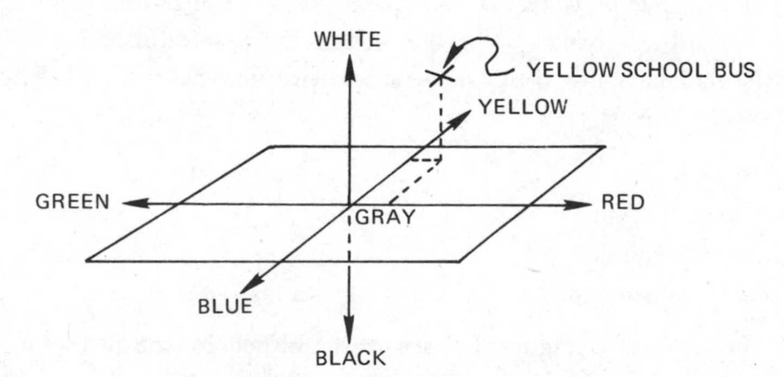


**Figure 1.6.** The three dimensions of opponent-colors space. This diagrams the experience of chromaticity as redness or greenness and yellowness or blueness, and shows the location of the yellow school bus color in opponent-colors space.

to-gray-to-white "axis" of the opponent-colors solid is as in the other systems. The red-to-green dimension runs from right to left, and the blue-to-yellow from front to back. The arrangement resembles that of the Munsell Color System, except that the dimensions are rectangular rather than cylindrical. In this system, the yellow school bus would be seen as having a fairly high lightness (white-to-black) attribute, a high yellow (as opposed to blue) attribute, and a slight red (as opposed to green) attribute. The opponent-colors dimensions are considered today to correspond most closely to the actual visual signals going from eye to brain.

Table 1.1 is a summary of the different views taken of three dimensional space by designers, color formulators, and color technologists.

## Geometric Attributes

Colors can be identified by three-dimensional coordinate systems, even though the systems of the artist, the product technologist, and the color scientist all differ. But the geometric attributes such as gloss, haze and translucency are more difficult to organize. Unlike color attributes, these are associated with the spatial distribution of light by the object and cannot be uniquely defined by any organized coordinate arrangement.

Fortunately, some meaningful simplification and organization of geometric appearance attributes is possible if we consider a relatively flat, uniform, area of surface. The light leaving that surface is characterized as having been either reflected or transmitted by the object. Reflected light leaves the object from the same side as is illuminated. Transmitted light leaves the opposite side after passage through the specimen.

*Diffuse and Specular Light Distributions.* Both transmitted and reflected light can be further divided, somewhat arbitrarily, into diffused light and not-diffused (specular) light, so that we end up with four main kinds of light distribution from objects:

1. Diffuse reflection
2. Specular reflection
3. Diffuse transmission
4. Specular transmission

The four objects in Figure 1.7 are distinguished by the fact that each is yellow because of a different type of light distribution. The yellow color of the plastic box is seen by diffuse reflection. We call the reflection from the yellow box diffuse because even if we move the light source or the eye the yellow color

**TABLE 1.1**
SYSTEMS OF THREE-DIMENSIONAL COLOR SPACE

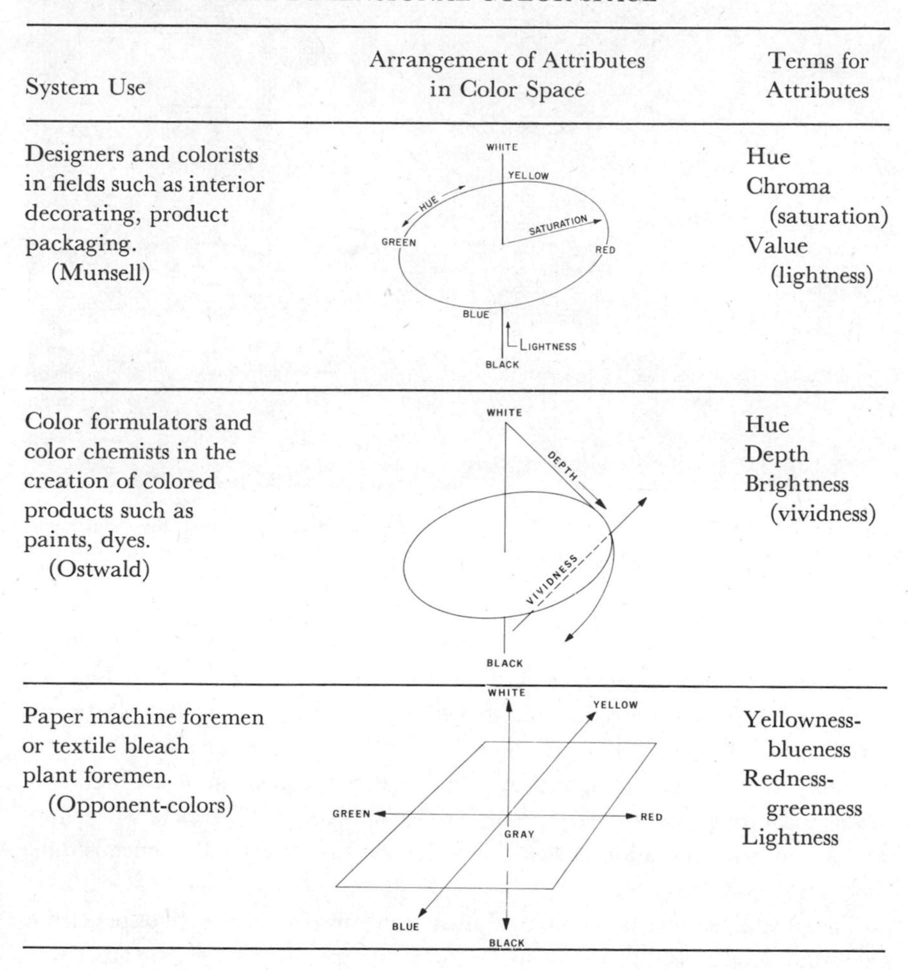


| System Use | Arrangement of Attributes in Color Space | Terms for Attributes |
|---|---|---|
| Designers and colorists in fields such as interior decorating, product packaging. (Munsell) | | Hue<br>Chroma (saturation)<br>Value (lightness) |
| Color formulators and color chemists in the creation of colored products such as paints, dyes. (Ostwald) | | Hue<br>Depth<br>Brightness (vividness) |
| Paper machine foremen or textile bleach plant foremen. (Opponent-colors) | | Yellowness-blueness<br>Redness-greenness<br>Lightness |

looks about the same. In other words, the yellow light from the box is reflecting in all directions, or is diffused. Objects whose color is seen by diffuse reflection constitute a major fraction of those analyzed for appearance in science and industry.

With the brass candleholder, yellow light is projected by specular reflection, the dominant light distribution. Specular reflection is reflection as from a mirror,

**Figure 1.7.** Four classes of objects. The plastic box shows diffuse reflection, the candleholder specular reflection; the plastic tumbler transmits light diffusely, the bottle of clear oil specularly.

which means that it is directed instead of being diffused. The yellow box also remits some specular reflection, but it is white since the box is a shiny plastic material. Virtually all of the yellow reflection from the brass candleholder is specular. If we move the light source or our eye, the position of the highlights seems to move on the object. Specular reflection from yellow metals is yellow, and it is in this coloration of specular reflection that shiny yellow metals differ from shiny nonmetals.

The plastic tumbler is yellow by diffuse transmission because light penetrates the yellow object, is diffused within it, and emerges, in part, on the other side. The diffusely transmitted light leaves the object surface in all directions. Diffuse transmission, at low levels, is often called haziness, or turbidity.

The bottle of oil is yellow by specular transmission, which means that the light passes straight through the oil altered only in color.

The foregoing classification of light distributions makes us aware of where we look and what we need to measure when we evaluate color or the geometric attributes of object appearance. It is particularly useful in our investigation of geometric attributes since it illustrates so clearly how dependent are gloss, haze, and the like on the way the light is distributed by the object.

## CLASSIFICATION OF OBJECTS

The classification of light distributions serves as a basis for placing objects into four groups according to the main, or dominant, way in which each object distributes the light that falls upon it. These groups with their dominant light distributions are

| | |
|---|---|
| 1. Opaque nonmetals | Diffuse reflection |
| 2. Metallic surfaces | Specular reflection |
| 3. Translucent objects | Diffuse transmission |
| 4. Transparent objects | Specular transmission |

This division of objects, which is illustrated in Table 1.2, is based on some simplifying assumptions. Not all objects fit clearly into one of the four categories—some fit not at all, and others seem to fall somewhere in between. Specular and diffuse components of reflected and transmitted light are seldom completely separable. The table ignores changes in object color resulting from a change in illuminant or in the manner of observation. An indication is given, however, of the complexity of real-life optical situations in the last column of Table 1.2. Here we see that gloss cannot be simply dismissed as "gloss." It really consists of at least six different appearance attributes, each of which is

**TABLE 1.2**

GEOMETRIC AND COLOR ATTRIBUTES OF OBJECT APPEARANCE

| OBJECT CLASS | IDEALIZED DISTRIBUTION OF LIGHT | DOMINANT LIGHT DISTRIBUTION | COLOR ATTRIBUTES (SEEN BY DOM. LIGHT DISTR.) | OTHER LIGHT DISTRIBUTION | GEOMETRIC ATTRIBUTES SEEN BY OTHER LIGHT DISTR. |
|---|---|---|---|---|---|
| OPAQUE NON-METALS<br>PLASTIC BOX | 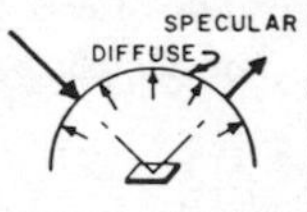 SPECULAR<br>DIFFUSE | DIFFUSE REFLECTION | LIGHTNESS AND CHROMATICNESS HUE AND SATURATION | SPECULAR REFLECTION | GLOSSINESS<br>SHININESS<br>LUSTER<br>DISTINCTNESS-OF-REFLECTED-IMAGE<br>SHEEN<br>SURFACE UNIFORMITY<br>DIRECTIONALITY |
| OPAQUE METALS<br>BRASS CANDLE HOLDER | 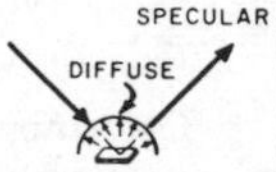 SPECULAR<br>DIFFUSE | SPECULAR REFLECTION | GLOSSINESS AND CHROMATICNESS | DIFFUSE REFLECTION | REFLECTION HAZE |
| TRANSLUCENT<br>PLASTIC TUMBLER | 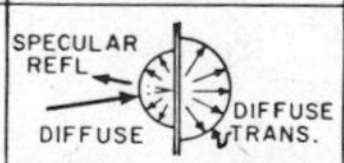 SPECULAR REFL.<br>DIFFUSE<br>DIFFUSE TRANS. | DIFFUSE TRANSMISSION | TRANSLUCENCY AND CHROMATICNESS | DIFFUSE REFLECTION | REFLECTION LIGHTNESS |
| | | | | SPECULAR REFLECTION | GLOSSINESS |
| TRANSPARENT<br>BOTTLE OF OIL | 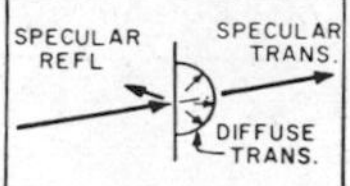 SPECULAR REFL<br>SPECULAR TRANS.<br>DIFFUSE TRANS. | SPECULAR TRANSMISSION | CLARITY AND CHROMATICNESS | SPECULAR REFLECTION | GLOSSINESS |
| | | | | DIFFUSE TRANSMISSION | TRANSMISSION HAZE OR TURBIDITY |

distinctively different from the others, and requires a different technique for its measurement.

Even with all of these reservations, however, Table 1.2 is very useful in our investigation of appearance attributes and how to measure them. Note that with all four groups the perception of color is caused by the dominant light distribution. Note, also, that color consists in each case of the two chromatic (hue and saturation, or red-greenness and blue-yellowness) attributes and a third luminous attribute which changes as the manner of light distribution changes. The lightness seen in the diffusely reflected color of nonmetal objects becomes glossiness in the brass candleholder, translucency in the plastic tumbler, and clarity in the clear yellow liquid. This luminous attribute, simultaneously a

**TABLE 1.3**
THE MOST IMPORTANT APPEARANCE ATTRIBUTES OF OBJECTS[a]

| Class of Object | Color Attributes | Geometric Attributes |
|---|---|---|
| Diffusing surfaces ("surface colors") | Color by diffuse reflection:<br>Lightness<br>Hue<br>Saturation | Glossiness by specular reflection |
| Metallic surfaces | Color by specular reflection:<br>Glossiness<br>Hue<br>Saturation | Reflection haze by diffuse reflection |
| Translucent specimens | Color by diffuse transmission:<br>Translucency<br>Hue<br>Saturation | Glossiness by specular reflection |
| Transparent specimens | Color by specular transmission:<br>Clarity<br>Hue<br>Saturation | Transmission haze, or turbidity by diffuse transmission |

[a] Summary of Table 1.2.

color attribute and a geometric attribute, is determined by the dominant projection of light by the object.

Most objects distribute light by projections in addition to the dominant projection. The attributes seen by these less-dominant light projections contribute significantly to the overall appearance of the object. Thus the glossiness of the box, which is carried by the less dominant specular reflection, tells the observer that he would find the box smooth to the touch. Similarly, any transmission haze present in the clear yellow liquid would be seen by diffuse transmission of light, a less dominant projection, but an important one in assessing the clearness of the yellow liquid.

## SUMMARY

In this chapter, we started with the proposition that sight, vision, and the way things look are an integral part of life. In order to begin to approach techniques for appearance evaluations in commerce and industry, we have identified the two ways we classify or "see" light: (1) by its color; (2) by where it comes from—its spatial distribution.

How we see that light can also be classified into three modes of appearance—aperture, illuminant, and object modes. The object mode is the one of primary interest to us, since we are treating here the subject of the appearance of objects.

Further we found it useful to divide all objects into four groups depending on the manner in which each object distributes the light that falls upon it. We were then able to identify appearance attributes with specific distributions of light. Table 1.3 is a summary of the most important appearance attributes of objects.

Recognition of the relationship between appearance and specific optical phenomena is the first step toward designing instruments that can measure what we see.

# CHAPTER TWO

# LIGHT AND LIGHT SOURCES

The human eye-brain mechanism makes astoundingly astute and rapid evaluations of object appearance all day long, every day. We take this for granted and seldom question the validity of the judgments made. However, there is a need in industry and science for these judgments to be convertible to numerical measurements. Therefore we must analyze this human process in order to learn how to build instruments that evaluate object appearances as humans do.

There are always three ingredients or elements in the object observing situation: a light source, an object, and a human observer. Figure 2.1 shows these three ingredients of appearance recognition. Let us consider each of them, and the part it plays in the process, beginning with the light source.

## WHAT IS LIGHT?

Most of us are familiar with the way a prism can be used to break up white light into different colors, or the way water droplets in the sky do the same thing to the sunlight, creating a rainbow. We use the word "spectrum" to describe the result of this separation of light into its brightly colored components.

Through many ingenious experiments scientists have shown that all light exhibits a wave motion. The prism or rainbow separates light according to the length of these waves (wavelength). The lengths of light waves are very short. In fact, the distance traveled by light as it passes from the front to the rear of our eyes is equal to about 50,000 wavelengths. The unit for measuring light waves, called the nanometer (nm), is $10^{-9}$ meters (about 25 million nanometers per inch). The visible spectrum contains wavelengths between 380 and 770 nm. The violet end is variously said to be from 380 to 400 nm; the red end, from 700 to 770 nm. The other hues have wavelengths somewhere in between. Wavelengths between 400 and 475 nm are usually called blue; near 500 nm, green-blue. The range from 500 to about 570 nm is called green, while the short span from 570 to 590 nm is normally said to be yellow. The rest of the

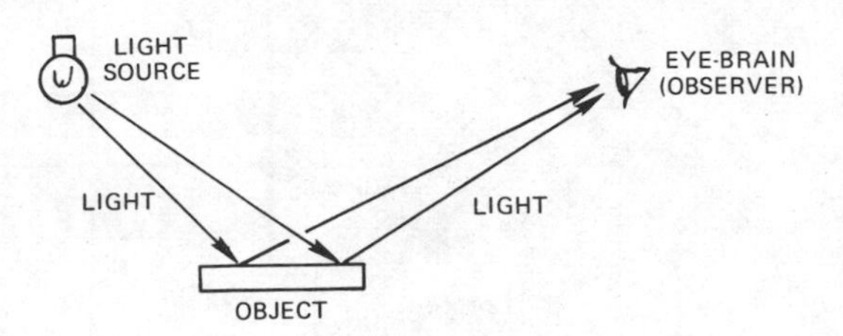


**Figure 2.1.** Ingredients of appearance recognition.

visible spectrum from 590 to 770 nm produces colors ranging from orange to deep red.

The visible light waves are a comparatively tiny part of a much longer array of waves called the electromagnetic spectrum, shown in Figure 2.2. The feature that distinguishes light waves from the other electromagnetic waves is visibility. Physically they are the same as X-rays or radio waves, differing from them only in wavelength. All of these waves, whether within the visible spectrum or not, are called radiant energy. We therefore define light properly as visible radiant energy.

## Luminosity

We introduce here one simple but important aspect of the relationships of wavelength to vision. Even within the range of visible wavelengths, some wavelengths can be seen more easily than others. Luminosity is the property of light by which we define how easily we can see it. The luminosity of various wavelengths differs; the eye is most sensitive to light at about 550 nm and is insensitive to radiation outside the visible spectrum. Thus the eye is more sensitive to a certain amount of energy at 550 nm than it is to the same amount of energy at 650 nm. The sensitivity of the eye to all wavelengths has been determined and can be simply summarized in the form of a graph. This graph is called the luminosity function for the human eye and is shown in Figure 2.3.

## Describing Light Spectrally

When we study light in reference to its wavelength we say that we are studying the spectral characteristics of the light. For example, the luminosity function of the human observer, which indicates our changing visual response to the same amount of energy at different wavelengths, is defined by the spectral curve shown in Figure 2.3. Looking at a spectrum, we say that blue light is spectrally different from yellow. It is the spectral properties of sources, objects, and observers that are responsible for all perceptions of color.

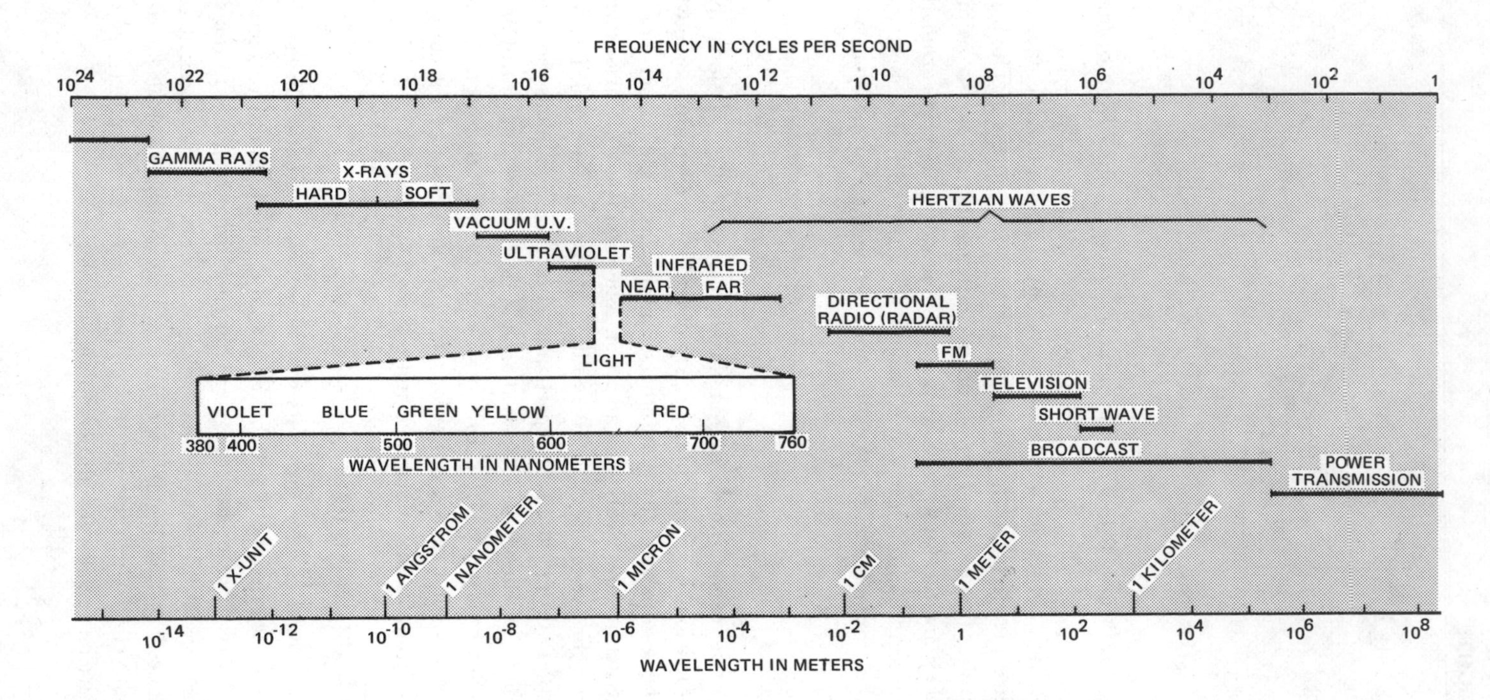


**Figure 2.2.** The radiant-energy (electromagnetic) spectrum. The visible portion of the spectrum is expanded to show the hues associated with different wavelengths of light. Reproduced from the *IES Lighting Handbook,* courtesy of the Illuminating Engineering Society.

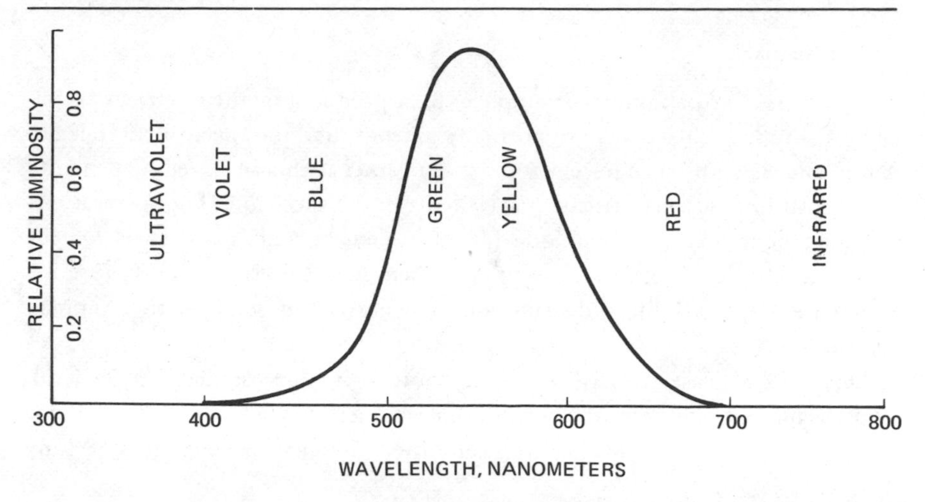


**Figure 2.3.** The luminosity function of the human eye. We see light in the yellow-green portion of the spectrum around 550 nm, much more easily than we see it elsewhere in the spectrum.

## LIGHT SOURCES

Light is created in one of two ways: by heating objects to incandescence (a red hot poker or an ordinary light bulb), or by atoms or molecules being subjected to excitation—as by an electrical discharge (an arc in air or in neon gas).

### Incandescent Sources

An incandescent source is a heated material. As its temperature goes up more light is produced, and the material changes color from red at low temperatures to yellow, and finally to nearly white as the temperature is increased, just as we have seen a poker change color as it is heated. The spectral characteristics of the light from any incandescent light source relate primarily to the actual temperature of the heated object, whether it is a tungsten lamp filament, a poker, or some other material. Thus, the colors of incandescent light sources have traditionally been identified by the temperatures of completely radiating (blackbody) sources having the same colors. (Blackbodies when hot enough are not black, but are light sources.) Figure 2.4 shows the relative spectral distribution (or the energy emitted at each wavelength relative to that at 560 nm) for a blackbody at different actual temperatures. The correlated color temperature of an incandescent source is defined by the temperature at which a blackbody operates to produce a visual match with the incandescent source. Reddish color sources have lower correlated color temperatures than white or blue sources.

## Other Sources

Neon lamps afford a familiar example of light produced by the elcctrical excitation of atoms. Such sources are normally gaseous, and the spectral curve is not continuous as with incandescent sources but rather is characterized by "spikes" or spectral lines at very specific wavelengths in the spectrum. The spectral distribution of this energy is unique for each element and is the basis for the spectrographic analyses so widely used in chemistry and astronomy. Figure 2.5 shows the "spectral line" distribution of energy from some of the common gaseous sources.

One special case related to light sources is fluorescence. In general, fluorescent materials are those that convert incident radiant energy of one group of wavelenths into energy of other, longer, wavelengths. Many

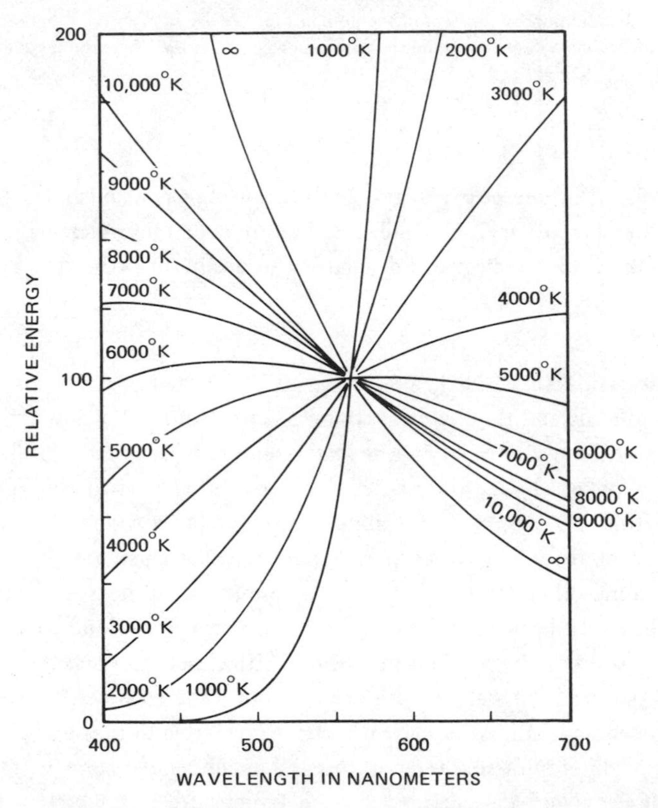


**Figure 2.4.** Relative spectral-energy distribution curves for blackbodies of different color temperatures. These curves have been normalized to 560 nm. Reproduced from *IES Lighting Handbook,* courtesy of the Illuminating Engineering Society.

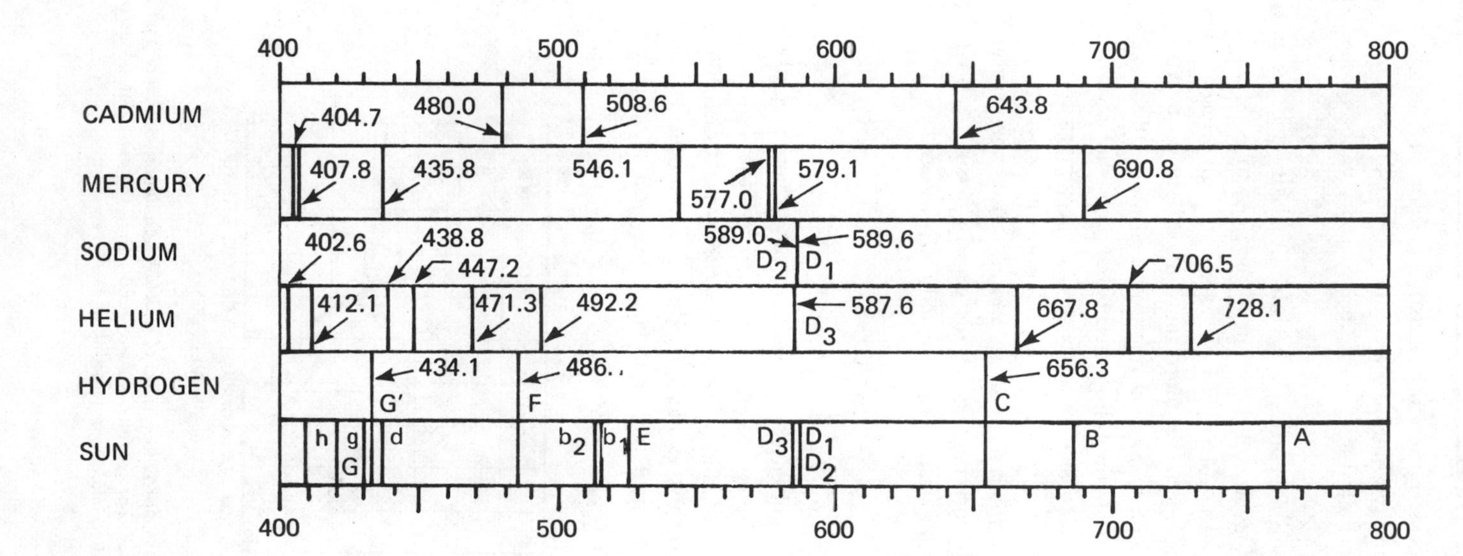


**Figure 2.5.** Spectral lines of certain gaseous sources. Spectral curves of gaseous sources are characterized by spectral lines at specific wavelengths rather than by a continuous emission of energy throughout the spectrum. From *Handbook of Applied Instrumentation* by D. M. Considine and S. D. Ross. Copyright 1964, McGraw-Hill Book Company. Used with permission of McGraw-Hill Book Company.

fluorescent materials possess the characteristic of absorbing radiant energy in the nonvisible region and reradiating it as light in the visible region.

In the common household fluorescent lamp, mercury vapor radiates spectral lines in both the visible and ultraviolet portions of the spectrum. Some of the spectral lines in the visible portion are transmitted through the translucent coatings of fluorescent powder on the inside of the glass tube. At the same time, the nonvisible ultraviolet spectral lines excite the fluorescent powder to generate additional and spectrally more continuous light in the visible region. The combination of mercury spectral lines and continuous fluorescent-powder spectral energy of a cool-white fluorescent lamp is shown in Figure 2.6. Fluorescent lamps are identified by correlated color temperatures, as are incandescent sources. However, there is no relationship between the correlated color temperature of a fluorescent lamp and its actual temperature, since a blue-white light (high correlated color temperature) can be produced by a fluorescent lamp at a comparatively low actual temperature.

## ILLUMINANTS

The illumination on an object may come from one or from several sources acting together. Often the combination is designated as "the illuminant." In colorimetry the illuminant is usually specified by its spectral distribution of light energy—as in a table of values measuring the sun's output at a specific time and date. When measurements are being made of object color, either by visual

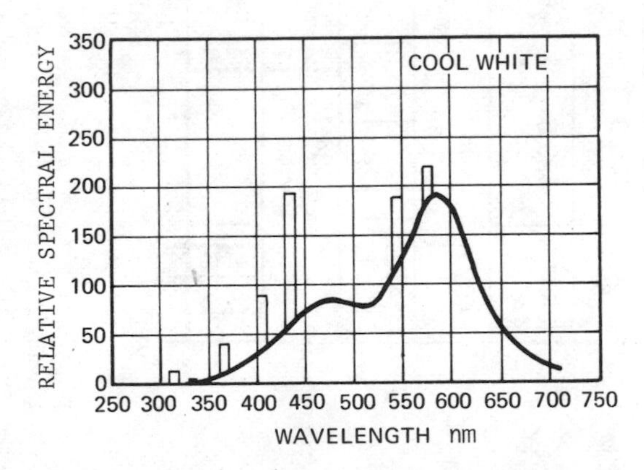


**Figure 2.6.** Spectral-energy distribution curve of a Cool-white fluorescent lamp. The spikes are mercury-vapor spectral lines. The continuous curve results from the excitation of the fluorescent coating of the tube by the ultraviolet spectral lines of the mercury vapor.

evaluation or by instrument, a particular illuminant is specified as part of the standardization of the viewing conditions.

## SUMMARY

Light is defined as visually evaluated radiant energy of wavelengths from about 380 to about 770 nm. Different wavelengths have different colors, and some wavelengths are visibly more intense than others. The eye's varying response to the same amount of energy at different wavelengths is represented by the luminosity curve. When we are concerned with wavelength-dependent properties, we use the word "spectral." Spectral curves can describe the amount of light or radiation at each wavelength, as in Figure 2.4, or our response to it, as in the luminosity curve (Figure 2.3).

Light can be produced by heating objects to incandescence or by the excitation of atoms and molecules. One special case is fluorescence, where light is converted from one spectral band to another.

A completely radiating source, called a blackbody radiator, can be used as a reference standard of identification of the color of practical incandescent light sources. The correlated color temperature of a light source is the temperature of a blackbody radiator visually closest to the appearance of the actual light source.

# CHAPTER THREE

# INTERACTION OF OBJECTS WITH LIGHT

In Chapter 1 we indicated that the appearance attributes of an object are related to the ways in which the object modifies the light that strikes it. The light can be modified spatially resulting in diffuse and specular reflection and/or diffuse and specular transmission. The light can also be modified spectrally (in color). This brings us to the second ingredient of the observing situation, the object, and what it does to the light.

## WHAT OBJECTS DO TO LIGHT

The four major things that can happen to light when it encounters an object are

1. Specular reflection at the skin of the object (associated with gloss)
2. Scattering within the material (associated with diffuse reflection and sometimes with diffuse transmission)
3. Absorption within the material (largely responsible for color)
4. Specular transmission directly through the object, if it is more or less transparent (associated with clarity)

To see these major processes in action, let us return to the yellow objects shown in Figure 1.7. Consider first the box of yellow pigmented plastic. Since we now know that the wavelengths of light are very small, we find we have to consider the microscopic detail of the box to understand what happens to the incident light. Figure 3.1 represents how the cross section of the box might look under a microscope.

When a beam of light encounters the skin of this object a small portion of it is reflected and does not penetrate. The actual amount reflected is dependent upon the smoothness of the surface, the refractive index of the material, and the angle at which the beam strikes the surface. This reflected light is what we see as specular reflection and is responsible for the glossy appearance of the box.

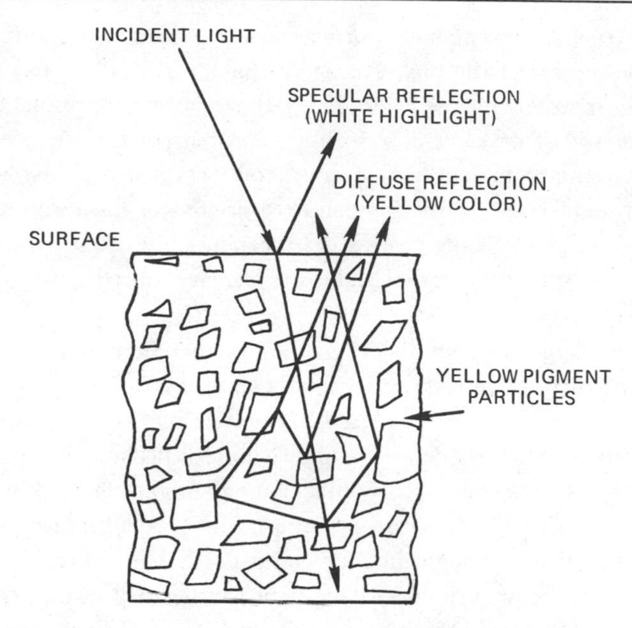


**Figure 3.1.** Microscopic view of light striking a yellow plastic box. Diffuse reflection from an opaque, nonmetallic object is the result of many reflections by the particles inside the body of the object. These yellow particles selectively absorb some of the wavelengths of light, causing our eventual perception of the diffuse reflection as being yellow.

With normal nonmetallic materials this reflected light is not changed in color. The highlights of the box look white, the color of the light source.

If the surface of the box had been rough, the light reflected at the surface would have been scattered in all directions. It would contribute colorless reflected light to the observed diffuse reflection. Thus the shiny yellow box appears darker and more saturated in color than would the same box with a rougher surface. Conversely, a matte object appears to be lighter and less saturated in color than does a shiny object of the same composition.

The major portion of the incident light beam penetrates the surface skin of the box and enters the body of the yellow box. This beam, called the refracted beam because it was changed in direction somewhat by moving from one medium (air) to another (internal body of the box), now encounters the numerous surfaces of the pigment particles that make the box yellow. It is partially reflected at each surface it encounters. This process of multiple reflection and refraction thoroughly diffuses the light and returns much of it to the surface of the box, where it leaves in all directions. This is called diffusely reflected light, and the process of diffusing the light is called scattering.

The reflections at the pigment particles are responsible for diffusion but not for the yellow color of the box. Passage of light through the yellow pigment particles is responsible for the color. The yellow appearance results because the coloring material removes or absorbs blue light but permits the green, yellow, and red wavelengths to continue. If green colorants had been used, then light absorption would occur in the blue and red regions of the spectrum, and the box would look green. This process of absorbing only certain wavelengths of light is called selective absorption and is the primary source of most of the color we see in everyday life.

The processes of reflection and absorption in combination with the features of object construction are responsible for the appearances of almost all objects. For example, consider the other objects in Figure 1.7. The brass candleholder's very shiny appearance is due to the almost total reflection that occurs at the skin. Metals are characterized by much higher first-surface (skin) reflection than nonmetals. This is a primary optical difference between metals and nonmetals. Penetration of light into a metal is negligible. Where metals appear colored (as with the brass holder), blue light is reflected less efficiently than is the rest of the spectrum. This selective reflection causes the reflected light to be yellowish.

The translucent yellow tumbler is structurally much like the box. The lack of a glossy reflection means that the first surface of the tumbler is rough, sending what would be the specular beam off in all directions. Thus there can be diffusion of light by reflection from rough surfaces as well as diffusion by internal pigment or fiber-particle scattering. The translucent tumbler also contains some yellow pigment. However, there are not enough multiple reflections within the tumbler to turn back all of the unabsorbed light even though the light that does pass through the cup is well-diffused. Thus it both reflects and transmits yellow light diffusely.

The glass bottle of transparent yellow oil has an outside glossy surface. There is no diffuse reflection because there are no particles of pigment within either the oil or the glass to scatter the penetrating light. The transmitted light is free from diffusion but is yellow because the oil removes or absorbs light of blue wavelengths.

Of the four optical types illustrated in Figure 1.7, the first, comprising diffusely reflecting objects, is the most important for appearance measurement. Probably 90% of the specimens measured for color or other appearance attributes fall in this first category. Table 3.1 identifies ten industries regularly concerned with analysis of appearance. Specific objects from these fields are listed according to optical type. The majority of objects and materials falls into

**TABLE 3.1**
INDUSTRIAL PRODUCTS CLASSIFIED BY TYPE OF OBJECT

| | Opaque Objects | | Light-Transmitting Objects | |
|---|---|---|---|---|
| Industry | Nonmetallic Objects (color seen by diffuse reflection) | Metallic Surfaces (color seen by specular reflection) | Translucent Objects (color seen by diffuse transmission) | Clear Objects and Materials (color seen by specular transmission) |
| Ceramic | Tiles, whiteware, porcelain enamels, bricks, cast cements, roofing granules | — | Translucent glasses | Transparent glasses |
| Paints | Most painted surfaces | — | — | Varnish and clear-lacquer films |
| Plastics | Opaque plastic objects and films | Metallized plastics | Translucent plastics | Transparent plastic films and objects |
| Paper | Most papers and pulps | — | Translucent papers | — |
| Printing | Most printing on paper | Clear ink printed on metal | — | — |
| Textiles | Most textiles | — | Translucent textiles | — |
| Pigments, dye, resins, and oils (raw materials) | Most specimens tested are incorporated into products above | — | Translucent test specimens | Clear resins and oils; dyes and clear pigments in solution |
| Cleaning and polishing materials and processes | Most cleaned and polished objects | Polished metals | — | — |
| Foods | Majority of foods | — | Translucent foods | Transparent beverages |
| Metals | Metals with diffusing metal oxide coatings | Most metals | — | — |

the first column. Only in plastics are there specimens of all four geometric types.

## THE OPTICAL SIMILARITIES OF DIFFERENT PRODUCTS

The optical behavior of light is much the same in many paints, plastics, papers, textiles, ceramic products, and foods. This is because all of these products have nonmetal surfaces that are shiny if optically smooth, lower in gloss if they are rough. Below this surface is a granular structure of pigment and/or fiber immersed in resin and/or air that scatters light diffusely. When some or all of the pigment absorbs light selectively, or when colored dyes are attached to fibers or pigments, there is absorption and therefore color. The similarity of the optical behavior of nine different diffusely reflecting products that are regularly measured for color and other appearance attributes is shown in Table 3.2.

## THE RELATION BETWEEN ABSORPTION AND SCATTERING

Selective absorption, which is largely responsible for color, takes place during the passage of the light through materials. Scattering occurs where light encounters interfaces between pigment and resin, fiber and air, and so on. Normally when particle sizes are made smaller less light is absorbed during passage through each particle (resulting in less color being apparent) and, at the same time, total particle surface becomes greater. The increase of particle surface leads to increased light scattering, or diffusion, since reflection occurs at the particle surfaces. The overall result of decreasing particle size, then, is less saturation and more lightness.

Figure 3.2 shows a striking demonstration of this change in absorption and diffusion with change in particle size. There are four vials all containing pieces of the same broken green bottle glass, but of decreasing particle size. Where the fragments are large, light travels extensively within the green glass and is selectively absorbed, producing the green color. The large fragments appear green and quite dark. Where the same glass is reduced to a fine powder, light is reflected every few thousandths of an inch. There is not as much opportunity for selective absorption to occur, and the powder appears to be nearly white. The effect of the frequent reflections predominates, and though only about 4% of the incident light is reflected at each encounter with a glass surface, the cumulative result is a diffuse reflectance of 70, 80 or even 90%. It is necessary to have small pigment particles to attain high diffuse reflectance. For color, on the other hand, the particles must either be large or have strong light absorption

**TABLE 3.2**

INGREDIENTS AND PROCESSES AFFECTING PRODUCT APPEARANCE

| Product | Light-Scattering (Diffusing) Ingredients | Light-Absorbing (Coloring) Ingredients | Skin-Gloss-Producing Ingredients or Processes |
|---|---|---|---|
| Ceramics | | | |
| Glass and Porcelain Enamel | $TiO_2$ and other white pigments | Colored inorganic oxides & pigments in glass | Glass, molds, and hardening conditions |
| Whiteware Tile, China | Clay, other white pigments | Overlying films of colored oxides in glass | The glass/pigment ratio and firing conditions |
| Paints | $TiO_2$ and other white pigments | Colored pigments both organic and inorganic | The resin/pigment ratio and the paint surface structure |
| Plastics | $TiO_2$ and other pigments | Colored pigments and dyes | The gloss of calender rolls or mold surfaces |
| Paper | | | |
| Uncoated | Cellulose wood fibers, white pigment sometimes added | Dyes on the fibers | Fiber smoothness and calendering |
| Coated | Cellulose fibers with clay in coating on surfaces | Dyes on fibers; colored pigments in coatings | Coating smoothness and degree of calendering |
| Printing Ink | Relies mainly on underlying paper for light scattering, but some white pigments are used | Dyes and colored pigments | Resin/pigment ratio and ink hold out from paper |
| Textiles | Textile fibers (pigments sometimes incorporated into man-made fibers) | Dyes, occasionally colored pigments used in man-made fibers | Fabric calendering and surface treatment |
| Food | Food pigments and fibers | Food pigments | Varies with type of food |

**Figure 3.2.** Four vials of the same green bottle glass, of different particle sizes. The vial on the left contains the largest particles. As the particle size decreases progressively in the three vials to the right, scattering power, and therefore lightness, increases.

for the short distances of travel in them. It should be noted also that there is a point, as particle size is reduced, at which the particles begin to lose their power to scatter light. Fred Stieg, in a paper given before the New York Society for Paint Technology, commented on this matter:

"Various estimates have placed the most effective particle diameter at approximately ½ the wavelength of the light involved. When, therefore, the diameter of a pigment particle which is being reduced in size becomes less than ½ of the shortest wavelength of visible light . . . it begins to disappear as it loses its ability to produce visible interference with the passage of light rays [Stieg, 1962]."

The relations involving absorption and scattering are very important in appearance technology. Knowledge of the absorption and scattering coefficients of specific materials enables one to predict the concentrations of component colorants needed to produce a given color. Computer programs utilizing these data are used to approach a match with a target color. Techniques for color formulation by computer are generally based on mathematical models such as the one developed by Kubelka and Munk (1931) and described by Park and Stearns (1944), Judd and Wyszecki (1963), and others. The applications of these formulas are limited by the extents to which the various products of commerce conform to the simple models. The simplest models consider only changes in internal absorption. More complex ones deal with both absorption and scattering, but very few of the useful models attempt to deal separately with both diffuse and specular light distributions.

## PHYSICAL EVALUATION OF LIGHT FROM OBJECTS

Techniques for quantitatively evaluating the light reflected from or transmitted by objects involve measurements of flat, uniform areas of the object. When light falls upon such a surface, a portion of it is always reflected. If the object is not opaque, some light is transmitted. The distribution of the light reflected or transmitted may vary with the following characteristics:

1. Wavelength of the light involved
2. Direction of incidence of light on the object
3. Direction from which the object is viewed (or, direction in which light is taken for measurement)

There are two primary instruments used for the physical measurement of light as it leaves an object: the spectrophotometer and the goniophotometer. The spectrophotometer measures the amount of light from an object wavelength by wavelength, and thus its readings relate primarily to the color of the object. The goniophotometer measures the quantity of light emitted from the object in different directions. Since it gives values of light reflectance or transmittance angle by angle, it provides data about the geometric attributes of appearance. Both the spectrophotometer and the goniophotometer are described in greater detail in Part Three.

### Spectrophotometric Curves

Measurements of the fractions of incident light either reflected or transmitted at different wavelengths are spectrophotometric in nature. The results of such measurements are usually presented as curves in which percent reflectance or percent transmittance is shown at each wavelength. The relationship of the shape of the curve to the curves of known hues gives some indication of the perceived color of the object.

Seven spectrophotometric curves showing percent of reflected light plotted against wavelength are given in Figure 3.3. These represent white, black, gray, and four colored specimens. Note that the white is characterized by high reflectance throughout the spectrum, the black by very low reflectance

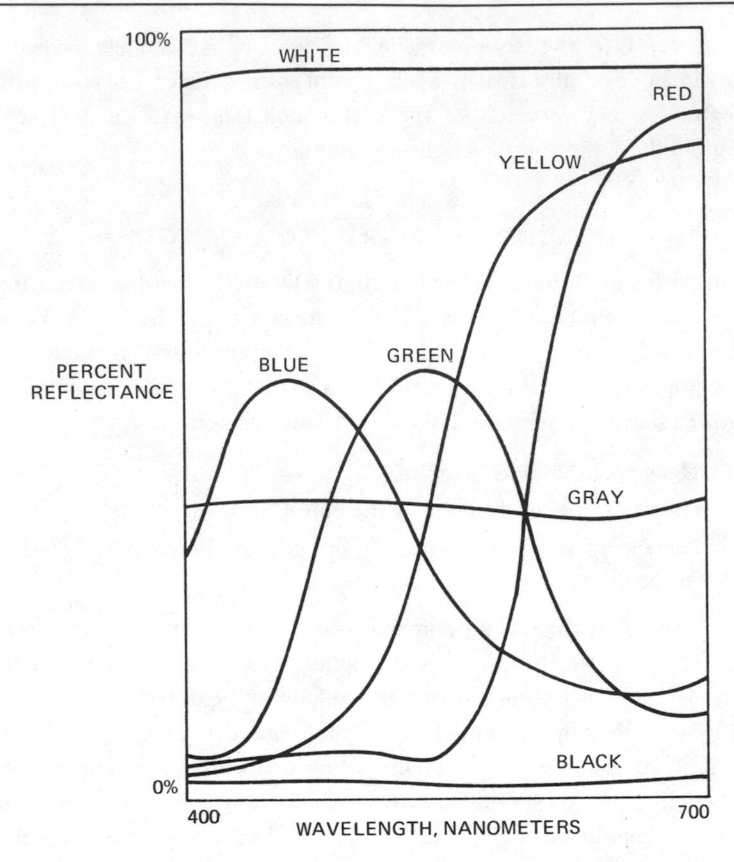


**Figure 3.3.** Spectrophotometric curves. Curves for white, gray, and black objects are characteristically uniform throughout the spectrum, but with high percentage reflectance for white, low for black, and medium for gray. Curves for colored objects normally show absorption in the areas of the spectrum not associated with the hue as we see it.

throughout the spectrum, and the gray by medium, uniform reflectance throughout the spectrum. The curves for the colored specimens are different in that percentages of reflectance change markedly from one part of the spectrum to another. The reflectance of the red specimen is high in the red end of the spectrum but low in the rest of the spectrum. Similarly, the green and blue specimens have their highest reflectance values in the green and blue parts of the spectrum, respectively. However, the curve labeled yellow is high in the red, yellow, and green parts of the spectrum and low only in that part called blue.

Spectrophotometric curves may change significantly with changes in the geometric arrangement of the light beams in the spectrophotometer. As we saw in Chapter 1, the color of a nonmetallic object is carried by the light diffusely reflected from the object. This light has penetrated the object and has been reflected and rereflected by particles within the object. If therein it underwent selective absorption by pigments or dyes, it would emerge as diffusely colored light. Thus, to measure color of nonmetal opaque objects, one uses a diffuse-reflectance spectrophotometer. However, if the spectrophotometer's geometric conditions of measurement include the specular, or first-surface, reflectance with the diffuse reflectance, the measured values of diffuse reflectance will be increased at all wavelengths by a constant amount, as is shown in Figure 3.4. This amount is the percentage of the incident light that is specularly reflected by the first surface, roughly 4% for a glossy surface. Where metallic pigments are present, spectral curves are affected even more strikingly by changes in geometry.

## Goniophotometric Curves

The amount of light reflected or transmitted by an object can vary as the direction of view is changed. Curves showing change of amount of reflectance or transmittance with change of angle of view are called goniophotometric curves. Such curves identify the properties of specimens responsible for gloss, haze, luster, and other geometric attributes. They are thus analogous in function to spectrophotometric curves, which identify the physical properties of specimens responsible for color. Figure 3.5 shows the geometric variables of goniophotometric measurements: the angles of incidence and viewing, each measured from perpendicular.

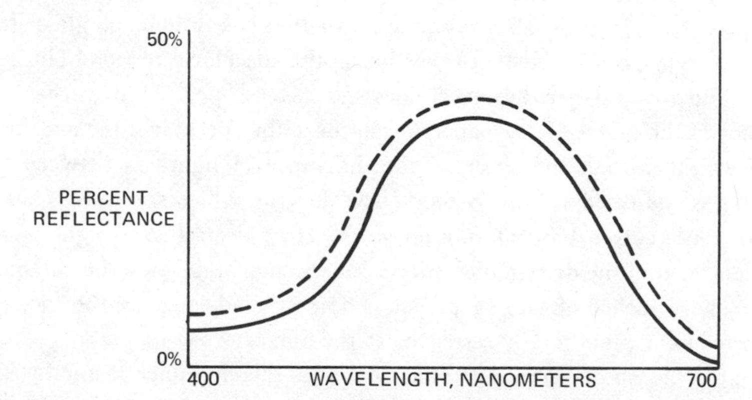


**Figure 3.4.** Two spectrophotometric curves of a green specimen. The dotted line resulted from the measurement including the specular reflectance; the solid line resulted when the specular was excluded.

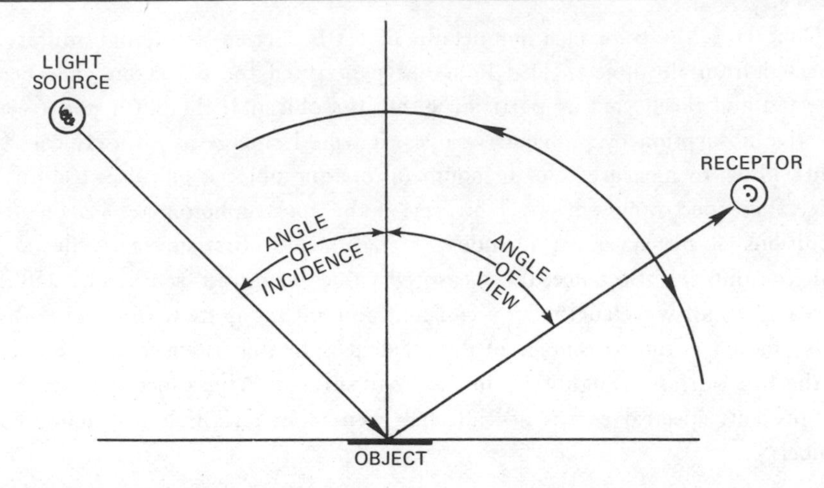


**Figure 3.5.** Geometric measurement conditions of a goniophotometer. The light source is fixed; the moving receptor measures the amount of light reflected at different viewing angles.

A goniophotometric curve is produced by fixing the light source at a specific angle and measuring the light reflected at different angles of view. Reciprocal conditions of fixed angle of view and variable illumination are also used. Figure 3.6 illustrates goniophotometric curves of four specimens. These curves show the amount of light reflected in various directions with the specimen illuminated at 45°. Note that the vertical reflectance-factor scale is logarithmic. The increase in light reflected in the −45° direction represents specularly reflected light and gives an indication of the gloss of the sample. One of the samples represented is a white matte paper that has little or no gloss. Its curve gives no evidence of a peak of reflection in the specular direction. The curve for the white glossy paper, however, does show such a peak. This curve is similar to that of the green glossy paper in and near the specular direction, but in directions other than the specular the difference in lightness between the two papers is indicated by the higher curve for the white specimen. The fourth curve represents a bare aluminum sheet. Here the reflected light is strongly concentrated in the direction of mirror, or specular, reflection, but there is some haze and evidence of lack of polish. These are indicated by the fact that the reflected light is not totally restricted to the mirror direction.

Figure 3.7 shows goniophotometric curves of reflectance factor for the four pieces of paper shown in Figure 3.8. Two of the papers are white (A, B), two are black (C, D), two are shiny (B, C), and two are dull (A, D). The curves in Figure 3.7 are for a fixed direction of incidence of 45°; the direction of specular

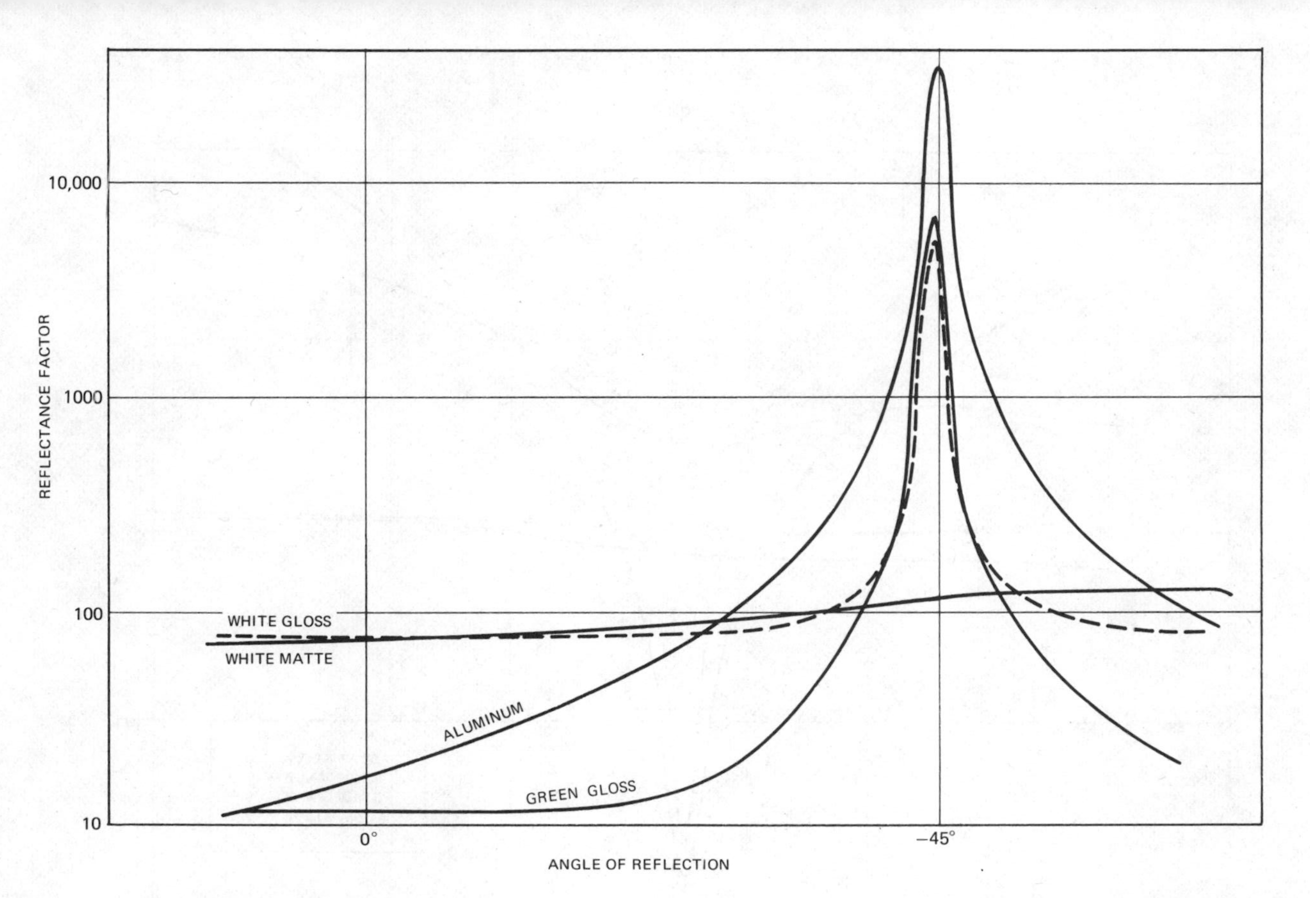


**Figure 3.6.** Goniophotometric curves for four papers. Four papers were illuminated at 45°, therefore the height of the curve at −45° is an indication of the paper's specular gloss. At angles other than the specular, the two white papers show a higher diffuse reflection than do the other two.

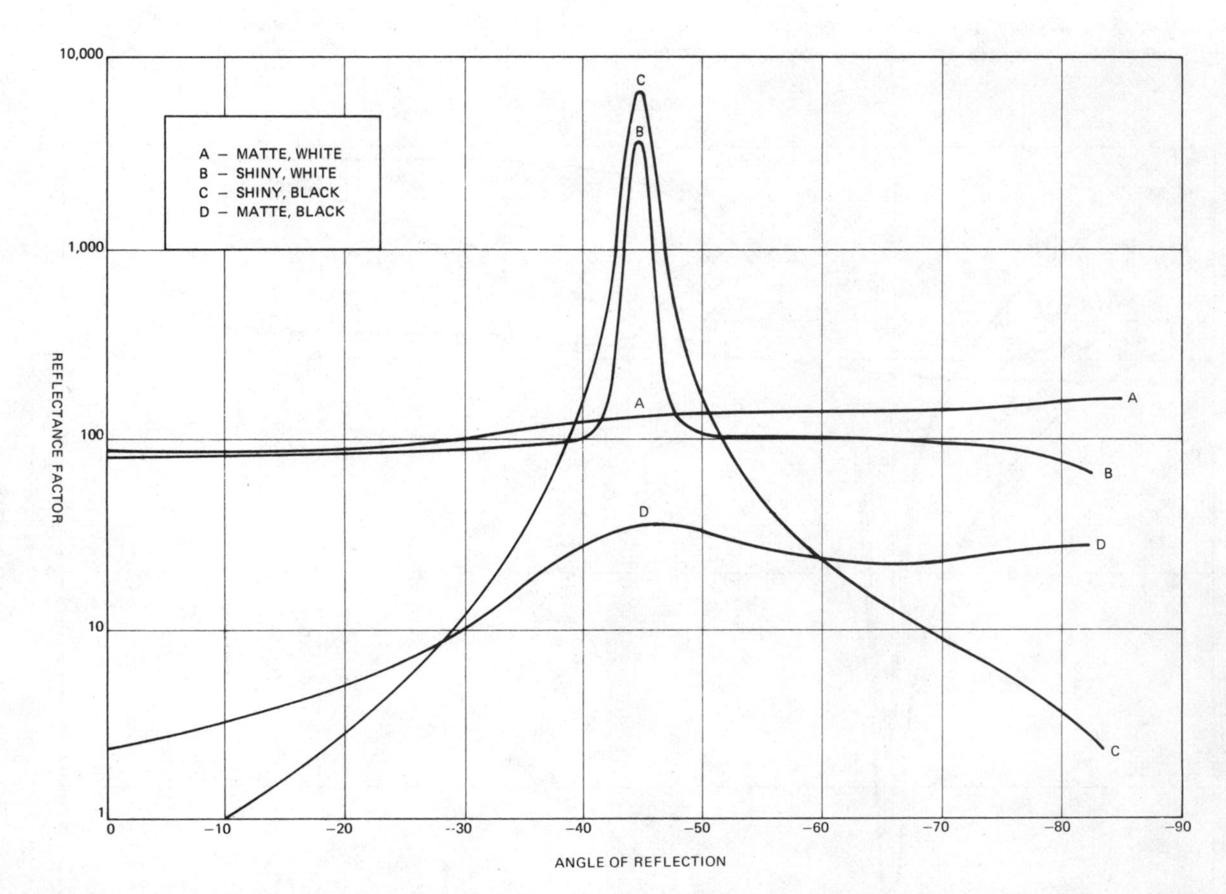


**Figure 3.7.** Goniophotometric curves for the four papers shown in Figure 3.8. Specular reflection is indicated at 45° viewing angle.

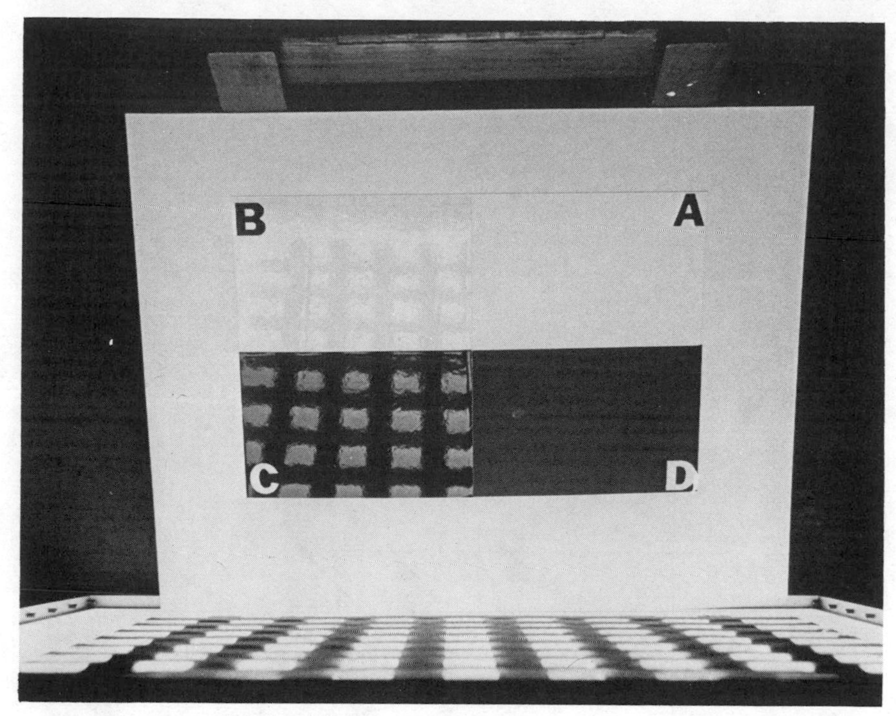


**Figure 3.8.** Four papers with surfaces reflecting a lighted grid in different manners. Papers A and B are white; C and D are black. A and D are matte; B and C are shiny. Goniophotometric curves are shown in Figure 3.7.

reflection is thus −45°. The glossy specimens B and C have curves with narrow, well-defined peaks in the direction of specular reflection; the matte specimens A and D show no such peaks in their curves. Note also the behavior of the curves near 0° (that is, perpendicular to the sample surface): the curves for the white papers A and B nearly coincide at about 80%, while the curves for the black papers C and D fall to below 3%. At this point, near the perpendicular, the curve is measuring diffuse reflection, thus the readings are typical of diffuse lightness, not gloss.

Figure 3.9 shows goniophotometric curves for the transmission of light through the four different plastic sheets shown in Figure 3.10. Two of these are transparent sheets of the type used in packaging. However, one of the sheets (2) is somewhat hazy, while the other (1) has good clarity. The haze is represented by the increased light intensity on either side of the central transmission peak.

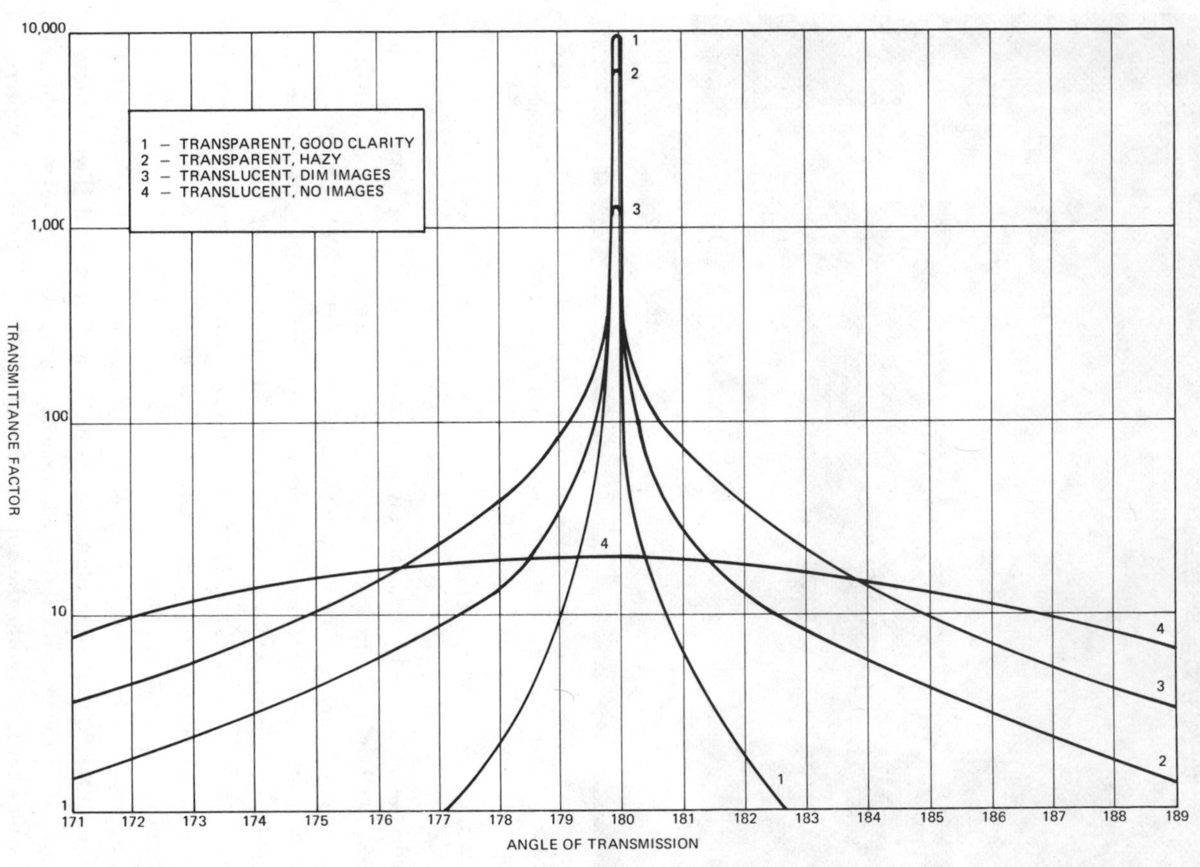


**Figure 3.9.** Goniophotometric curves for four plastic sheets shown in Figure 3.10. Specular angle of transmission is at 180°.

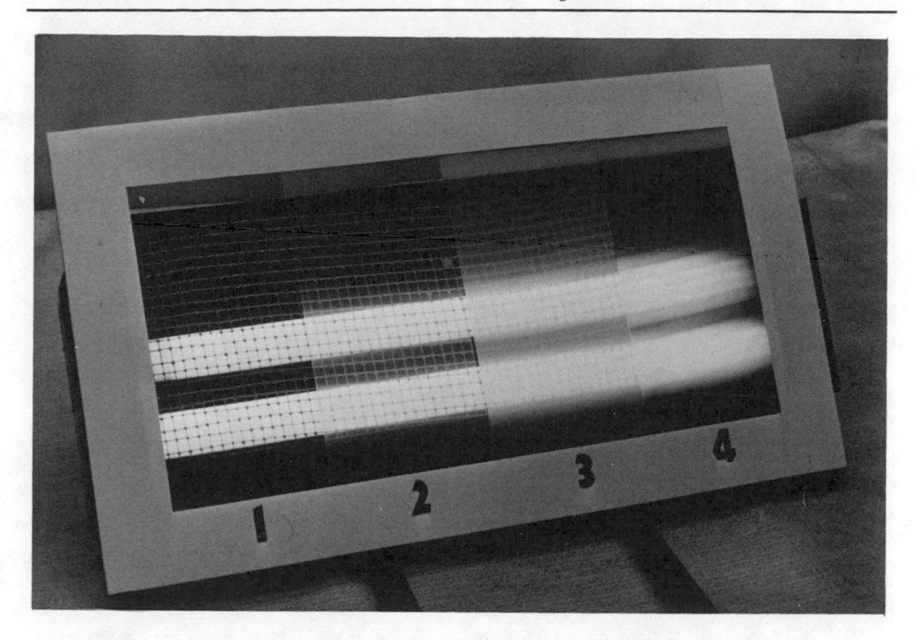


**Figure 3.10.** Four plastic sheets transmit light from a grid-covered fluorescent tube. Plastic sheets 1 and 2 are transparent; 2 transmits less well than 1. Sheets 3 and 4 are translucent; 3 transmits only a dim image of the lighted tube; 4 transmits no image at all. Goniophotometric curves are shown in Figure 3.9.

The two other plastic specimens represented in Figure 3.10 are translucent. One of these sheets (3) transmits dim images; the other (4) transmits no images.

The three curves in Figure 3.11 represent reflection from a single sheet of glossy photographic paper. The curves are for light incident on the paper at 45, 60, and 75°. Note how much the concentration of light reflected in the direction of specular reflection increases with specular angle. This is typical of most nonmetal surfaces.

We noted earlier that spectrophotometric curves may vary with geometric conditions of measurement. Conversely, the goniophotometric curve representing a specimen may change not only with the direction of the incident light but also with the spectral quality of this light. All the curves shown in this chapter were made with white light as a source and a spectral receiver with a response similar to the luminosity response of the eye.

Because spectrophotometric curves may vary with geometric condition, and goniophotometric curves may vary with directions of incidence, it becomes ap-

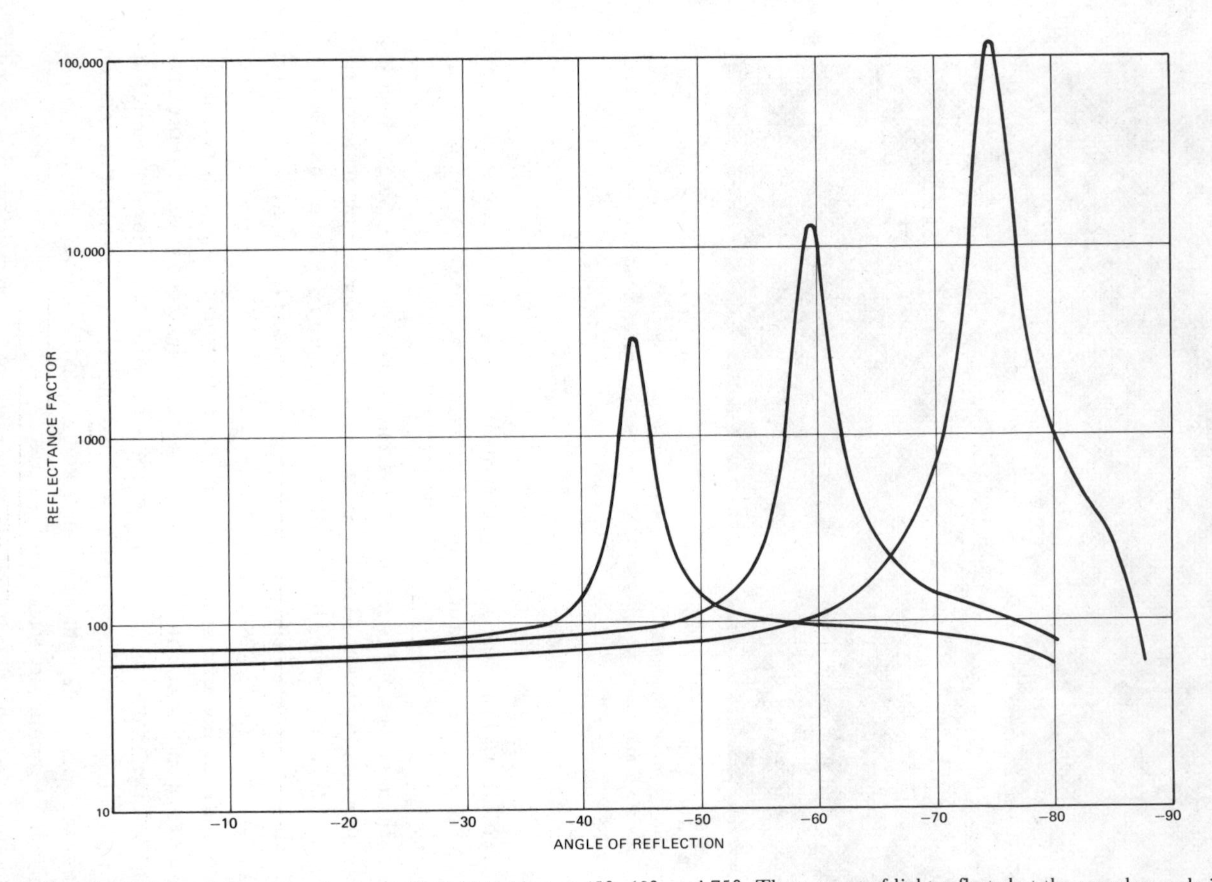


**Figure 3.11.** Goniophotometric curves of paper with light incident at 45°, 60°, and 75°. The amount of light reflected at the specular angle increases as the size of the angle increases.

parent that a whole family of curves is necessary in order to describe the physical light-distributing capacity of just one elemental specimen area. Complete spectral and goniophotometric analyses of light distributions by objects are much too cumbersome and complex to be attempted in any normal circumstances.

## SUMMARY

The light striking an object will be affected by interaction with the object in a number of different ways. The resulting light distributions give us our impressions of what that object looks like. Specular reflection, for example, makes the object look glossy or shiny. Metals are distinguished by stronger specular reflection than that from other materials, and smooth surfaces are always shinier than rough ones.

The diffuse reflection caused by the scattering of light within a material, where no selective absorption by dyes or pigments is invovled, causes a white appearance. White paper, for example, is white by diffuse reflection. This means it must contain many internal surfaces, each of which reflects some of the light. This is indeed the case, as paper is composed of many small fibers. It is white because the fibers absorb almost no light. Snow, talcum powder, milk, and all other white materials look white because they scatter light by multiple reflection but do not absorb it.

When absorption is the dominant process, dark colors result. If all wavelengths are absorbed, black results. Selective absorption of just certain wavelengths results in our perception of color.

In the majority of objects all these processes are operating; specular (shiny) reflection, diffuse reflection by scattering, and absorption. To make physical analyses of the combined results of these three processes, spectrophotometric and goniophotometric measurements are used. Spectrophotometric curves measure the reflection or transmission of light from objects by wavelength. Spectral curves thus relate to color. Goniophotometric curves describe the reflection or transmission of light from objects as a function of angle, and relate to the geometric attributes such as gloss and haze. Although these measurements do not provide conclusive values for *appearance,* they quantify the light-object interaction part of the observing situation.

# CHAPTER FOUR

# THE HUMAN OBSERVER AND VISUAL EVALUATION OF OBJECT APPEARANCE

The observer is the last stage in the three-part observing situation. The ultimate visual perception is a result of how the eye interprets the light reaching it from objects. In this chapter we discuss the basic concepts of human vision, the capabilities of the human observer, and the techniques and settings he uses for the visual appraisal of appearance differences.

## THE HUMAN OBSERVER

### Color Vision

We have seen that the eye is not equally sensitive to light at all wavelengths and that its spectral-intensity response can be described by the luminosity curve. We have also found, however, that we see light as more than just light-or-dark luminosity. We see colors, and we can arrange them into three-dimensional configurations.

Scientists have long known that the three-dimensionality of color means that our eyes must have at least three spectrally different receptors of light radiation. Direct means of measuring the three actual responses have recently been developed, but most of the important work in color vision was done before direct spectral analyses of the receptors were possible.

The history of the search to understand color vision really begins with Sir Isaac Newton (around 1700) who first demonstrated with prisms that white light, instead of containing no color, was made up of lights comprising all the spectral colors. Later investigators found that by using just three colored lights—one red, one green, and one blue—they could, by adding together varying amounts of these three lights, create yellow, violet, white, orange, and the many shades in between. Thomas Young, in 1807, suggested a three-component theory, which was elaborated by Helmholtz about 50 years later.

This theory assumed that the eye contained only three spectrally unique responses—one primarily red, another primarily green, and the third primarily blue.

In 1878 Ewald Hering provided additional insight by presenting an "analysis of sensations of color rather than of the stimuli required to evoke them [Judd, 1966]." He proposed that there are six independent colors—red, green, yellow, blue, white, and black. These colors are perceived by three opponent-colors systems, black-white, red-green, and yellow-blue. Thus, an observer sees chromaticity in terms of redness or greenness, and yellowness or blueness.

Through the early 1900s a large number of experiments were carried out to learn more about the precise nature of color perception. These studies confirmed beyond doubt the three-dimensionality of color vision and the plausibility of Hering's opponent-colors theory. For many years the Young-Helmholtz and Hering theories were considered to be in competition. Then Mueller proposed in his three-stage theory that the Young-Helmholtz concept of three types of color receptors in the retina of the eye was correct, but that responses from these three receptors were converted in the elaborate nerve-signal switching areas within the eye and optic nerve to opponent-colors signals such as Hering postulated (Mueller, 1930). Thus, the currently accepted theory says that both the Young-Helmholtz and the Hering theories are correct. They merely refer to different stages in the process of color vision. In 1949 Judd showed that the Mueller theory successfully explained the observed effects associated with different forms of color blindness (Judd, 1949). Commenting on the subject of recent advances made in this area, L. F. C. Friele writes:

"The most recent developments in physiological knowledge confirm the Mueller concepts in a striking way. These developments open the possibility for a more adequate description of the threshold ellipsoids in terms of retinal and neural signal transport. Very promising progress has been made in this respect by the Institute for Perception in the Netherlands [Friele, 1971, private correspondence]."

### Visual Sensitivity to Light Signals

The field of psychology provided another important piece to the puzzle of human vision. Psychologists, beginning with E. H. Weber (1795–1878), have shown that sensory responses are not linear with the amount of stimulus; that, in fact, with touch, taste, and sight we differentiate between amounts of a stimulus according to the ratio between them, or the percentage change from one to the other (Weber's law). In the understanding of vision, the most important application of this principle is in the way we see lightness. For example, if we look at three ceramic tiles in a row, a white one reflecting 80% of the incident light, a light gray one reflecting 40%, and a dark gray one reflecting 20%, the eye sees the lightness *difference* between adjacent tiles as about equal. This is because the change from 80 to 40% is 50%, and the change from 40 to 20% is also 50%. On a log-reflectance scale these equal ratios become equal increments. The name Fechner is associated with this fact. When expressed on the log basis the term Weber-Fechner law is often used. Figure 4.1 shows what the relationship between reflectance factor and perceived lightness would be if based on Weber's law. In actual practice the surroundings and adaptation modify this relationship, especially for dark surfaces.

## The Capabilities of the Human Observer

The eye-brain mechanism is incredibly sensitive and flexible. Not only can the eye differentiate among about 10 million juxtaposed surface colors, but also it can distinguish lines that are separated by as little as one minute of arc. The eye-brain combination is not only sensitive, but is also capable of high-speed data analysis and interpretation enabling other parts of the body to react quickly to visual stimuli. Consider the tremendous amount of information that flows through the eye when one drives at night in a crowded city. Despite low illumination levels and distractions of light and motion, one is able to sort out quickly the important information needed for driving and still occasionally observe and appreciate other things.

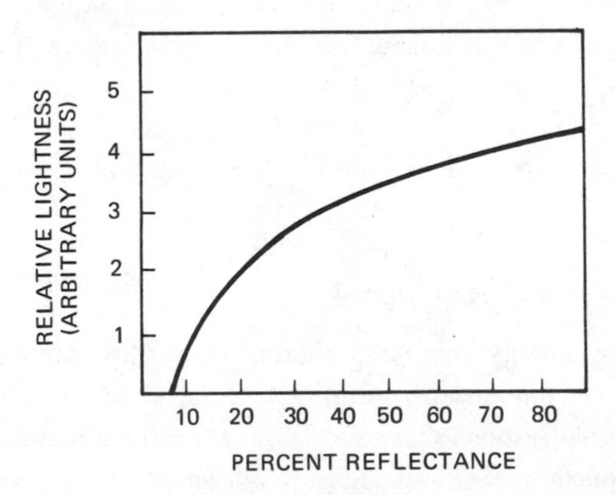


**Figure 4.1.** Logarithmic relationship between light level and perceived lightness.

Although the eye is a marvelously sensitive and discriminating sensor, it cannot make quantitative measurements. One can see very slight differences between colors, for instance, but only by making a side-by-side comparison. Under these conditions, 5 to 10 million different colors can be discriminated. When one looks at a colored object and then at a similar one a few minutes later, one cannot tell from memory whether the colors are different unless the difference is substantial. The eye can identify from memory only about 300 different colors. The same contrast between precision of discriminations by direct comparison and precision by memory applies to the geometric as well as to the color attributes.

When we first try to measure appearance, the extreme agility of the eye and brain sometimes seems to work against us. Because we can so readily perceive glossiness and color visually, we are generally unaware of what is actually involved in making the perception. Because the technology of appearance measurement is so far behind our abilities as human observers, it is essential that we take the time to look at objects visually and to understand what we are seeing and why objects look as they do before we thrust them into machines to measure them for color and gloss.

Valid psychological techniques for evaluating visual differences have been developed. There are accepted standardized viewing conditions that are used a great deal in industry, science, and technology. Before we discuss these viewing practices, we must examine some of the factors that can affect the observer's visual evaluation of object appearance.

## VISUAL EVALUATION OF APPEARANCE

To be useful in industry, visual evaluations of product appearance must be precise and reproducible. This requires that the method and physical arrangements of observation be the same for all evaluations. Because the human observer so readily compensates for differences in light level or spectral distribution (e.g., the color constancy phenomenon), we are often unaware of how significantly the results of visual evaluation will change with conditions of view. Standardized conditions of observation are absolutely necessary to obtain results that are comparable with previous inspections, whether by an instrument or by an inspector. This is admittedly a departure from the flexibility of viewing conditions possible in real life, where an object is viewed by an observer from the angle that best reveals the appearance characteristics of interest to him. The viewing conditions that affect appearance can be grouped into spectral conditions and geometric conditions.

## Spectral Conditions

With human observers, the spectral responses are built in, and we have to accept them as standardized. However, since the light coming from an object is due to the properties of the light source as well as the object, appearance will vary with a change in light source. For example, the appearance of redness in an object is due partly to selective absorption of wavelengths other than red by materials in the object. However, if red wavelengths are not present originally in the light that impinges upon the "red" object, the object will appear to be black (or nearly so). The effect of a light source on the color appearance of an object when compared to the object's color appearance under a standard source is called "color rendering."

## Geometric Conditions

Object appearance is directly related to the geometric conditions of observation, that is, the directions of illumination and view. For example, the glossy highlights of colored nonmetal objects (such as the plastic box in Figure 1.7) appear to be white. One does not look at the highlights to see color, but that is precisely where one looks to see gloss. Conversely, for observation of the color of the box a geometric arrangement is used where only the diffusely reflected light is seen, and not the specularly reflected light.

Figure 4.2. shows the simple geometries used for viewing gloss and color. To observe color, one should avoid seeing any specular reflection, since glossy highlights will mask the color. A diffuse angle of viewing, therefore, should be used, such as 0° when the light is incident at 45°. However, if one desires to assess glossiness, one should view the sample at the same, but opposite, angle at which the light is incident on the sample.

Since the specular-diffuse separation is somewhat arbitrary, we find that by viewing the diffuse reflection from different angles we often get a variation in

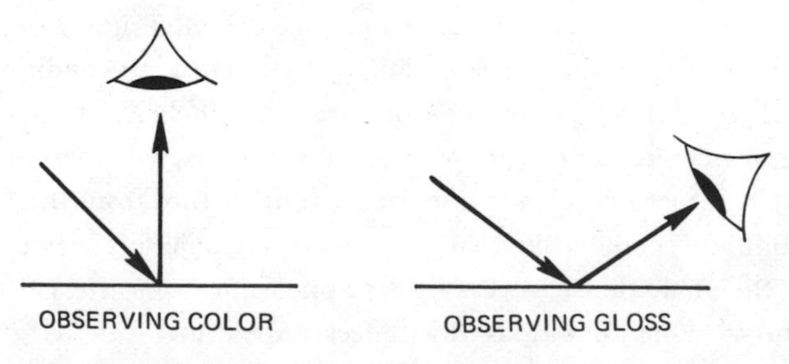


**Figure 4.2.** Geometries for viewing color and gloss of nonmetal surfaces.

color, and that by viewing specular reflection at different angles and with concentrated or diffused illumination we see differences in gloss. Thus for critical evaluations one needs to be more specific about the geometry of viewing than is provided by the single separation of diffuse and specular.

### Light Intensity

Visual discrimination is excellent over a wide range of illumination levels, but with too much or too little light visual sensitivity decreases. For discrimination between small color differences, it is recommended that at least 100 footcandles be used for light colors, and up to 200 or 300 footcandles illumination be used for darker colors. For color attributes, a light source of extended area is preferable to a concentrated source. A uniform extended light source for color inspection lessens the distracting effects of geometric attributes such as gloss, shape, and texture.

### Summary of Conditions to be Controlled

To make reliable visual evaluations of appearance and appearance difference we must control the following six things apart from the actual objects observed:

1. The spectral quality of the light source
2. The intensity of the light source
3. The angular size of the light source
4. The direction from which the light strikes the object (angle of incidence)
5. The direction from which the object is viewed (angle of viewing)
6. The background

## STANDARDIZED CONDITIONS FOR VISUAL EVALUATION OF COLOR

### Illuminants

For the detection of differences in color under diffuse illumination, both natural daylight and artificial simulated daylight are commonly used. A window facing north (to be free of direct sunshine) is the natural illuminant normally employed for visual color examination. Natural daylight, however, varies importantly in spectral quality with time of day, weather, direction of view, time of year, and geographical location. The recent trend in industrial testing, therefore, has been toward the use of artificial daylight because it can be standardized and is stable in quality.

In order to define the artificial light sources used in appearance evaluation, the Commission Internationale de l'Éclairage (CIE) established in 1931 "Standard" Illuminants, which have the characteristics of spectral closeness to natural light sources and reproducibility in the laboratory (CIE, 1931). Illuminant A defines light typical of that from an incandescent lamp, Illuminant B represents direct sunlight, and Illuminant C represents average daylight from the total sky. These are shown in Figure 4.3.

Within the last few years, there has developed an increased interest in the ultraviolet content of the illuminants used for visual evaluation. The major reason for this is an increase in the commerical use, chiefly in papers and textiles, of fluorescent materials that are activated by ultraviolet light. For the proper color examination of these materials it is necessary to control not only the visible but also the near-ultraviolet energy that falls on them.

In 1966 a fourth, D, series of illuminants was proposed (CIE, 1971). These not only more completely represent daylight than do Illuminants B and C, but also are defined for complete yellow to blue series of color temperatures. Figure 4.4 shows spectral distribution of typical daylight for correlated color temperatures. Note that the curves start at 300 nm, not 400 nm as in Figure 4.3. The D illuminants are usually identified by the first two digits of the color temperature, so that the 6500 K curve is known as D65.

Until the present time Illuminant C has enjoyed the greatest use in the

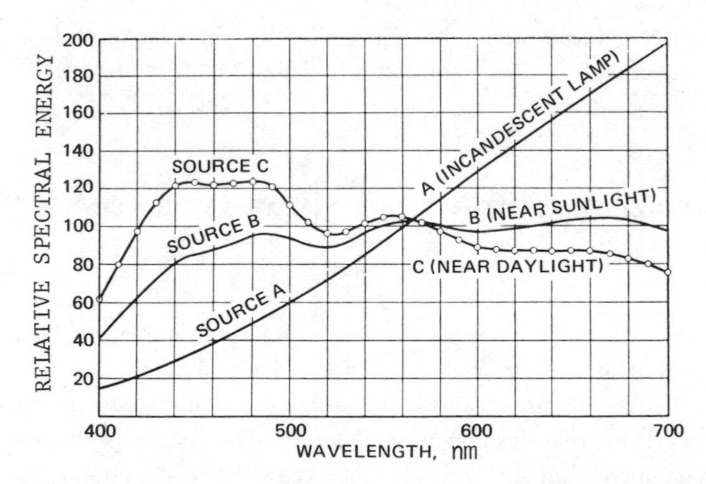


**Figure 4.3.** Spectral distribution of CIE sources A, B, and C. Illuminant A represents light from an incandescent lamp; Illuminant B represents direct sunlight. Illuminant C represents average daylight from the total sky. From *Color in Business, Science, and Industry,* by D. B. Judd and G. Wyszecki, John Wiley & Sons, 1963. Used with permission of *J. Opt. Soc. Am.*

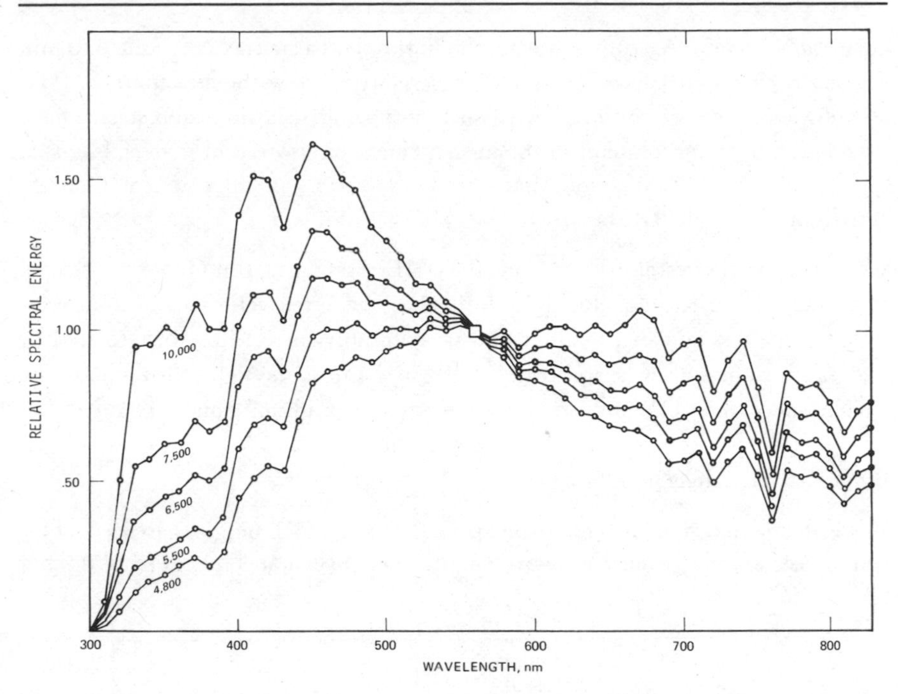


**Figure 4.4.** Family of reconstituted spectral energy distribution curves of typical daylight. From Judd, MacAdam, and Wyszecki, *J. Opt. Soc. Am.* **54,** 1964.

laboratory. Most color specifications are in terms of appearance under illuminant C. However, an increasing trend toward the D series of illuminants, particularly D65, is likely in the future, especially in the evaluation of fluorescing or "brightened" materials, where the spectral energy distributions of the light source in the 300–400 nm ultraviolet range contributes to color appearance.

The assumption is often made that a tungsten incandescent source, because it has less ultraviolet energy than has Illuminant D, cannot be used satisfactorily for instrumental color measurement of fluorescent materials. It has been found in actual practice that filter-modified incandescent sources are often quite satisfactory. The reason for this is that in the colorimetry of fluorescent whitening agents, the most important factor for the light source is the ratio of energy in the range of maximum absorption of the brighteners (350 to 400 nm) to energy in the range of maximum emittance (400 to 450 nm). Figure 4.5 illustrates this point by comparing the curve for Illuminant D65 with the curve of an incandescent quartz-iodine lamp whose curve has been normalized to the D65 curve at 450 nm. As can be seen, the ratios between the 380 and 450 nm regions of the two curves correlate reasonably well with one another. Tristimulus colorimeters equipped with such an incandescent tungsten source have been found to be quite useful in the measurement of fluorescent materials.

## Artificial Daylight Illuminants

Even before the establishment of the CIE Standard Illuminants, selected colorants were used in a blue glass filter that converted yellowish incandescent light to blue daylight in a manner that was not only correct in approximate hue and saturation, but was spectrally close to natural daylight throughout the visible spectrum. The industrial use of artificial illuminants for the visual examination of colors has increased greatly since 1931. Two artificial-illuminant units in general use are:

1. Filtered tungsten in which the special "daylight" blue glass is used. This blue glass absorbs long red wavelengths while regular blue glasses do not

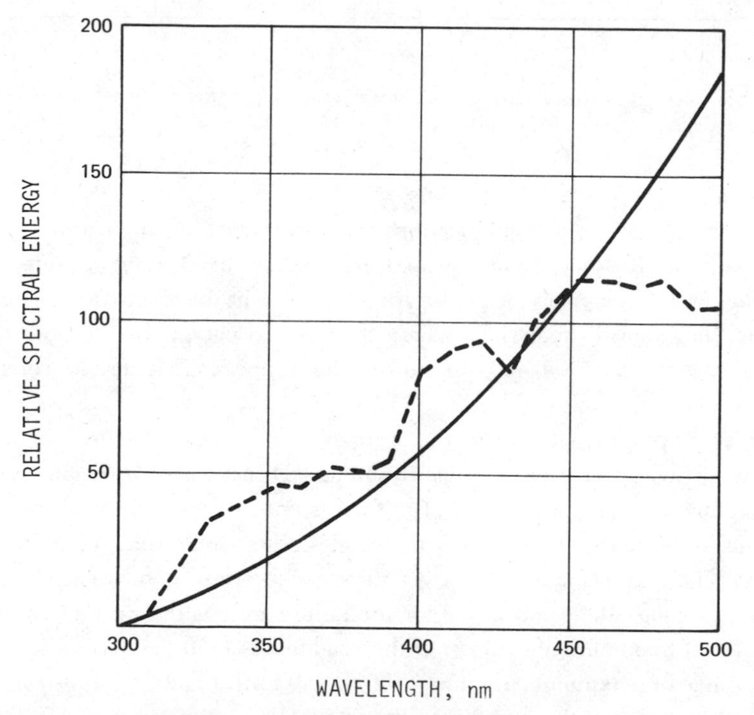


**Figure 4.5.** Comparison of curves normalized at 450 nm for tungsten lamp of approximately 3000 K (solid line) with Illuminant D65 (dotted line).

(Gage, 1933). These units normally include bare incandescent lamps to provide a second illuminant used for comparing color under daylight with color under lamplight.

2. Special fluorescent sources designed to simulate daylight spectrally. These units may also have yellowish incandescent lamps for supplementary and/or separate sample illumination. These make possible the evaluation of color matches under two light sources to assure that "matched" samples will appear the same under different lighting conditions. Further, the incandescent lamps add supplementary long-wave energy where the fluorescent tubes are deficient.

There is a current trend toward the use of standard cool-white fluorescent light as a third illuminant for examining products of commerce. It is spectrally quite different from both natural daylight and incandescent light and is used in industrial appearance testing because many products are first viewed by the consumer under a cool-white fluorescent illuminant. A shopper in a department store or supermarket frequently makes his choices under this illuminant.

Another daylight-simulating source now being tried for visual color examination is the high-pressure xenon arc. The bare arc is actually a little too purple and requires a pale green filter to provide a closer rendition of natural daylight. It has too much ultraviolet so that some has to be removed. The xenon arc has promise as a light source for visual grading of color, but its use is not yet well established. Chapter 12 describes a number of actual instrument light sources.

## Lighting Arrangement and Placement of Specimens

Figure 4.6 diagrams two of the lighting arrangements used to evaluate color and color difference. With the standard artificial daylighting booth, illumination on the specimen is about perpendicular to the specimen plane (0° illumination) and viewing is done at an angle of 45° (45° view). This arrangement is reciprocal to the standard condition for examination of colors by natural daylight (45° illumination and 0° viewing), but it gives the same results. The background against which the specimens are placed for view should be without distractions and similar to, but grayer than, the specimens.

All specimens to be compared should be placed side by side on the same plane. The closer together they are, the more the dividing line between the specimens tends to disappear and the more sensitive the eye becomes to small visual differences. When objects are examined in pairs, the positions of the two specimens should be interchanged and examined a second time to avoid bias in viewing them in only one order. When the sizes of the specimens being compared differ, or when a specimen has different colors in different areas, com-

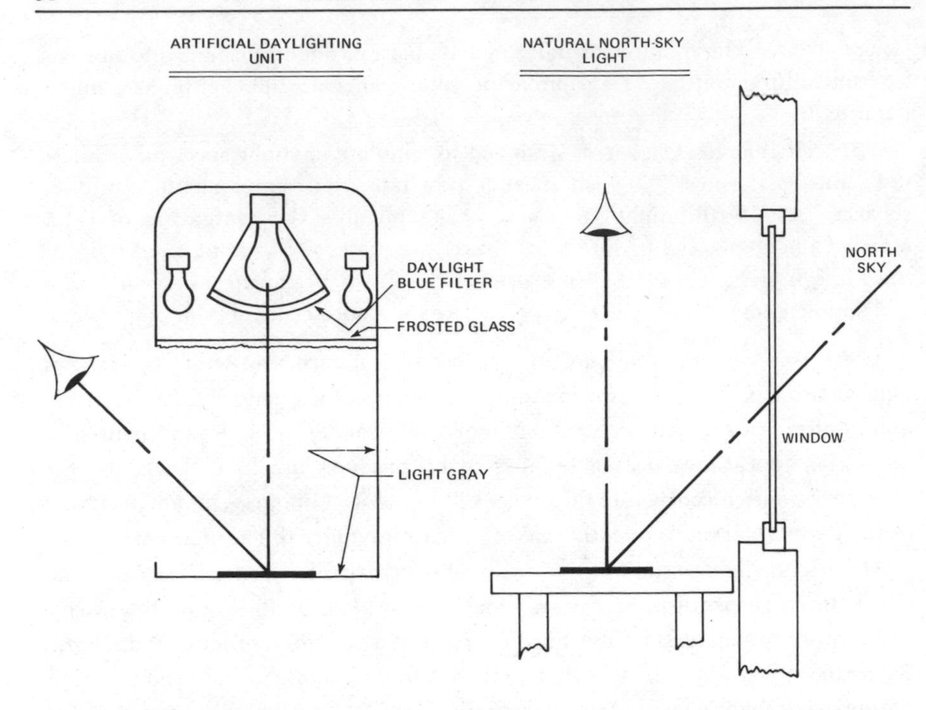


**Figure 4.6.** Observing conditions used for visual color examination. A geometry of 45° for light incident on the sample, and 0° viewing (or the reciprocal arrangement) is used for visual color evaluations.

parisons should be made between areas of like size and pattern. This is achieved by the use of masks, usually of gray paper with rectangular holes cut out. The specimens being compared are covered by the masks, allowing only the equal areas visible through the holes to be compared. Masks should always be used when one is seeking the small color chip of a collection that is closest to the color of a large sample. Separation of the specimen areas being compared by masks or other means tends to focus attention on chromatic differences between samples. By contrast, the bringing of specimen areas into direct contact with each other, so that the line between them is hard to see, enhances any lightness differences that may exist.

## STANDARDIZED CONDITIONS FOR VISUAL EVALUATION OF GLOSS

In examining an object for gloss, luster, or transparency, the spectral quality of the light is not as important as its geometric distribution. Standard conditions for the visual examination of geometric attributes are not generally established. However, the nature of the gloss phenomena to be observed will provide guidance in establishing the illuminating situation required for critical evaluations. Whereas illumination for color examination should be diffuse, that for gloss studies should be nondiffuse with strong, sharp patterns. A patterned source for appraisal of geometric attributes is shown in Figure 4.7.

With such a unit, low-gloss specimens are compared by examining the reflected light from lamp tubes at high (near-grazing) angles. On the other hand, the best comparison of high-gloss specimens is made by studying, in a near-perpendicular direction, the reflected image of the wire screen in front of the lamp. While gloss differences are seen at the specular angle of reflection, haze differences are seen at angles near to the specular direction, whether by reflection or transmission. Figure 4.8 shows techniques for viewing gloss.

A gloss difference between specimens may influence the judgment of color differences, even though gloss differences may be minimized by the diffuse surrounding which is used. The gloss can produce veiling reflections that alter the apparent contrast. Texture differences may also influence the judgment of color differences. The distracting effects of differences in specimen size, gloss or tex-

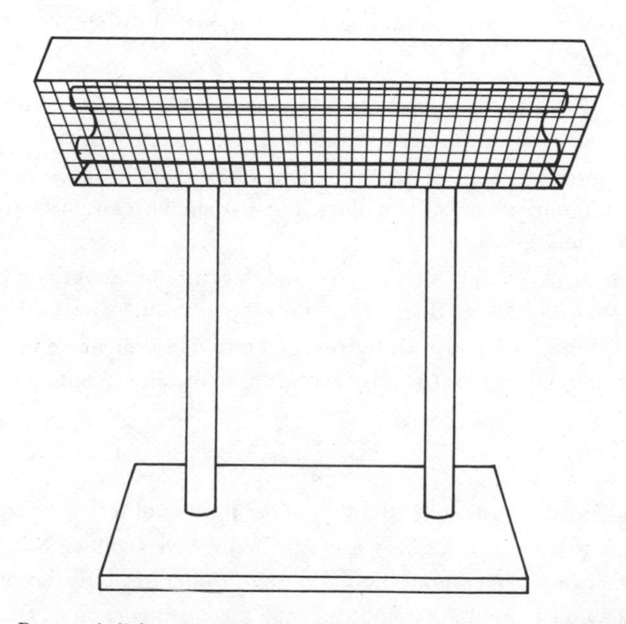

**Figure 4.7.** Patterned light source for examination of geometric attributes. An ordinary fluorescent lamp, with black velvet in back of the tubes and a wire screen in front of them, makes a good light source by which to evaluate gloss.

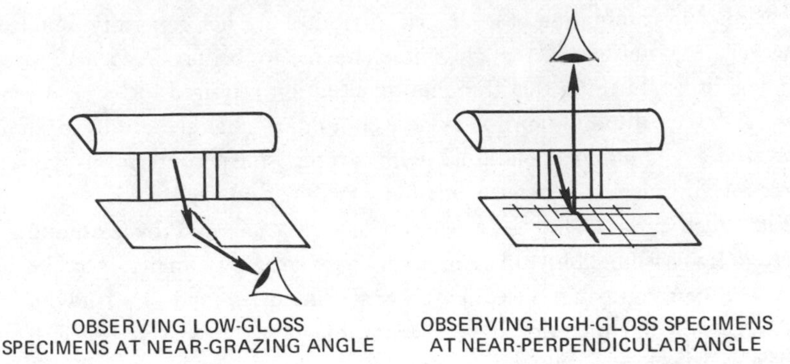


**Figure 4.8.** Geometries used for viewing different kinds of gloss. Low gloss is seen most easily at angles close to grazing; high-gloss specimens are best evaluated at angles close to perpendicular.

ture can be minimized by the use of masks, which helps to equate the areas being compared.

## GENERAL RULES FOR VISUAL EXAMINATION

In general, for achieving maximum sensitivity to visual differences, these rules can be suggested:

1. Place the specimens in immediate juxtaposition. As the visual dividing line separating objects grows thinner, visual differences become easier to see.
2. Keep the intensity of illumination high. Only at light levels approaching those of an outdoor overcast sky does the eye make comparisons with its maximum precision.
3. Have the background similar to and, if anything, grayer than the specimens so that it offers no distracting contrast with the visual task.
4. Have the illumination spectrally representative of that normally employed in critical commercial studies (usually actual or artificial daylight).

## SUMMARY

The final stage in the observing situation is the human observer, whose complicated reactions to the light he sees are still being investigated. Many theories concerning the eye-brain combination and how it operates have been proposed. Enough is known about the capabilities and the insufficiencies of the eye and the brain so that the most effective conditions for making visual evaluations can be employed. Such evaluations are primarily comparisons of similar specimens rather than individual identifications. Standardized conditions must be used in order that such evaluations be reproducible. The conditions that must be standardized are the light source (its intensity, angular size, and spectral distribution), and the observation geometry (the angle of incidence of the light and the angle of view).

More important than standardization as such, however, is the requirement to use the conditions appropriate and most effective for the attribute being examined. Color evaluations should use a diffuse source and diffuse geometry, with the light source controlled spectrally. Gloss evaluations require a well-defined (perhaps patterned) source, and judgments will vary significantly depending on whether the specimens are viewed at near grazing angles or at a more perpendicular angle.

# CHAPTER FIVE

# PSYCHOPHYSICAL SCALES FOR APPEARANCE MEASUREMENT

An analysis of the light reflected or transmitted by an object has enabled us to see how the eye and brain qualitatively convert the light to a perception of appearance. We began our analysis in the realm of physics as we investigated light and some of the things that happen when light interacts with an object. We moved into the psychological area with the introduction of an observer who perceived the light reflected or transmitted by the object and made certain judgments about the appearance of the object.

We now come to the crux of the matter of appearance measurement. How is it possible to put numbers on qualitites that exist only in the brain of an observer? "Beauty is in the eye of the beholder"—and also all color, gloss, haziness, and the myriad factors that contribute to ugliness and beauty are in the eye alone.

The attributes of an object's appearance that we wish to quantify seem to be a relatively unchanging aspect of that object. Therefore, it would appear to be feasible to put numbers on the attributes for the purpose of measuring them. Only investigation of the way in which observers respond to light stimuli is needed. Such investigations have revealed that the responses of observers with normal vision are also uniform and unchanging enough so that a "Standard Observer" can be quantified. On this basis appearance-measurement scales have been created and used successfully for the measurement of appearance attributes. The next section of the book is concerned with how we describe these attributes with numbers in order to measure them.

## PSYCHOPHYSICS AND APPEARANCE SCALES

Underlying the concept of all appearance-measurement scales is the science of psychophysics. Psychophysics is the branch of psychology that is concerned with human responses to physical stimuli, particularly the perception of physical magnitudes. As such, it is not concerned primarily with the mechanics of how the responses are generated, although theories in this area are often useful in helping to relate the physical and psychological events being studied.

Psychophysics has played a primary role in appearance measurement. Without the benefit of an adequate model of the actual visual process, psychophysical techniques have been used to generate scales for the measurement of appearance that closely duplicate the ultimate human response. These "appearance scales" relate the physical measurements of spectral and spatial light distributions described in Chapter 3 to the psychological evaluations of color and gloss. Our preceding analysis of how objects look indicated that the attributes an observer sees fall into two categories—those associated with color, and those, such as gloss, associated with the geometric or spatial distribution of the light by the object. The development of measurement scales follows this same division, and we have color scales and geometric scales.

1. Color scales grew out of experiments that determined the specific way human observers respond to spectral distributions of light. Chapters 8 and 9 show two distinct kinds of experiments that were employed to generate the color scales now accepted in science and industry. These experiments were used because of the two major requirements for a useful color scale.

First, the scale should identify colors in a way that assures that visually identical colors have the same numbers. "Color matching" experiments were used to supply these data.

Second, the scale units should relate meaningfully to the visual attributes of the color and to the magnitude and character of perceived color difference between any two colors. "Color spacing" experiments were used to generate these data.

2. Gloss scales and scales for other geometric attributes relate to our geometric perceptions, which do not have the tridimensional configurations of our color perception. The psychophysical experiments behind gloss and geometric scales have been based on direct comparisons of visual rankings of geometric attributes with reflectance and transmittance factors to find which geometry produces the best appearance differentiation for each type of specimen. Good correlation has resulted, and gloss scales are in wide use.

These two types of scales, based as they are on the fact that an observer sees the color of nonmetallic objects by diffuse reflection and sees gloss and other geometric attributes by specular reflection, understandably have different references as the basis for their measurements. When diffused light is measured (as in surface color measurements), it is measured in terms of the reflection by the perfect diffuse white under the same geometric conditions. A comparison is made of the diffuse reflectance or transmittance factor to that of the perfect diffuser. The term "reflectance factor" is now used to identify these comparisons with perfect white (CIE, 1970), and this scale is called a "factor scale." However, when a specular beam from a glossy material is measured (as in gloss measurements) it is the fractional relation of the specular beam to the incident light that is important, not its relation to ideal white. Therefore it is called a "fractional scale." (See Chapter 15.)

Because surface color is associated with diffuse distribution of light, surface color values are based on factor scales and are related to the perfectly diffusing or transmitting reference standard. Gloss, on the other hand, is measured on a fractional scale with a mirror as a standard. A good rule of thumb is that if, when visually evaluating the attribute of importance, the observer sees the ob-

**TABLE 5.1**
MAJOR SPECTRAL AND GEOMETRIC ATTRIBUTES OF OBJECT APPEARANCE

| Category | Spectral (Color) Attributes | Geometric (Luminous Intensity) Attributes |
|---|---|---|
| Physical analyses | Spectrophotometric Curves 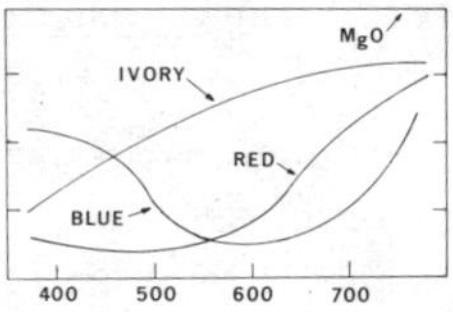 | Goniophotometric Curves 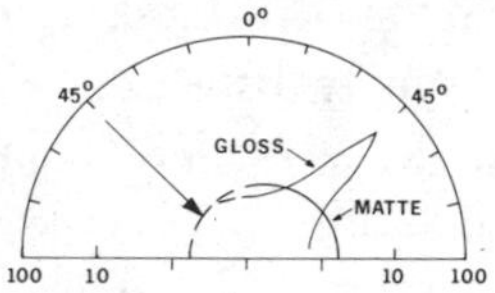 |
| Psychological (visual) | Color:<br>Hue (red, blue, etc.)<br>Saturation (chroma)<br>Lightness, or one of the other geometric attributes | Luminous Intensity:<br>Lightness (by reflection)<br>Glossiness<br>Lightness by diffuse transmission<br>Clarity by specular transmission |
| Basis of psychophysical measurement | Ability of human observer to match color of any object by adding proper amounts of three colored primary lights | Comparisons with ideal objects that either diffuse light perfectly or have perfect specular projection |
| Psychophysical quantity | Chromaticity (hue and saturation together) | Diffuse reflectance factor, or lightness<br>Gloss<br>Diffuse transmittance factor<br>Specular transmittance factor |

ject as a diffuser of light, he should use the factor scales for instrumental measurement. If the object appears to be an attenuator (reflector or filter) of a continuing beam of light, one of the fractional scales would be appropriate.

Table 5.1 summarizes the relationships between the spectral and the geometric appearance attributes of objects, and indicates the bases for the psychophysical scales that were developed for their measurement. Also shown are some of the psychophysical quantities measured by these scales.

The scales for the geometric attributes, such as gloss, are the subject of the next chapter. The scales for the measurement of color are covered in the remainder of Part Two.

## SUMMARY

In Part One of this book we moved from the physics of the makeup and interaction of light and object to psychology as we analyzed how the human observer perceives light. In Part Two, we operate in the area of psychophysics, as we investigate physical measurements that have a psychological basis. Psychophysical scales for measurement of appearance are based on visual evaluations of stimuli for which the physical properties are known.

# PART TWO

# THE DEVELOPMENT OF NUMERICAL SPECIFICATIONS FOR APPEARANCE ATTRIBUTES

# CHAPTER SIX

# SCALES FOR GLOSS AND OTHER GEOMETRIC ATTRIBUTES

The geometric attributes of objects such as gloss, haze, luster, and transparency are perceived as being distinct from color. Our perception of geometric attributes does not have the tridimensional limitation of our color perception. Because it is not possible to produce an equivalent gloss stimulus synthetically (as color stimuli are matched by mixtures of three standard lights), no Standard Observer for gloss has ever been developed. Instead, to formulate each geometric scale, a number of specimens are ranked according to their visual attributes. The geometric distributions of light are then measured under different conditions. When the correlations between visual ranking and the measurements are optimized, a scale is selected.

With this in mind, let us turn to a consideration of the types of visual gloss encountered, and then to evolution of the instrumental scales designed to measure them.

## GLOSS

Gloss is the attribute of surfaces that causes them to have a shiny or lustrous appearance. It is generally associated with specular reflection by the surfaces of objects. However, specular reflection can vary from one surface to another as follows:

1. The fraction of light reflected in the specular direction
2. The manner and extent to which light is spread to either side of this specular direction
3. The change of fractional reflectance as specular angle changes

The author first became aware of these variations of specular reflection after he designed, in 1934, a glossmeter to compare the capacities of materials to reflect light specularly at 45° from the perpendicular. In 1935 he was asked to evaluate a series of white porcelain enameled panels on his newly designed (and heretofore quite successful) glossmeter (Hunter, 1934). The panels exhibited important, readily discernible gloss differences. However, values of relative specular reflectance as measured on the glossmeter did not correlate with the visual rankings. Two of these panels are shown in Figure 6.1. The primary distinguishing feature between them is the sharpness, or distinctness, of the reflected images. Analysis of the problem revealed that the image-reflecting capacity of a surface was not what the specular glossmeter was measuring. The viewing-beam field angle on the glossmeter was about 8°, whereas it would have had to have been about 0.5° to successfully differentiate between a distinct image and a blurred one.

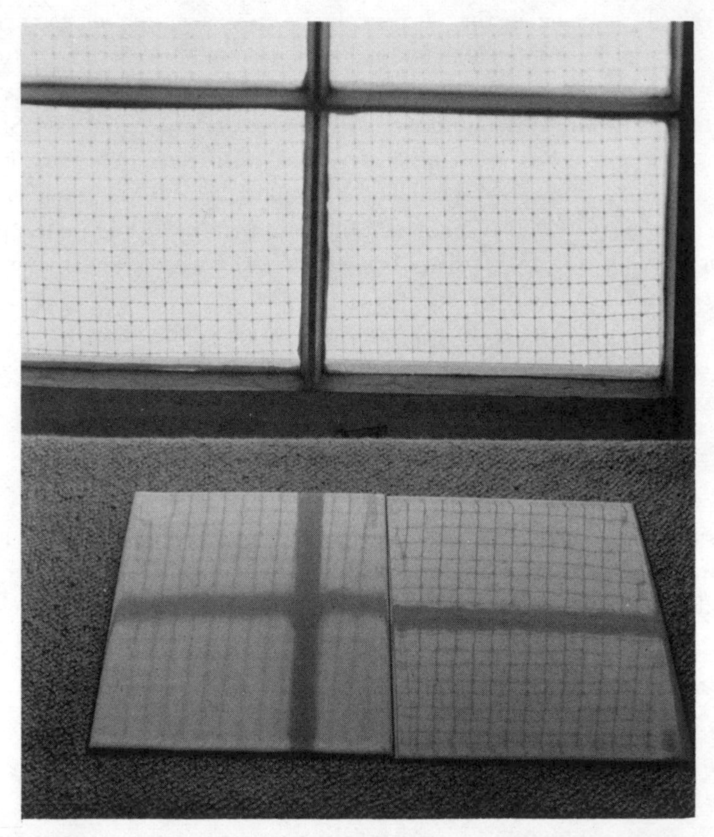

**Figure 6.1.** Two white porcelain enameled panels exhibiting a difference in distinctness-of-image gloss. These panels demonstrated to the author that "gloss" could be more than the simple specular gloss that his 1934 glossmeter was designed to measure.

**Figure 6.2.** Difference in shininess, or specular gloss, of two bowls.

During the remainder of the decade the author studied many different materials and their glossiness rankings. During the course of these investigations it became apparent that there were at least six different visual criteria by which glossiness rankings are made. Table 6.1 summarizes the result of that study (Hunter, 1937).

For those unfamiliar with evaluating gloss visually, some practice is necessary in order to be able to separate these attributes. Examples of different kinds of gloss are shown in the following group of figures, in an attempt to make these gloss attributes visually meaningful.

Figure 6.2 is a picture of two bowls showing a difference in shininess, or specular gloss, which is readily seen in the curved surfaces of the bowls. Figure 6.3 is of two chromium-plated bumper bars showing a difference in distinctness-of-reflected-image gloss. At the present time there are no optical glossmeters that will measure curved surfaces for differences of the types shown. Commercially, the metal-plating differences are quite important for they correlate closely with the amount of hand labor expended in buffing the specimens prior to plating them.

**TABLE 6.1**
SIX TYPES OF GLOSS

| Types of Gloss | Visual Evaluation | Reflectance Function | Types of Surfaces |
|---|---|---|---|
| Specular gloss | Shininess, brilliance of highlights | 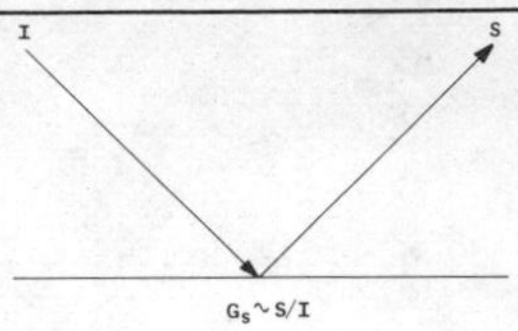 I, S; $G_s \sim S/I$ | Medium-gloss surfaces of book paper, paint, plastics, etc. |
| Sheen | Shininess at grazing angles | 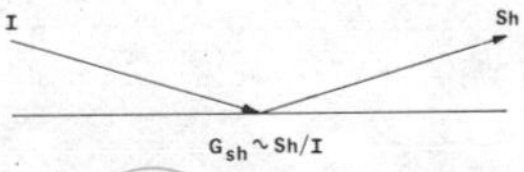 I, Sh; $G_{sh} \sim Sh/I$ | Low-gloss surfaces of paint, paper, etc. |
| Contrast gloss or luster | Contrast between specularly reflecting areas and other areas | 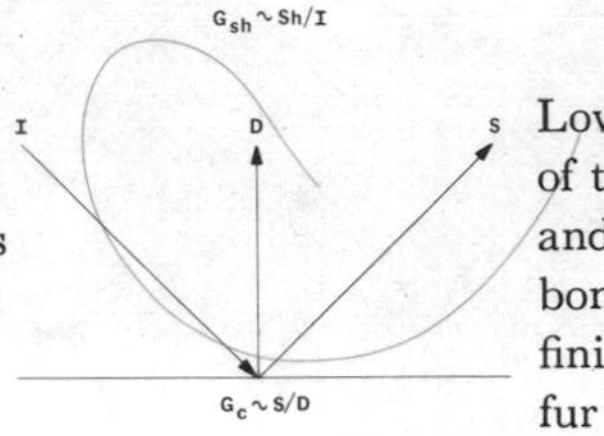 I, D, S; $G_c \sim S/D$ | Low-gloss surfaces of textile fiber, yarn and cloth, newsprint, bond paper, diffuse-finish metals, hair, fur etc. |
| Absence-of-bloom gloss | Absence of haze, or milky appearance, adjacent to reflected highlights | 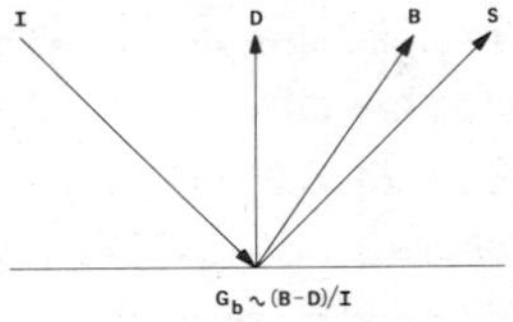 I, D, B, S; $G_b \sim (B-D)/I$ | High- and semigloss surfaces in which reflected highlights may be seen |
| Distinctness-of-image gloss | Distinctness and sharpness of mirror images | 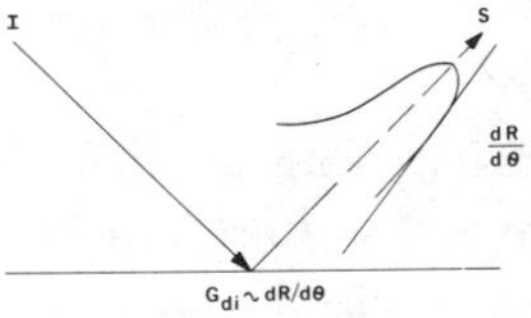 I, S, $\frac{dR}{d\theta}$; $G_{di} \sim dR/d\theta$ | High-gloss surfaces of all types in which mirror images may be seen |
| Surface-uniformity gloss | Surface uniformity, freedom from visible nonuniformities such as texture | Not a function of reflectance | Medium-to-high-gloss surfaces of all types |

**Figure 6.3.** Difference in distinctness-of-reflected-image gloss of two bumper bars.

Two white rayon taffetas showing a difference in contrast gloss, or luster, as it is usually called in the textile field, are pictured in Figure 6.4. Note that in this case the visual evidence of luster is the contrast between the highlight-reflecting areas and the adjacent darker areas. It is the apparent "darkness" of the adjacent area that we see as luster.

Figure 6.5 shows two plastic wall tiles exhibiting a difference in reflection haze. Absence of haze (bloom) is seen as a form of gloss. Visually, haze and bloom both cause a milky appearance adjacent to the specular highlight. In the paint and metals industries, bloom has a slightly different meaning from haze, but both are evaluated visually in the same manner. In these industries this appearance is called bloom if it can be wiped off, and haze if it cannot be wiped off.

Sheen is the low gloss of matte surfaces that must be viewed at an almost grazing angle such as 85° to be seen. A difference in the sheen of two low-gloss painted metal panels is shown in Figure 6.6.

Figure 6.7 shows two panels that display a difference in directionality as they each reflect a light bulb. Directionality is usually the result of buffing or machine marks, extrusion ridges, and the like, and is evidenced by a spread of specular reflection in the horizontal direction, but not perpendicularly.

Figure 6.8 shows two aluminum panels contrasting a textured surface on the left with a smooth surface on the right. Texture is seen as surface nonuniformity and detracts from glossy appearance. Surface uniformity, or depth of finish, is the freedom of a surface from texture and other markings with which the eye can identify the position of the surface.

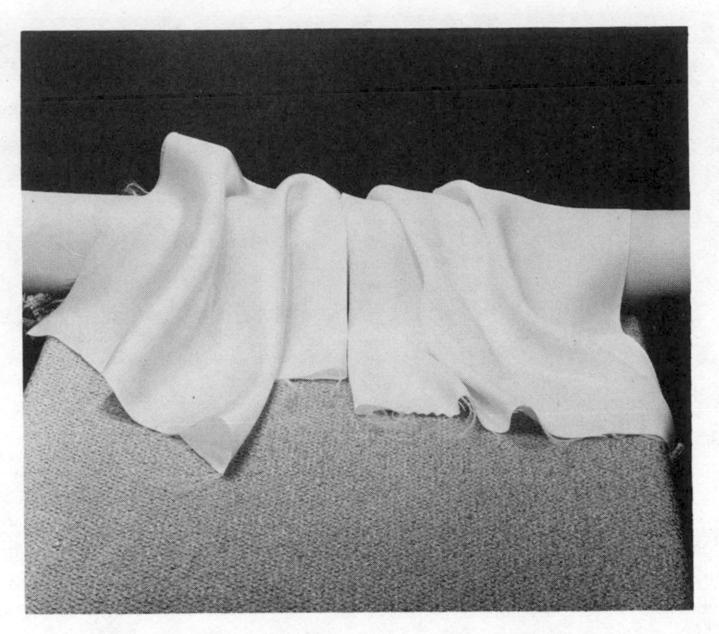

**Figure 6.4.** Difference in contrast gloss, or luster, of two fabrics.

**Figure 6.5.** Difference in reflection haze of two plastic wall tiles.

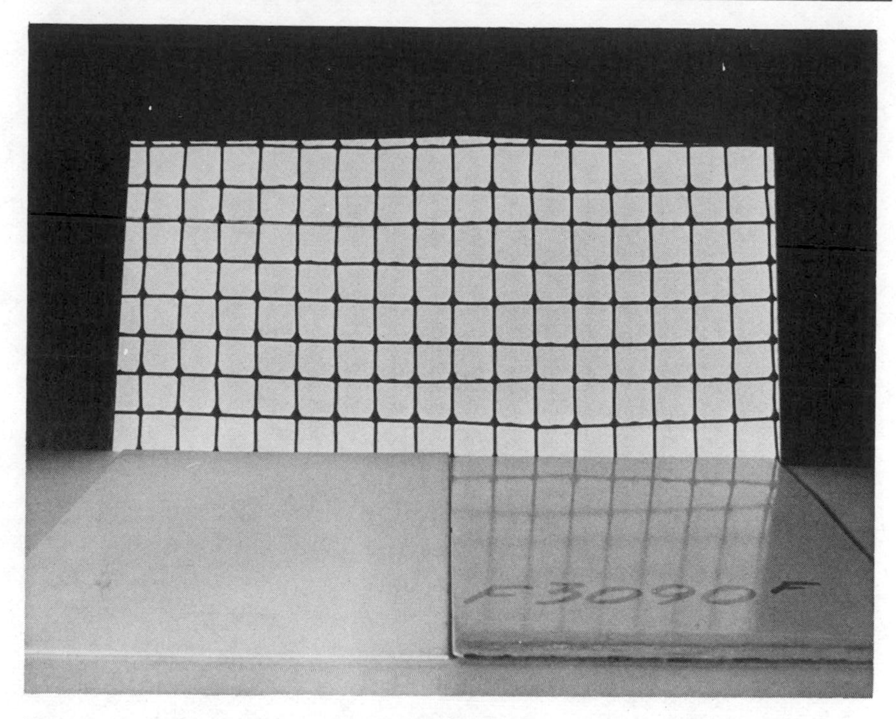


**Figure 6.6.** Difference in sheen of two painted metal panels.

## ESTABLISHED METHODS TO MEASURE GLOSS

There is a variety of established methods of gloss measurement because it is necessary to measure different aspects of reflection in order to duplicate as far as possible the different visual gloss grading procedures. Even where the same aspect of reflection (for example, specular gloss) may be used to grade two types of materials, the best geometric conditions for differentiating specimens of one of the types of materials may differ from those that are best for the other.

The first established method for gloss measurement (Ingersoll, 1914) was based on the principle of polarization of light that occurs at some angles in specular reflection as opposed to the lack of polarization in diffusely reflected light. The Ingersoll glarimeter had a specular geometry with incident and viewing angles at 57½°. Gloss was evaluated by a contrast method that, through the use of a polarizing element, "subtracted" the specular component from total reflection.

About 1930 Pfund pointed out that specular shininess is the basic (objective) evidence of gloss but that actual surface glossy appearance (subjective) relates to

**Figure 6.7.** Difference in directionality of two panels reflecting a light bulb.

**Figure 6.8.** Difference in texture of two aluminum panels.

the contrast between specular shininess and the diffuse lightness of the surrounding surface area (now called "contrast gloss" and "luster"). If black and white surfaces of the same shininess are compared visually, the black will appear to be glossier because there is greater contrast of a specular highlight with black surroundings than with white surroundings. This proposal by Pfund was the first suggestion that more than one method of measurement is needed to analyze gloss (Pfund, 1930).

Hunter's investigations of gloss in the early 1930s employed a specular geometry at 45°, as did the first photoelectric gloss methods. But visual evaluations of a large number of samples (primarily paint) eventually demonstrated that the 60° angle provided the best overall estimates (Hunter and Judd, 1939). The 60° method suggested in 1939 was subsequently adopted by the American Society for Testing and Materials in ASTM Method D523. It enjoys greatest use on paints, plastics, waxes, and floor coverings, and is more widely used than any other gloss test procedure. The old 45° method is now used primarily on glazed ceramics and polyethylene and other plastic films.

Incorporated into ASTM Method D523 in 1951 as alternate procedures were a 20° specular test for evaluating high-gloss finishes, developed earlier at the duPont Company (Horning and Morse, 1947) and an 85° method for evaluating low-gloss, matte surfaces. The latter sheen method was developed in 1938 by J. W. Ayers of the C. K. Williams Company (Ayers, 1938; and Hunter, 1952), and was used subsequently to test matte camouflage finishes used by the Ordinance Department. It is now used for measuring interior flat wall paints, and low-gloss exterior painted aluminum siding.

A two-parameter method for 60° specular gloss designed to distinguish image-forming gloss surfaces from non-image-forming gloss surfaces of nonmetals was adopted as ASTM Method D1471 in 1969 (Nimeroff, 1957). It was designed to cover those cases in which measurements made by the ASTM Method D523 60° gloss method did not correlate with gloss appearance. Method D1471 specifies a second, 2°, receiver aperture to be used in conjunction with the aperture of Method D523. An evaluation is then made of the amounts of light received by each of the apertures.

In 1937 the paper industry adopted a 75° specular-gloss method because the angle gave the best separation of coated book papers (Institute of Paper Chemistry, 1937; Hunter, TAPPI, 1958). This method was adopted in 1951 by the Technical Association of Pulp and Paper Industries as TAPPI Method T480. For the evaluation of waxed, lacquered, and cast-coated high-gloss papers at 20°, TAPPI Method T653 was adopted in 1958 with window angular sizes different from those of the 20° ASTM paint method (Hunter and Lofland, 1956).

Notice that with all of these specular-gloss methods, the higher-angle methods are used to distinguish between lower-gloss specimens, with 20° working best on high-gloss materials, and 75° or 85° working best on low-gloss materials.

Figure 6.9 illustrates roughly the differences between values of gloss measured on four of the specular-gloss scales just discussed. The vertical axis shows numerical values of gloss on each of the four scales. The horizontal axis is intended to show a visually uniform spacing of gloss from matte to high

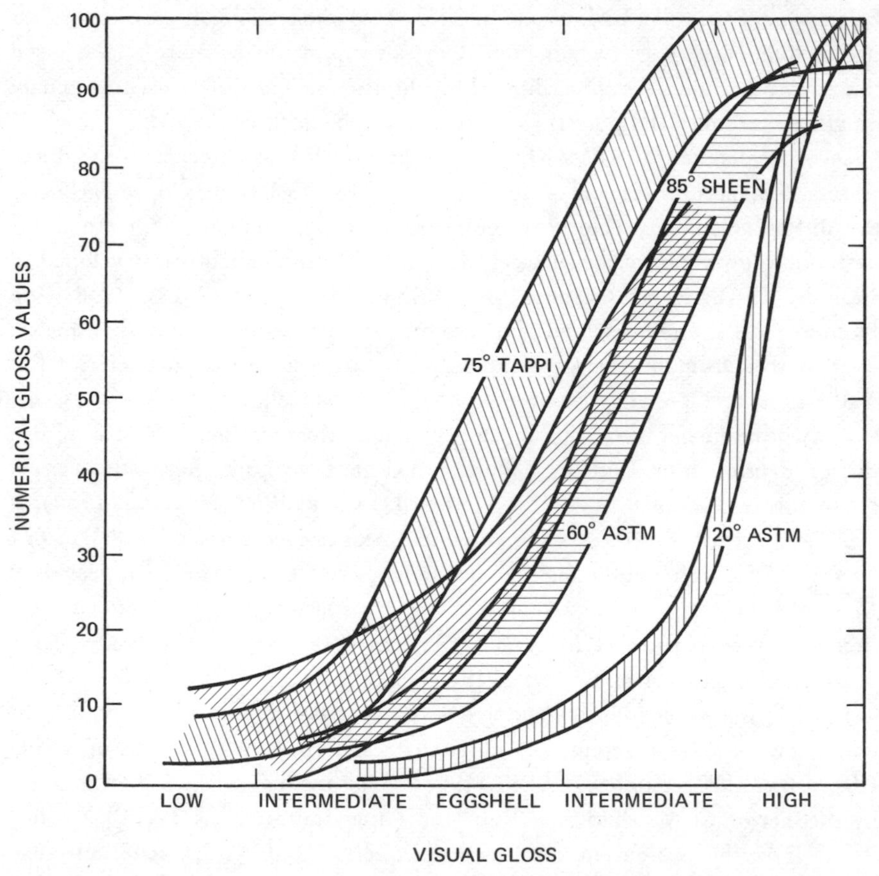


**Figure 6.9.** Difference between values of four numerical gloss scales as plotted against visual ratings of gloss.

gloss. Note that through much of the gloss range the 75° test method gives higher values than the 85° sheen method. This is contrary to what we would expect and is due to the fact that the receiver field angle of the 85° method is much smaller than that of the 75° gloss test, allowing less light to enter and giving numerically smaller values. (See Table A.1.)

As we have seen, each specific method of gloss measurement has been designed to provide correlation between instrumental values and visual appraisals of glossiness of a specific class of surfaces. Although some scales, such as the ASTM 60° specular-gloss scale developed in 1939, have been used successfully for gloss measurements of wide varieties of materials, it is frequently true that a unique product requires its own unique technique for gloss measurement. Some materials require more than one gloss scale to evaluate properly the appearance attributes of surfaces that are of practical importance. Refrigerator and appliance enamels are measured for both haze and distinctness-of-reflected-image gloss to assure that they are suitable in quality and have been applied properly.

Metals owe their appearance chiefly to the manner in which light is reflected in and near the directions of specular reflection. Since specular reflection is dominant, it is not surprising that gloss appearance attributes and methods of analysis are useful for metallic appearance evaluation. The same distinctions that have been made concerning the glossy appearance of nonmetals can also be made of metals (Hunter, 1970). In 1959 Barkman suggested a method for the evaluation of the "brightness" of aluminum surfaces based on the separation of diffuse and specular reflectance (Barkman, 1959, 1970). The Barkman method is published as ASTM Method E429.

A proposal for more complete gloss evaluation of metallic appearance involves measurement of distinctness-of-image, two kinds of haze, and diffuseness. With a narrow incident beam at 30°, evaluations are made at four near-specular angles: 0.3° from specular for distinctness of image, 2° from specular for bloom, 5° from specular for haze, and 15° from specular for diffuseness. Figure 6.10 is a diagram showing the angles of these beam axes suggested for metal appearance evaluation (Hunter, 1970; Christie, 1970; ASTM E430).

Table A.1 lists the gloss measurement procedures presently in use. For each of these scales, the table gives a diffuse correction factor for white. This is supplied to answer the question, "If the vinyl fabric I'm measuring for gloss is white and not black, what effect does the increase of diffuse reflectance due to its whiteness have on its measured gloss value?" The adjustment to the gloss value is made by multiplying the diffuse-reflectance-factor value of the vinyl fabric by the diffuse correction factor shown in the table, for the appropriate

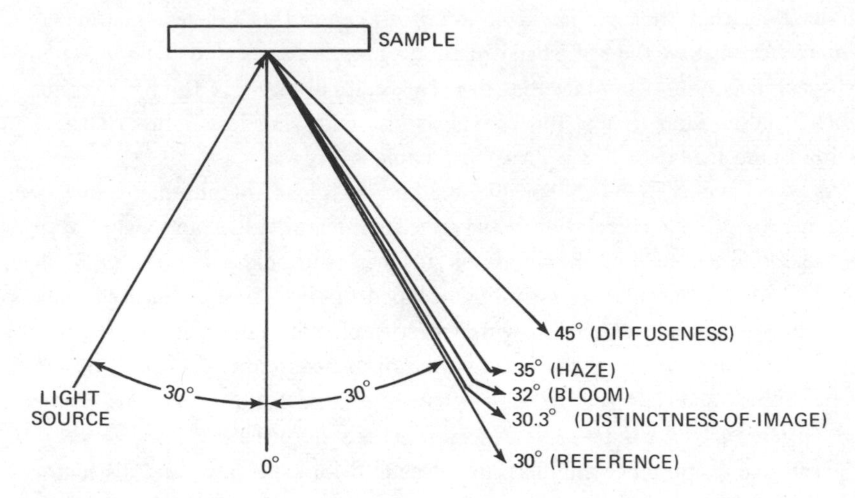


**Figure 6.10.** Four angles used in the appearance measurement of metals. For metallic surfaces, distinctness-of-image, haze or bloom are often of more critical interest than is specular gloss alone.

gloss scale, and then adding the result to the measured gloss value of the vinyl fabric.

## LIMITATIONS OF INSTRUMENTAL METHODS FOR MEASURING GLOSS

In thinking about gloss scales, we must keep in mind the relatively limited abilities of instruments when compared to the human eye. For example, the attribute of distinctness of reflected image seen in Figure 6.1 is difficult to measure photometrically because the human eye has a higher resolving power than do most reflectance measuring instruments. The normal human eye can see two lines as being separate when they are only 0.01° apart. Photometric measurements of reflectance, on the other hand, are generally limited to receptor field angles about 100 times as large, although recent developments allow glossmeters to achieve resolutions of about 0.1° in some situations.

The requirement for flat test surfaces for narrow-angle measurements of high gloss is very important because an instrument cannot differentiate between a low reading due to poor image-reflecting quality of the specimen and a low reading caused by surface curvature that has directed the specularly reflected beam improperly. The eye, however, can readily distinguish between these effects. Specimen nonflatness distorts, but does not destroy, the visual images reflected in its surface. Inaccurate positioning of a surface in a high-gloss instrument creates the same sort of measurement problem.

## TRANSPARENT MATERIALS

Transparent materials are transmission analogs of metals in that the specular component is dominant. With these materials the matter of interest is specular transmission rather than specular reflection. Again, as with metals, the analogy to the various types of gloss seems to have at least some application. Haze, an attribute of plastic and glass objects, and turbidity, the analogous optical attribute of liquids, command major attention and are frequently measured. The standard ASTM test for haze of plastic film involves measurement of light diffused by more than 4° from the specular direction. There is also a standard test for plastic films, which is analogous to the distinctness-of-image test for reflecting surfaces. It is called a test of clarity and measures loss of any light spreading 0.1° or more from the specular direction. With transparent films analogies can almost certainly be found for surface uniformity (freedom from texture) and contrast or diffuseness attributes. So far, however, interest in these attributes has not been great enough to bring about the development of scales for their measurement.

## OPACITY

Opacity is an important appearance attribute applying to the diffuse rather than to the specular aspect of appearance. Specifically, opacity is the property of a material that hides what is behind it. Opacity is important for paints, paper, and other (usually white) diffusing films normally seen as covers over a substrate or background. When this background is nonuniform, as with black printing on white paper, the effectiveness of the covering film in hiding the patterned background becomes important. Normally a value of opacity is measured by the ratio of the reflectance factor obtained with a black backing for the film to the same factor obtained for the same film with a white backing. This "contrast ratio method" is described in TAPPI Opacity Method T425 and ASTM Method D589 for paper, and ASTM and Method D2805 for paint.

The approaches to making and using opacity measurements improved markedly when the Kubelka-Munk equations for diffuse laminar materials were first proposed. Judd and others who followed prepared a number of charts similar to the one shown in Figure 6.11 showing how opacity and reflectance of the product changed with product thickness and scattering power (Judd and Wyszecki, 1963). With graphs such as these, it is possible to determine from a measurement at one film thickness what thickness change is required to obtain a film of the same material having a different specified opacity. The following method can be used with Figure 6.11 to determine the scattering power *SX*, the scattering coefficient *S*, reflectance $R_0$, and absorption coefficient *K* of a sample

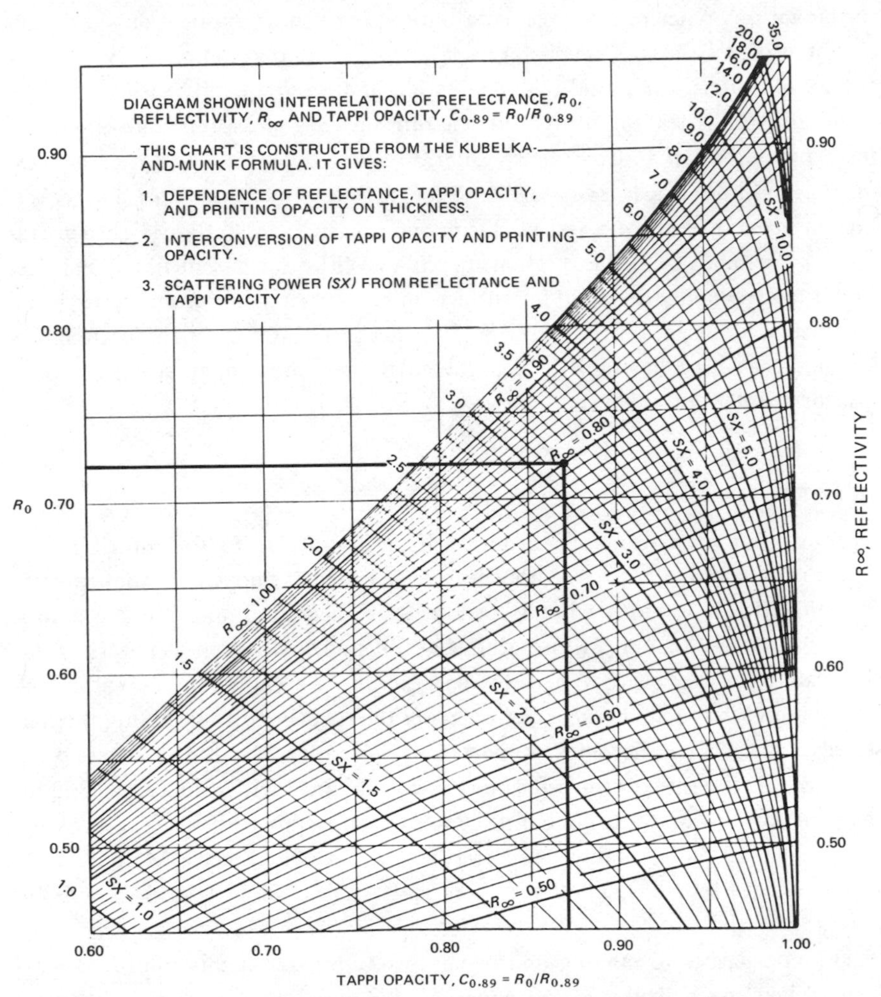


**Figure 6.11.** Reflectance-opacity chart for white backing of 89% reflectance. From *Color in Business, Science, and Industry* by D. B. Judd and G. Wyszecki, John Wiley & Sons, 1963. Used with permission of *J. Opt. Soc. Am.* See text for explanation.

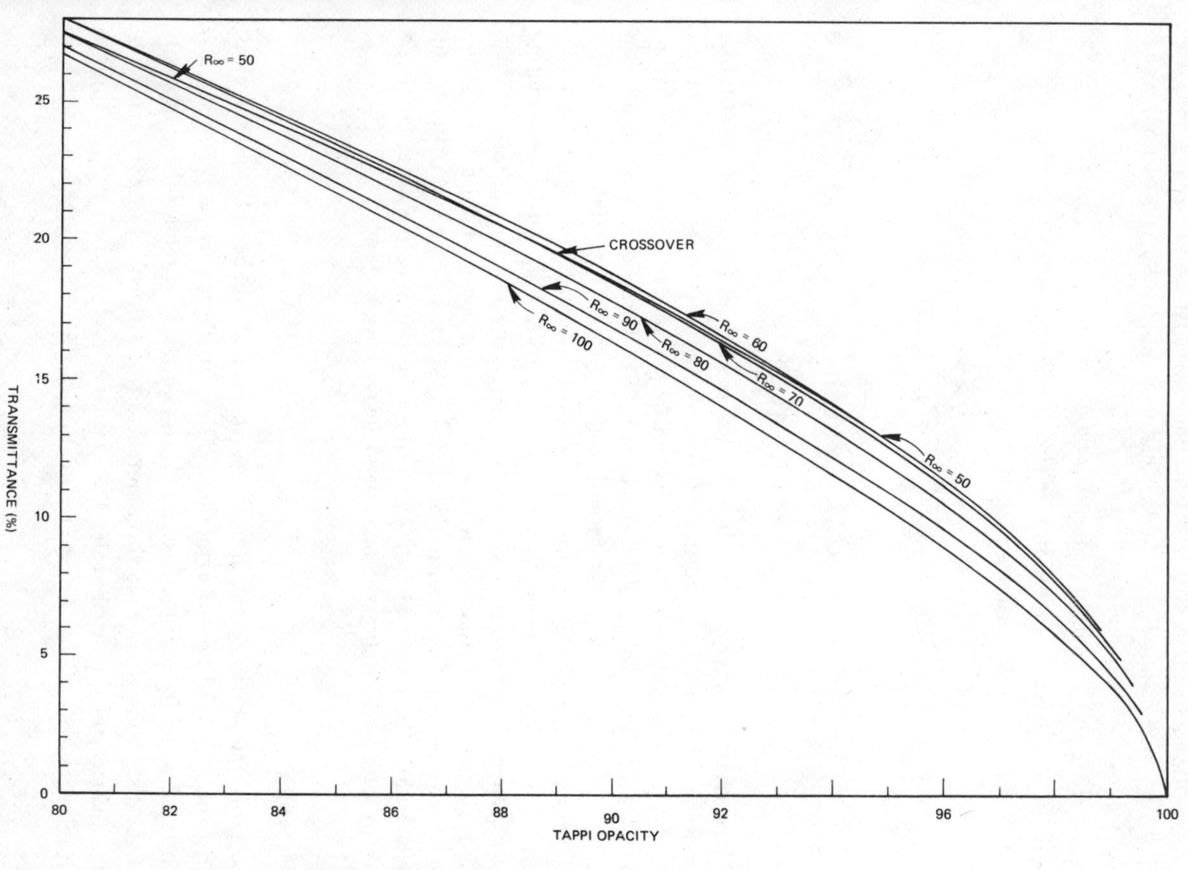


**Figure 6.12.** Relation between TAPPI opacity $C_{0.89}$ and transmittance.

of white paper with a Tappi Opacity $C_{0.89} = 0.87$, a brightness $R\infty = 0.80$, and a basis weight $X = 50$ lbs..

1. Find the intersection of the vertical line corresponding to $C_{0.89} = 0.87$ and curved line of constant $R\infty = 0.80$. Read $SX = 3.2$ from curved lines of $SX$ (values shown along line of constant $R\infty = 1.00$).
2. $SX = 3.2$ and $X = 50$; therefore $S = 0.064$.
3. Read $R_0 = 0.72$ at point of intersection, using left ordinate values of $R_0$.
4. Using the Kubelka-Munk equation,

$$\frac{K}{S} = \frac{(1 - R\infty)^2}{2R\infty}$$

$$\frac{K}{0.064} = \frac{(1 - 0.80)^2}{2\,(0.80)}$$

$$K = 0.0016$$

More recently, attention has been called to the advantages of substituting measurements of total-transmittance factor of a film for opacity. The contrast ratios of practical interest to the paint and paper industry are usually higher than 90%. Economically, a 1% contrast-ratio difference may be quite important. It is difficult to measure such contrast ratios with the high accuracy and precision needed to determine, for instance, whether additional $TiO_2$ pigment is necessary to obtain the specified opacity for a lightweight printing paper. When transmittance measurements are substituted for contrast-ratio determinations, the precision requirements drop markedly. Figure 6.12 shows several curves relating transmittance to TAPPI opacity (TAPPI Useful Method 550). In fractional increments the transmittance measurement is six times as sensitive to changes in opacity as the contrast method.

## SUMMARY

The geometric attributes of appearance involve a number of types of gloss, clarity, opacity, and light distributions. These attributes are perceived visually as being distinct from color. Because there is no way to reduce the complexity of light distributions down to three numbers (as with color) or even ten numbers, a variety of geometric scales have evolved, each to meet unique application requirements. Table A.1 is a tabulation of the gloss methods presently in use.

# CHAPTER SEVEN

# THE CIE STANDARD OBSERVER

The scientific basis for the measurement of color is the existence of three different color-response mechanisms in the human eye. The spectral responses of these light-receiving devices to different wavelengths is now fairly well known. In order to identify a color stimulus by numbers, the responses by the color-sensing devices in the eye to different wavelengths of light have been standardized and incorporated into the CIE Standard Observer for colorimetry. The Standard Observer, like a standard illuminant, is a table of numbers. It is designed to be representative of a normal observer, but its responses do not refer to any specific observer.

## LUMINOSITY

One aspect of the response of the three light-receiving devices in the human eye has to do with perception of luminosity. Luminosity refers to the relative visual-intensity response to different wavelengths of light. The luminosity function (see Chapter 2 and Figure 2.3) was developed from a study of 52 observers (Gibson and Tyndall, 1923), which supplemented the findings of earlier investigators (Coblentz and Emerson, 1918–1919). It was adopted by the CIE in 1924 as a standard function of the eye at normal levels of illumination (CIE, 1924). Table A.2 shows the luminosity curve of Figure 2.3 in tabular form.

To obtain these numbers for the luminosities of different wavelengths, an experiment was conducted in which human observers visually adjusted the intensities of different wavelengths until they appeared equally luminous. Determining the actual energy levels of these equally luminous stimuli made it possible to compute the relative efficiencies of our eyes in converting that energy to light-intensity sensation. This degree of efficiency defines luminosity. In practice, the precision of such experiments is poor if they are conducted in a manner that requires the observer to assess the luminous difference between two wavelengths of very different color. The chromaticity difference distracts the observer, resulting in imprecise estimates.

To minimize this difficulty, two experimental techniques that produce consistent results have been used. In the cascade method the observer sees a screen illuminated on his left by one wavelength and on his right by another. The wavelengths are spectrally close together so that the chromatic difference is small. In this circumstance the observer is able to adjust the intensity of the second wavelength so that a minimal difference results. The observer then uses the second wavelength as a standard, comparing it with a new third wavelength still farther away from the first. This step-by-step process is continued through the entire spectrum, and the energy levels in each case are recorded.

Flicker photometry, on the other hand, uses a screen illuminated alternately by two wavelengths in rapid succession. The observer adjusts the relative energies to minimize the visual flicker. With this arrangement the sensation of flicker predominates over the chromaticity difference, and the two wavelengths may vary greatly in chromaticity (red and green, for example) without destroying the ability of the observer to adjust for minimum flicker. Thus with flicker photometry a single reference light may be used for comparison with all wavelengths.

## ADDITIVE AND SUBTRACTIVE COLOR MIXING

In addition to its response to luminosity, the human eye's evaluation of chromaticity had to be investigated in order to derive the CIE Standard Observer. This was accomplished through an experiment involving the mixture of colored lights. Before describing this experiment, note should be taken that there are two different types of color mixing.

1. Subtractive (product) mixing occurs when colored substances are combined so that each in turn absorbs light. An example of subtractive mixing is the passage of a beam of light through two colored filters, one placed on top of the other, or the mixture of two or more dyes in a solution to be used to color white fibers. Mixing of two or more paints in a paint box also gives subtractive color mixture. Since each component absorbs light, subtractive mixtures are always darker than the components separately. Only through spectral analyses involving light-absorption properties and light-scattering properties at all wavelengths can one predict color appearance due to subtractive mixtures of known coloring components.
2. Additive (light) mixing occurs when two or more colored lights are added as by shining them on the same surface. Additive mixtures are always lighter than any of the individual components since each light adds luminous energy to the final mixture. Color appearance resulting from an additive mixture can be predicted from the color attributes of the lights and the amounts of each that

are used in the mixture. It is because colors can be matched visually by additive mixture using only three primary lights that numerical color specifications are feasible. All tristimulus color specifications are ultimately identifications of the amounts of three primary lights required by the Standard Observer to match the unknown color stimulus.

## ANALYZING THE WAVELENGTH RESPONSE CHARACTERISTICS OF OBSERVERS

To develop the Standard Observer, which is the basis for all instrumental color measurement, normal observers were asked to perform an experiment in which they matched visually the colors of lights of the individual spectrum wavelengths by mixing together the lights from three colored primaries. Figure 7.1 illustrates the way in which this was done. The observer looked at a white screen about 2° in diameter. The bottom half of this field was illuminated by the light to be matched. As the experiment progressed this light was changed in wavelength. The upper half of the field was illuminated by an additive mixture of three colored lights that shone simultaneously on it.

Usually in such experiments, lights of single wavelengths in the blue, green, and red parts of the spectrum, respectively, are used as primaries. In fact any lights may be used so long as no one of them can be produced by an additive

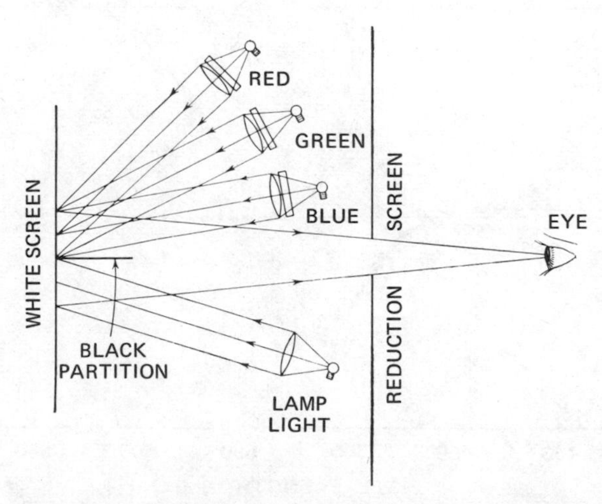


**Figure 7.1.** Visual arrangement for additive colorimetry. Observed through an aperture, the bottom half of the screen is illuminated by a light matched by mixing three lights together on the other half of the screen. From *Color in Business, Science, and Industry,* by D. B. Judd and G. Wyszecki, John Wiley & Sons, 1963. Used with permission of *J. Opt. Soc. Am.*

mixture of the other two. The observer adjusts the intensity of each of the three lights until both halves of the screen are identical in color appearance. The energies of each of the three lights are then recorded, and the experimenter progresses to the next test wavelength.

Historically the most important color-matching experiments were carried out by W. D. Wright in 1928 and J. Guild in 1931. Seven observers were analyzed in Guild's experiments and ten observers in Wright's. Figures 7.2 and 7.3 show the relative amounts of the three primary lights needed to match each wavelength of light as demonstrated by the observers in the two sets of experiments.

It was not actually possible for the observers in the Wright and Guild experiments to match the colors of all the spectrum wavelengths with the three primary lights. To match a yellow wavelength, for example, large amounts of red and green primary lights were used. The mixture was still visually less yellow than the test wavelength. The experimental solution to this dilemma was to add blue primary light to the yellow test wavelength until both halves of the field matched. This addition of one of the primaries to the opposite side was

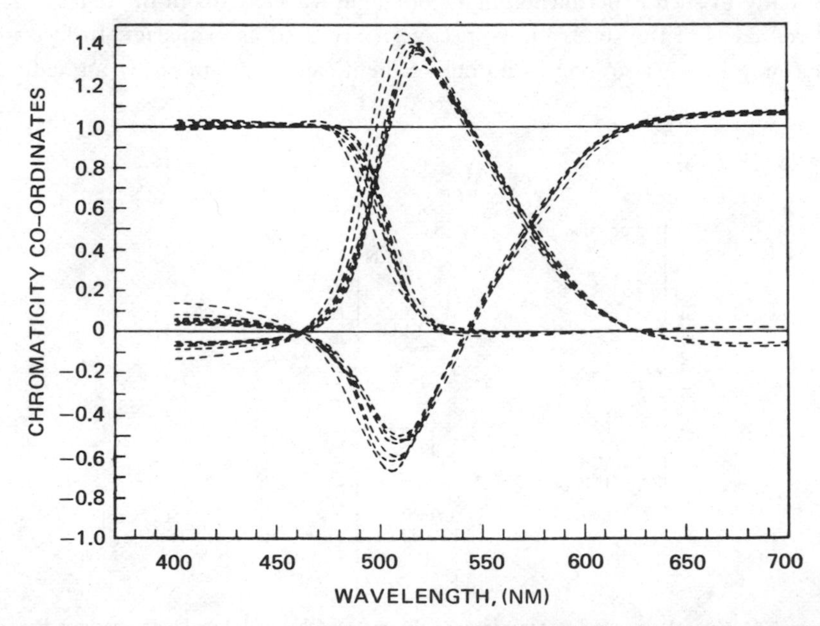


**Figure 7.2.** Relative amounts of three primaries used in Guild's experiment to match the spectrum colors. Guild's source-plus-filter primaries were 460 nm, 550 nm and 630 nm. From *The Measurement of Colour* by W. D. Wright, Alger Hilger Ltd., London, Van Nostrand Reinhold Co. in the U.S., 1969.

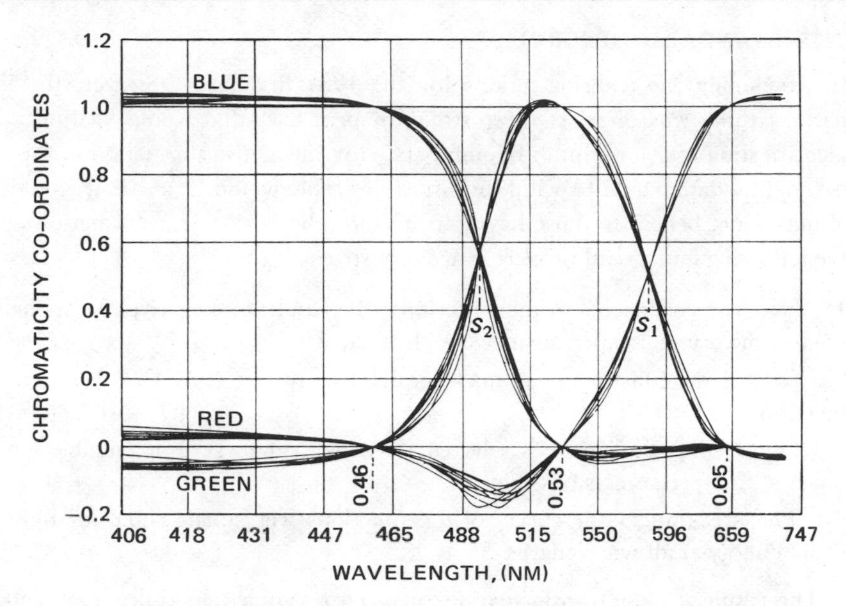


**Figure 7.3.** Relative amounts of three primaries used in Wright's experiment to match the spectrum colors. The three primaries used in this experiment were wavelength lights of 460 nm, 530 nm and 650 nm. From *The Measurement of Colour* by W. D. Wright, Alger Hilger Ltd., London, Van Nostrand Reinhold Co. in the U.S., 1969.

treated as adding a negative light to the mixture. Negative amounts of primaries are represented in Figures 7.2 and 7.3 wherever the curves drop below the zero value.

## TRANSFORMATIONS

Note that the curves obtained by Wright and Guild in their two experiments are substantially different—far more different than can be attributed to differences among observers. The reason is that Wright and Guild did not use the same sets of three primary wavelengths in their laboratory experiments.

J. C. Maxwell (1860) and H. E. Ives (1915) had shown that the color mixture functions obtained by using different sets of primaries can always be related mathematically by a set of equations called a linear transformation. Transformation techniques (Judd and Wyszecki, 1963) made it possible to convert the experimental curves of Wright and Guild into more usable functions, and the 1931 CIE Standard Observer resulted from such a transformation (CIE, 1931).

## THE STANDARD OBSERVER

In developing the transformations for thc 1931 Standard Observer, the important thing was to select three standard primaries that would facilitate the identification of color stimuli by numbers. Any linear transformation of the experimental data would be valid, but most desirable would be one that would be of maximum benefit to those having to use it. Therefore a transformation with the following four useful properties was constructed:

1. Negative coefficients in the equations (the result of "negative" lights being used in the original experiment) were eliminated.
2. One of the functions was made equivalent to the 1924 CIE Luminosity function.
3. One of the functions was selected to be as near to zero as possible for as much of the spectrum as possible.
4. The areas under the curves of the functions were made equal for light of equal energy at all wavelengths.

The result of these transformations was a set of functions called $\bar{x}$, $\bar{y}$, and $\bar{z}$, since they did not represent any real colored primaries. Table A.3 specifies, in the columns labeled "Tristimulus Specifications of the Equal Energy Spectrum," the exact functions adopted in 1931. The $\bar{x}$, $\bar{y}$, and $\bar{z}$ values of each wavelength are called the CIE tristimulus values for that wavelength.

Note that in the table there are no negative numbers, the $\bar{y}$ function is the same as the luminosity function shown in Table A.2, the $\bar{z}$ function is zero after wavelength 625, and the column totals (the integrals, or areas under the curves) are equal.

The 1931 CIE Standard Observer can also be shown as a graph, as in Figure 7.4.

### Using the Standard Observer Functions

The usefulness of the Standard Observer is that it provides a standard means for converting any spectral curve into three numbers called tristimulus values, $X$, $Y$, and $Z$, that properly identify the color of the object or light source in terms of the mixture of the primary lights that match it visually. Two different spectral curves that reduce to the same $X$, $Y$, and $Z$ values are curves for objects or sources that appear visually to be the same color.

The simplest spectral curve to convert is one with light at only one wavelength. In this case, $X$, $Y$, and $Z$ are equal to $\bar{x}$, $\bar{y}$, and $\bar{z}$ at that wavelength. Spectral curves of more than one wavelength require that we consider the

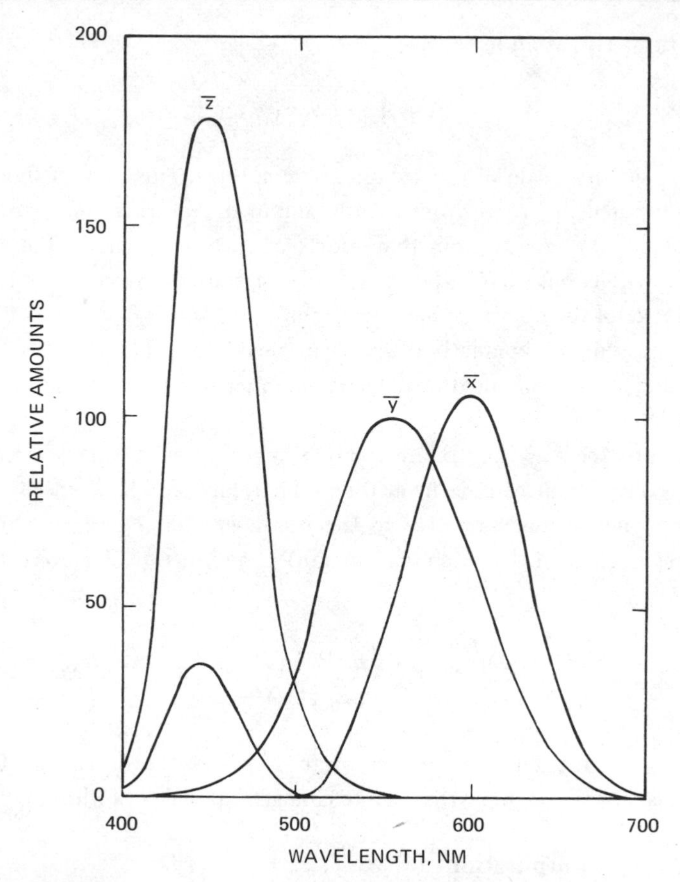


**Figure 7.4.** The color matching response functions of the 1931 CIE 2° Standard Observer. The curves indicate the relative amounts of three imaginary red, green, and blue lights mixed to match each wavelength test color. From *Principles of Color Technology* by F. W. Billmeyer, Jr. and M. Saltzman, John Wiley & Sons, 1966.

contribution of each wavelength for each primary. At a particular wavelength, if $x_\lambda$ is the amount of contribution toward $X$ of one unit of that wavelength, then it must be multiplied by the amount of that wavelength actually present. This amount we get by measuring the spectral energy $S_\lambda$ at that wavelength. This contribution toward $X$ of one particular wavelength, then, will be $S_\lambda \bar{x}_\lambda$. Contributions made by each wavelength are then combined by addition so that we consider all of the light at all wavelengths for each primary:

$$S_{\lambda 1}\bar{x}_{\lambda 1} + S_{\lambda 2}\bar{x}_{\lambda 2} + S_{\lambda 3}\bar{x}_{\lambda 3} + \cdots + S_{\lambda n}\bar{x}_{\lambda n}$$

or, more properly, as an integral

$$X = \int_{400}^{700} S_\lambda \bar{x}_\lambda \, d\lambda$$

Depending on the mode of the stimulus, the spectral energy $S$ of the stimulus will be computed differently. For an illuminant or aperture color, stimulus $S_\lambda$ will equal $E_\lambda$, the energy from that source at each wavelength. For reflecting objects, $S_\lambda$ will equal $E_\lambda R_\lambda$, the energy of the source reduced by the percent reflectance $R_\lambda$ of the object at that wavelength.

In the preceding explanation we have considered only the $X$ tristimulus value; $Y$ and $Z$ are computed in the same manner using, of course, the $\bar{y}$ and $\bar{z}$ Standard Observer functions.

Objects are seen as they relate to their surroundings, not in terms of the absolute level of light coming from them. Therefore, $X$, $Y$, $Z$ specifications of objects are always made relative to the luminosity of a perfect white object (reflectance equal to 1.0 at each wavelength). The functions for objects thus become

$$X = \frac{\int E_\lambda R_\lambda \bar{x}_\lambda \, d\lambda}{\int E_\lambda \bar{y}_\lambda \, d\lambda} \qquad Y = \frac{\int E_\lambda R_\lambda \bar{y}_\lambda \, d\lambda}{\int E_\lambda \bar{y}_\lambda \, d\lambda} \qquad Z = \frac{\int E_\lambda R_\lambda \bar{z}_\lambda \, d\lambda}{\int E_\lambda \bar{y}_\lambda \, d\lambda}$$

By definition, the $Y$ for the perfect white is always 1.0 (or 100%). The magnitudes of $X$ and $Z$ for the perfect white change with color of illuminant.

## Object Color Computations for *X, Y, Z*

To convert spectral reflectance (or transmittance) curves to $X$, $Y$, $Z$, three methods of integration are available:

1. Weighted-ordinate integration
2. Selected-ordinate integration
3. Instrument (automatic) integration

*The Weighted-Ordinate Integration Method for Computing X, Y, Z Values.* Most color computations are carried out to determine appearance under one of the Standard Illuminants A, B, C, or one of the D's (see Chapter 4). In order to simplify computation, Standard Observer tables have been prepared in which the $\bar{x}$, $\bar{y}$, and $\bar{z}$ functions already have been multiplied at each wavelength by the spectral curves of these illuminants. Table A.4 shows these values for CIE Illuminants A, B, C, and D65. The tables are normalized so that the $\bar{y}E$ total is a multiple of 100, making the division by the $\bar{y}$ integral a matter of moving a decimal point. Note that the column totals (or integrals) are no longer equal, since the functions have been modified for the illuminant. The weighted-ordinate method computes $E_\lambda R_\lambda \bar{x}_\lambda$, $E_\lambda R_\lambda \bar{y}_\lambda$ and $E_\lambda R_\lambda \bar{z}_\lambda$ at each wavelength, using Table A.4 and the reflectance curve of the object. These components are then added, and the three totals are divided by the $\bar{y}E$ column total. Table A.5 is an example of a weighted-ordinate calculation, using the yellow school-bus color (see Figure 7.5) as an example.

*The Selected-Ordinate Integration Method.* For this type of computation, the Standard Observer functions become three lists of selected wavelengths, one for each response function $\bar{x}$, $\bar{y}$, and $\bar{z}$. The intervals between selected wavelengths are small where the particular Standard Observer function is high and large where it is not. Figure 7.6 shows the wavelength spacing used for the $\bar{y}$ function and 30 selected coordinates for computation of $Y$. To compute $X$, $Y$, and $Z$ values by the selected-ordinate method, the measured reflectance from the spectral curve of the object is taken at each of the selected-ordinate wavelengths for the desired illuminant. These reflectances are then added together and divided by the number of ordinates in the table.

Table A.6 shows lists of 30 selected ordinates for illuminants A, B, and C. Computations in the selected-ordinate method must account for the change in $X$ and $Z$ weighting with change in illuminant by multiplying the $X$ and $Z$ totals by constants specific for the illuminant involved. These constant factors are given at the bottom of the table.

Table A.7 illustrates the derivation of $X$, $Y$, $Z$ values using the selected-ordinate method. The example used is the yellow school-bus color, the reflectance curve of which is shown in Figure 7.5.

*Instrument Automatic Integration.* Two distinctly different techniques exist for obtaining $X$, $Y$, and $Z$ values without manual calculation. Some spectrophotometers have tristimulus integrators that compute $X$, $Y$, and $Z$ while the curve is being drawn. In another wholly different type of instrument called a photoelectric tristimulus colorimeter, photodetector-filter arrangements are used optically to duplicate as closely as feasible the $\bar{x}$, $\bar{y}$, and $\bar{z}$ functions. When these photodetector-filter arrangements receive light from colored objects, they give signals closely proportional to $X$, $Y$, and $Z$.

## METAMERISM

Metamerism is present when two objects having the same color appearance nevertheless have different spectral curves. The layman recognizes metamerism when two objects that match under one illuminant fail to match under a

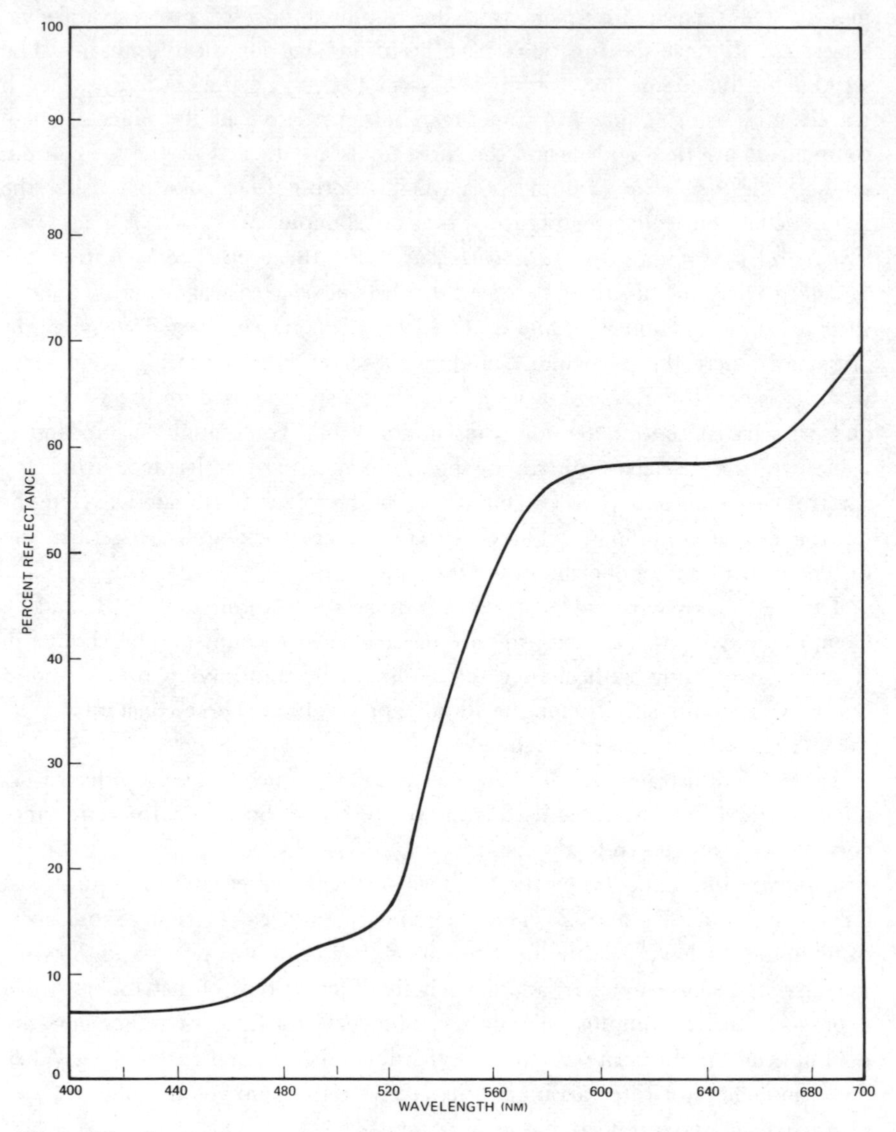


**Figure 7.5.** Spectrophotometric curve of yellow school bus. See Tables A.5 and A.7.

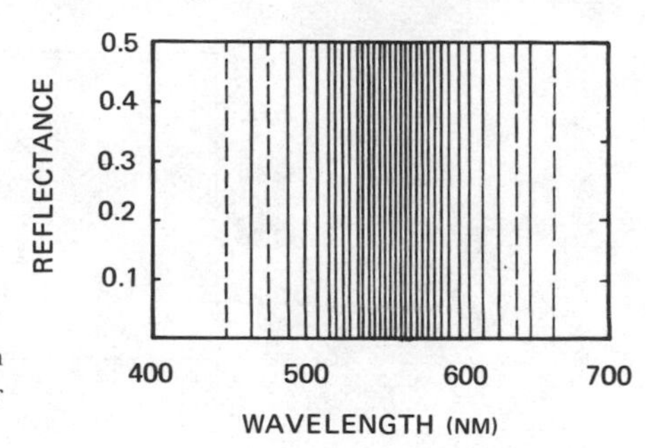


**Figure 7.6.** Template showing wavelength spacings used in selected ordinate method for integrating $Y$ tristimulus value.

second. From the viewpoint of colorimetry, metamerism occurs when $X$, $Y$, and $Z$ of two specimens match under the first illuminant but not the second. Metamerism is possible because the Standard Observer computations reduce an entire spectral curve to only three numbers. It is possible, and often happens, that two objects with different spectral curves will reduce to the same tristimulus specifications and will therefore be a visual match under one illuminant. However under a different illuminant ($E_\lambda$ above) the spectral curves will change so that the tristimulus values $X$, $Y$, and $Z$ may no longer match, and the two objects will not, therefore, be a visual match. Figure 7.7 shows curves of pairs of samples that exhibit metamerism.

## CIE CHROMATICITY

The tristimulus values $X$, $Y$, and $Z$ are limited in value as color specifications because they correlate poorly with any arrangement of visual attributes such as those described in Chapter 1. While $Y$ correlates with lightness, $X$ and $Z$ by themselves do not correlate with hue, saturation, depth, vividness, redness-greenness, yellowness-blueness, or with any visually meaningful attribute of color appearance. When the Standard Observer was established, the CIE recommended a chromaticity system to identify those aspects of color appearance separate from lightness. For this purpose the Commission proposed that chromaticity coordinates $x$, $y$, and $z$ (also called trichromatic coefficients) be defined as

$$x = \frac{X}{X+Y+Z} \qquad y = \frac{Y}{X+Y+Z} \qquad z = \frac{Z}{X+Y+Z}$$

Since the sum of $x$, $y$, and $z$ will always be unity, only two of the chromaticity

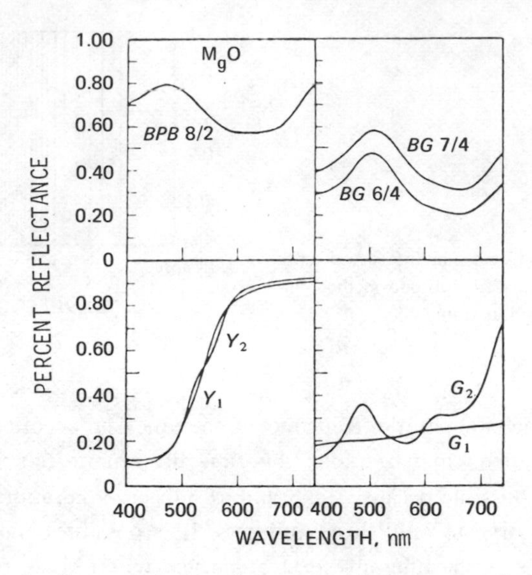


**Figure 7.7** Spectral reflectance of pairs of samples exhibiting various degrees of metamerism. The upper two pairs differ considerably in color, but show little or no metamerism. The lower two pairs are near matches: the left pair is moderately metameric and the right pair is strongly metameric.

coordinates (the CIE recommended use of $x$ and $y$) are needed to specify chromaticity. Neither $x$, $y$, nor $z$ correlates with any of the meaningful attributes of color appearance. Taken together and incorporated into a chromaticity diagram, however, relationships of $x$ and $y$ with color appearance are developed.

## The Chromaticity Diagram

For a visual display of chromaticities and their relationships, the CIE recommended a graph using $x$ and $y$ as axes. In such a graph the trichromatic coefficients of the spectrum colors (Table A.3) form a horseshoe-shaped curve within which the $(x, y)$ chromaticities of all color stimuli fall. This chromaticity diagram is shown in Figure 7.8. The straight line that connects both ends of the spectrum closes the diagram. All real chromaticities fall within the enclosed area, and every point inside represents a distinct chromaticity. For example, the chromaticity of a white or neutral stimulus is found at point E (equal energy source) in the diagram. The upper area of the diagram contains the various chromaticities that we would call green. Blues and violets fall in the lower left region, while reds lie in the lower right-hand section. Between the green and the red, on the right-hand side, fall the yellows.

An interesting property of the chromaticity diagram is that the chromaticities of all colors resulting from the additive mixture of two colored lights fall on the straight line connecting the two lights. Thus, the chromaticities of all additive mixtures of Illuminant C (point C in the diagram) and the spectral color at 520 nm fall on the straight line shown in Figure 7.8.

The fact that *Y, x, y* is superior to *X, Y, Z* as a specification of color ap-

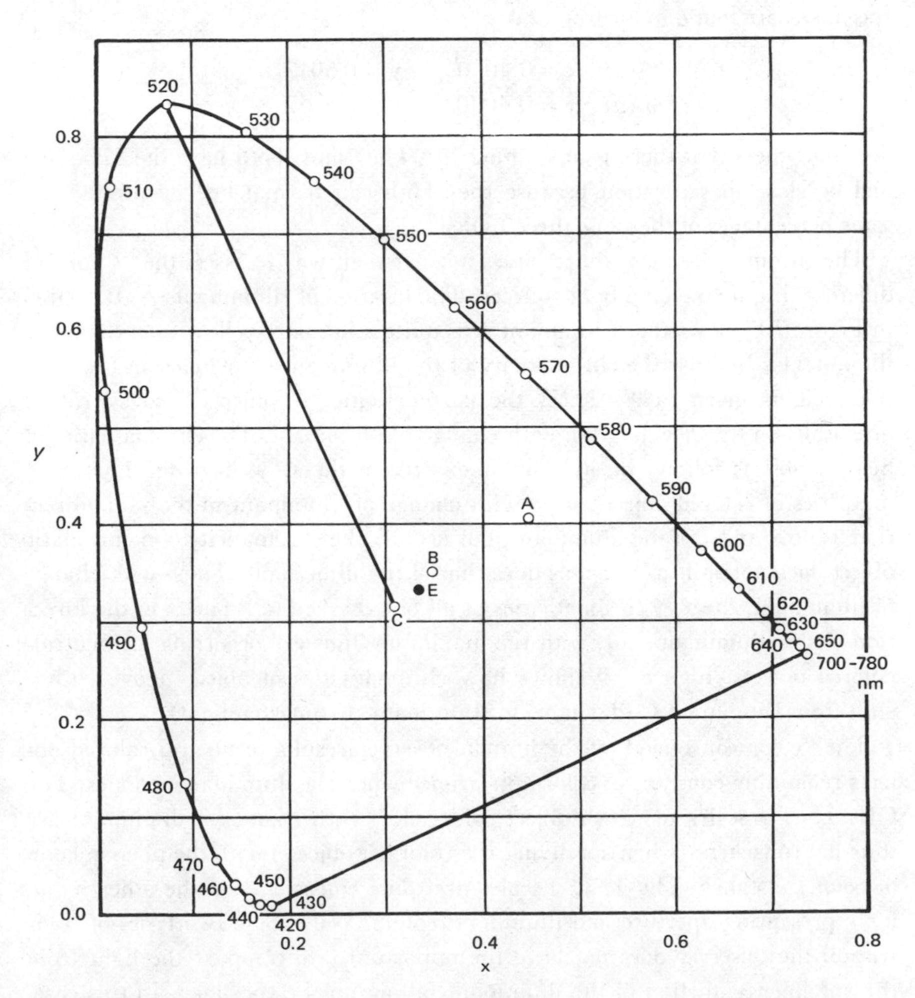


**Figure 7.8.** The CIE *x, y* Chromaticity Diagram, showing locations of Illuminants A, B, C, and E (equal energy). The chromaticity of any mixture of Illuminant C and the spectral color at wavelength 520 will fall on the straight line connecting them on the diagram.

pearance was suggested by Hardy, who illustrated his point with an example similar to the one that follows (Hardy, 1936). Suppose one has two objects for which color measurements are as follows:

$$(a)\quad X = 48.36 \qquad Y = 60.14 \qquad Z = 11.50$$
$$(b)\quad X = 53.20 \qquad Y = 66.15 \qquad Z = 12.65$$

One knows from the foregoing that (*b*) is lighter than (*a*) because $Y$ is 66.15 for (*b*), only 60.14 for (*a*). Not until chromaticities are computed, however, and the specimens are found to measure:

$$(a)\quad x = 0.4030 \qquad y = 0.5012$$
$$(b)\quad x = 0.4030 \qquad y = 0.5012$$

does one guess that there is any similarity. They must both have the same hue and be close in saturation because they both can be matched by mixing the same percentages of the same three lights.

The chromaticity diagram provides a convenient way to assess the "color" of different illuminants and light sources. The location of Illuminants A, B, and C in Figure 7.8 shows that Illuminant A is quite a bit more yellow than the other illuminants. Because the chromaticity of the illuminant is included in the relative weights given to $X$, $Y$, $Z$, the chromaticities of objects "move" rather drastically in the $x$, $y$ diagram with change of illuminant. The chromaticities of neutral objects follow the illuminant exactly in the $x$, $y$ diagram. The chromaticities of colored objects move with change of illuminant in the same direction as do $x$ and $y$ of the illuminant, but usually the chromaticity of a chromatic object does not shift as much as does that of the illuminant. Thus, with change of illuminant, the $x$, $y$ chromaticities of all objects tend to "move" in the direction of the illuminant shift, with the maximum "moves" occurring for neutral-colored objects. Figure 7.9 shows how chromaticities of objects move with a shift from Illuminant C (daylight) to Illuminant A (tungsten light).

The "color constancy" of the human observer results in neutral-colored objects remaining constant in color appearance when the illuminant changes. The CIE $Y$, $x$, $y$ scales in which object colors shift with change of illuminant are thus not consistent (when specifying the colors of objects) with the phenomenon of color constancy. The $Y$, $x$, $y$ scales are quite consistent, on the other hand, for representing aperture and illuminant colors. With these two types of color stimuli the observer does not have the opportunity to compare the light from the specimen with that of the illuminant falling on the specimen. In this case, color appearance does change with change of illuminant.

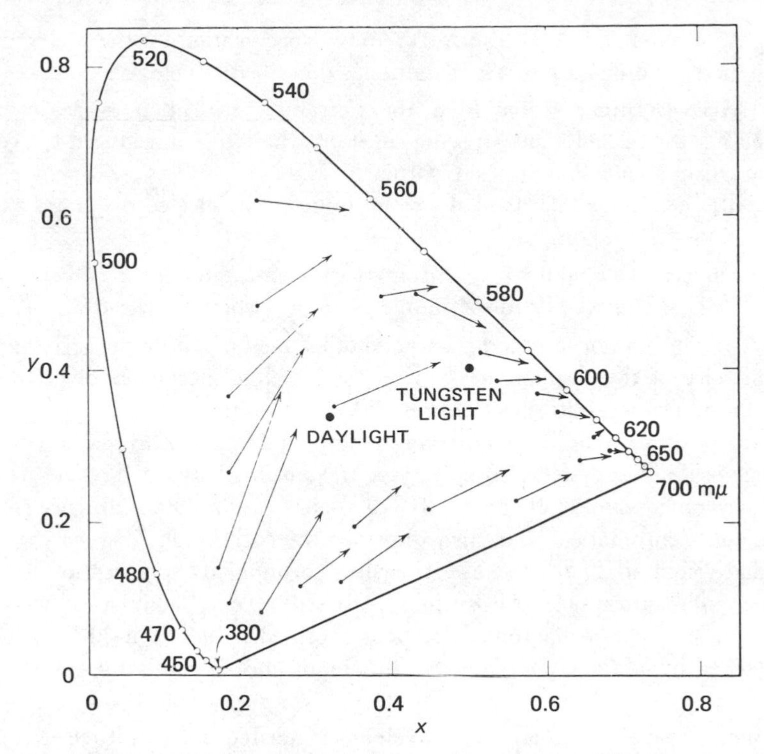


**Figure 7.9.** Typical chromaticity changes in objects with a change in Illuminant from C to A. Each vector shows the change in chromaticity which was required to maintain a visual match between the two halves of a tristimulus colorimeter when one part of the central area of the retina was adapted to daylight and an adjacent part of the same retina was adapted to tungsten light. From Color in *Business, Science, and Industry* by D. B. Judd and G. Wyszecki, John Wiley & Sons, 1963. Data obtained by D. L. MacAdam, and used with permission of the *J. Opt. Soc. Am.*

## Adapting CIE Specifications to Object Colors

Two methods are used to adapt CIE scales for object-color specifications by referring to the chromaticity of the illuminant as neutral, so that measures of color result that are consistent with the phenomenon of color constancy.

1. Percent specification of $X$, $Y$, $Z$. Percent specification compensates for the shifting chromaticity of any illuminant by computing $X$, $Y$, $Z$ each relative to the values for a perfect white under any illuminant. Percent specifications are written $X_{\%}$, $Y_{\%}$, $Z_{\%}$.

In a weighted-ordinate calculation this has the effect of equalizing the areas under the $\bar{x}$, $\bar{y}$, and $\bar{z}$ curves for any illuminant used. It is accomplished by dividing $X$, $Y$, and $Z$ each by the $X$, $Y$, and $Z$ specification for the perfect white rather than dividing only by the $Y$ values as described (Table A.5).

In a selected-ordinate calculation the percent specification is achieved by eliminating the $X$ and $Z$ multiplying constants that were introduced to account for the chromaticity shift in the first place. The values of $X_{\%}$, $Y_{\%}$, and $Z_{\%}$ are simply the column totals divided by the number of selected ordinates (Table A.7).

It is incorrect to compute $x$, $y$ chromaticity coordinates from *percent* values ($X_{\%}$, $Y_{\%}$, $Z_{\%}$). Only $X$, $Y$, and $Z$ should be used to compute $x$ and $y$.

2. Dominant wavelength and purity. Another method of compensating for the chromaticity of the illuminant is to use scales that specify an object's chromaticity in terms of its position relative to the illuminant, rather than its absolute position. The dominant wavelength and purity chromaticity system was one of the first systems for specifying the chromaticity of objects other than by their $x$, $y$ values. It not only compensates for the influence of the illuminant's chromaticity, but also improves the correlation between the numbers and visual attributes because it permits chromaticity specification in terms of hue and saturation. The system is based on the additive-color-mixing properties of the $x$, $y$ diagram. A color is specified by describing how it would be matched by additively mixing the illuminant and light of some single wavelength.

Dominant wavelength is the wavelength needed for mixture with the illuminant. In general it identifies the hue of the object's color.

Purity is the percentage contribution of the dominant wavelength to the mixture. Thus, 1.00 is the purity of all spectral colors and 0 is the purity of the illuminant. Purity correlates with saturation.

In order to derive dominant wavelength and purity for the yellow school-bus color, for example, the position of the Illuminant C–bus-color combination is located by plotting on the $x$, $y$ diagram its $x$, $y$ values as computed in Table A.5 (P in Figure 7.10). Illuminant C is plotted using its $x$, $y$ values as shown in Table A.4. A straight line drawn through the Illuminant C point and the object point P intersects the spectrum locus at point S or 582 nm. The dominant wavelength for the color is thus 582 nm. Excitation purity, commonly called purity, is the ratio of the distance between illuminant and object points to the distance between illuminant and spectrum locus. It can be expressed as 100 (CP/CS), and for the school-bus yellow color P is about 70%.

For a purple color such as M in Figure 7.10, the line through C and M intercepts the spectrum locus (not the purple line) at 540 nm. Being on the opposite side of C, this is called the complementary dominant wavelength of M.

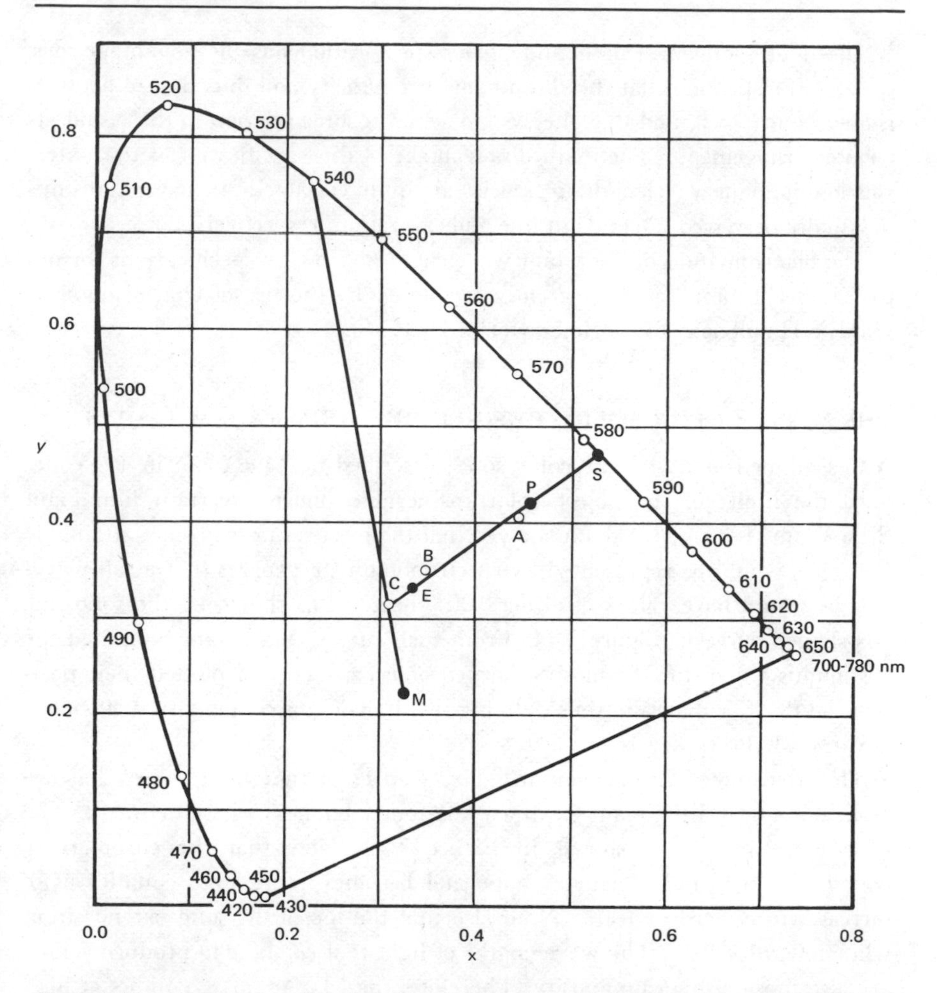


**Figure 7.10.** Dominant wavelength and purity plotted on the CIE *x, y* Chromaticity Diagram. The dominant wavelength for point *P* under illuminant C is found by drawing a straight line from the illuminant C point through P to the spectrum locus, where it intersects at 582 nm, the dominant wavelength. Excitation purity is the percentage defined by C*P*/C*S*, the percentage the distance from illuminant C to *P* is of the total distance from illuminant C to spectrum locus.

Note that dominant wavelength is not a physical property, but is obtained from a chromaticity diagram after the entire spectral curve has been reduced to *Y, x, y*. The specification has nothing to do with any peaks that may occur in the spectral curve of the object. Dominant wavelength is a psychophysical specification.

The dominant wavelength and purity specification has the advantage over the *x, y* specification that the illuminant chromaticity and thus that of neutrals is always at P = 0, and that the system achieves some relation to the visual-attribute arrangements. The main disadvantage of this specification is that intervals in dominant wavelength and purity are quite variable in their relationships to visually perceived intervals in hue and saturation, respectively.

The determination of dominant wavelength and purity specifications for objects under Illuminant C is greatly facilitated by the use of charts found in Hardy's Handbook of Colorimetry (Hardy, 1936).

## THE Y, *x, y* COLOR SOLID, DEFINED BY THE MacADAM LIMITS

A three-dimensional *Y, x, y* color solid, described by MacAdam in 1935, defined the limits of the object colors achievable under a certain illuminant (MacAdam, 1935). MacAdam showed that the most saturated colors attainable in objects would be represented by spectrophotometric curves that at all wavelengths would have values of either 100% or zero, as shown by the solid-line box-shaped curve in Figure 7.11. From such curves, MacAdam computed the tristimulus values for the most saturated object colors, and plotted their positions on the *x, y* diagram, thus defining the limits of maximum visual efficiency at various lightness levels.

The coordinate arrangement in *Y, x, y* space of the object colors that are realizable under Illuminant C, and whose outer boundaries are defined by the MacAdam limits, are shown in Figure 7.12. Note that the chromaticity boundaries in *x* and *y* change shape and become progressively smaller as *Y* increases from zero to 100%. Note also that the top of the solid extends from white toward yellow. The wavelengths of light that combine to produce yellow are also those of high luminosity. Therefore, the MacAdam *x, y* limits at high

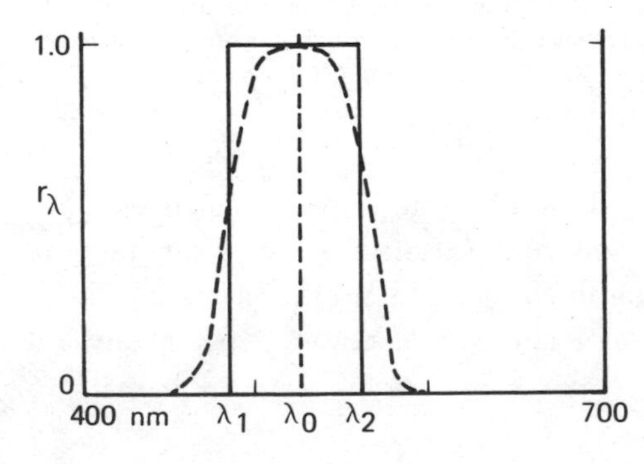


**Figure 7.11.** Spectrophotometric curves for actual (dotted line) and ideal (solid line) green samples having dominant wavelength about 525 nm and luminous efficiency about 0.50. Box shaped curve has maximum attainable purity. Used by permission of the *Journal of the Optical Society of America*, **25**, 1935.

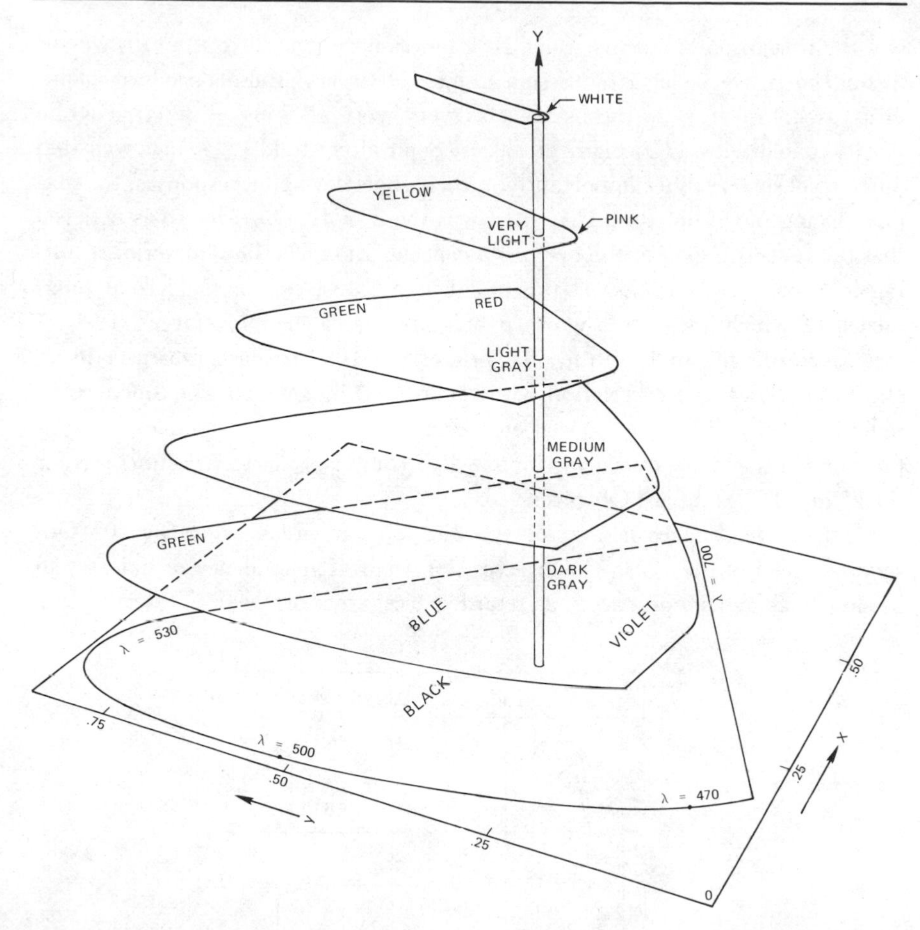


**Figure 7.12.** The CIE *Y, x, y* color solid, with outer boundaries defined by the MacAdam limits. All achievable object colors lie within the limits of this color solid.

levels of *Y* encompass primarily white and yellow colors. Thus the solid defined by the MacAdam limits serves to delineate which product colors, for example, are practically obtainable and those which are impossible to achieve, such as a saturated blue with a high *Y* value.

## THE 10° STANDARD OBSERVER

The 1931 CIE Standard Observer was based on an experiment that used a 2° field of view for determination of average human response. Jacobsen (1948) was the first to point out that the CIE $\bar{z}$ function does not give as much weight to the shortwave violet as do actual human observers grading products of industry with their eyes. Jacobsen's observers were working with large white panels coated with $TiO_2$ pigment pastes. Soon after, Judd (1949) showed that different observers differ importantly in their shortwave blue responses, the majority being more violet sensitive than was the 2° CIE Observer. It is believed that the central region of the eye has a concentration of yellow material resulting in poor violet response of this small area. Most commercial visual judgments, of which Jacobsen's white panels are typical, employ larger fields of view. Accordingly, in 1960 the CIE proposed a 10° Standard Observer in an effort to obtain better correlation with commercial judgments. The functions finally adopted in 1964 were based on a combination of the independent evaluations of Stiles and Speranskaya. Figure 7.13 compares the $\bar{x}$, $\bar{y}$, $\bar{z}$ functions for the 2° and 10° Standard Observers.

Most of the changes in a color specification caused by using the 10° Observer instead of the 1931 2° Observer are small. The tendency of industry to be slow to change its practices has retarded acceptance of the 10° Observer.

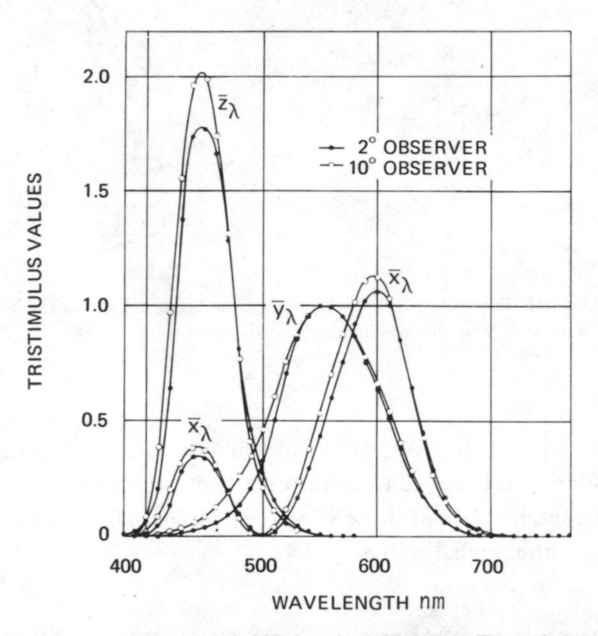


**Figure 7.13.** Comparison of 2° and 10° CIE Standard Observers. The 10° Observer gives more weight to the shorter wavelengths and is believed to more adequately represent the object-color response function of human observers. From *Color in Business, Science, and Industry,* by D. B. Judd and G. Wyszecki, John Wiley & Sons, 1963. Used with permission of *J. Opt. Soc. Am.*

## SUMMARY OF TERMS

$\bar{x}, \bar{y}, \bar{z}$ The three standard observer functions. They specify the relative response of the CIE Standard Observer to each wavelength.

$X, Y, Z$ CIE tristimulus value specifications for the color of a particular spectral distribution. $X, Y, Z$ are integrals of $\bar{x}, \bar{y}, \bar{z}$, each multiplied by the spectral distribution of the color stimulus.

$x, y, z$ CIE chromaticity coordinates (trichromatic coefficients) for any color stimulus:

$$x = \frac{X}{X+Y+Z} \qquad y = \frac{Y}{X+Y+Z} \qquad z = \frac{Z}{X+Y+Z}$$

$Y, x, y$ Specification in a form designating luminosity $Y$ and chromaticity $x, y$.

$X_{\%}, Y_{\%}, Z_{\%}$ Tristimulus values expressed as percent of $X$, $Y$, and $Z$ of the standard white surface.

Although nomenclature has not yet been standardized, subscripts of 2° and 10° for any of the above specifications may be used to indicate which Standard Observer is employed. No subscript implies the 1931 2° Observer.

## SUMMARY

The system of specification of color by $X$, $Y$, $Z$ tristimulus values is based on the 1931 CIE Standard Observer. This table of $\bar{x}$, $\bar{y}$, and $\bar{z}$ functions was derived from additive light mixing experiments and is used to specify color in terms of $X$, $Y$, and $Z$ tristimulus values for all stimulus modes. The weighted-ordinate and selected-ordinate methods are two ways of computing $X$, $Y$, $Z$. Spectrophotometers equipped with integrators and tristimulus colorimeters are used more generally to arrive at the tristimulus values without laborious manual calculation. Chromaticity of stimuli can be specified in terms of $x$ and $y$, and color can be given in terms of $Y$, $x$, $y$. In order to specify color in the object mode, percentage specifications may be used to account for color constancy—or the $x$, $y$ values can be converted into dominant wavelength and purity.

# CHAPTER EIGHT

# UNIFORM COLOR SCALES

Within two or three years of the development and international acceptance of the 1931 CIE color scales, there appeared the first of a series of specialized scales now generally called "Uniform color scales" (UCS). Each of these scales was unique, although all were related to the CIE system as a base. New color scales have continued to appear right to the present time, so that there are now 20 to 30 such uniform color scales* described in the literature.

In one sense it seems strange that so soon after the international adoption of a standard system for the measurement of color alternate systems started to appear. However, the reasons become apparent when one considers the basic nature of the CIE Standard Observer. It was the result of many years of research and study here and abroad. It was designed to provide an accurate and useful tool for the identification of color stimuli. In the basic experiment used to obtain the data for the Standard Observer, only the color of the light being matched was analyzed. The observer used the aperture mode of observation, viewing the test lights through a peephole of 2° diameter.

Thus a scientifically controlled experiment provided the data needed to represent an average human observer. However the CIE scales have three major inadequacies from the point of view of one who is using them to measure colors or color differences of products:

1. The CIE scales were not intended for, nor are they convenient for, identifying the colors of objects, because they were designed for color stimuli of all modes: apertures, illuminants, objects. With objects, the CIE color specifications always refer to combinations of illuminant and objects, not to the object itself, the point of interest.
2. The CIE dimensions, $X$, $Z$, $x$, and $y$, are difficult to relate to perceived dimensions of colors of objects. Only $Y$, with its high correlation to luminance and object lightness, is easily interpreted.
3. CIE scales are not uniform in the spacing of colors as related to visual dif-

* In Table A.8 and occasionally in this chapter, scales are identified by means of subscripts. Where identification is clear without them, subscripts have been omitted.

ferences. The relative sizes of differences between colors as plotted in various parts of the CIE *Y, x, y* color space do not correspond to the sizes of the differences as seen by the human eye. This is understandable when one recalls that the basic experiment was one in which observers were asked only to match one test light with a mixture of three other colored lights, not to estimate differences. Figure 8.1 illustrates this nonuniformity by showing a vertical section through the *Y, x, y* color solid, with ellipses of uniform visual-difference radii. The major axes of the ellipses lengthen in the horizontal direction as one goes to the dark part of the solid. An ideal uniform color solid would plot as circles of equal diameter throughout.

The development of alternate scales soon after the adoption of the CIE system, therefore, was due to the need for object appearance scales, as opposed to the aperture color-matching scales of the CIE. The Standard Observer pro-

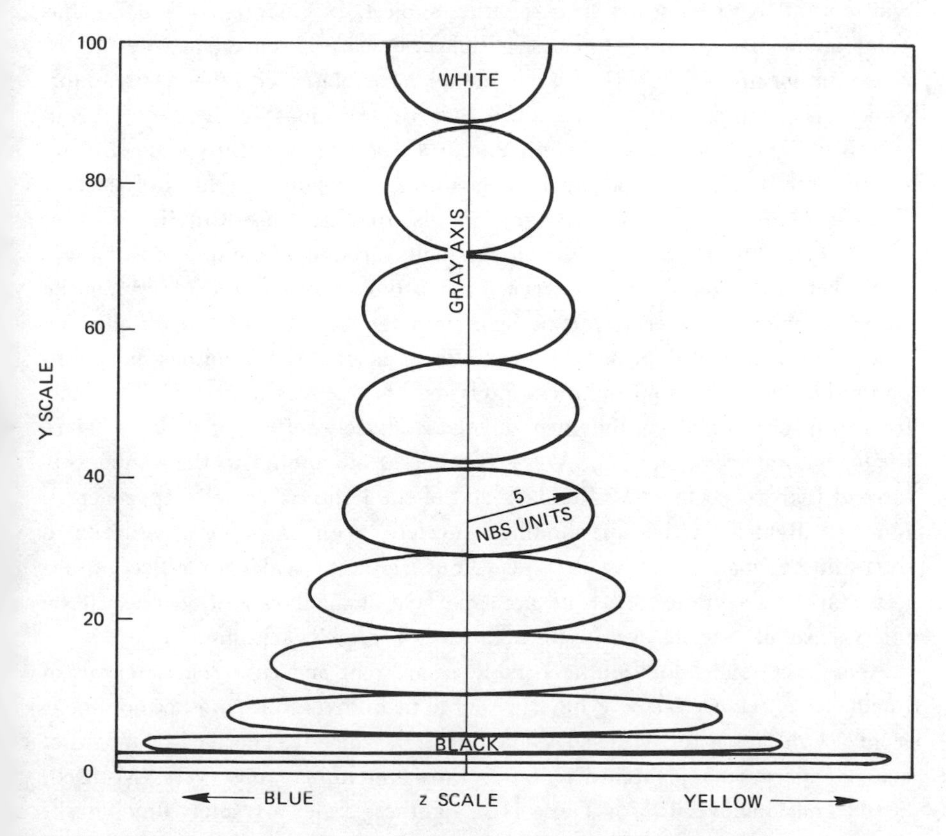


**Figure 8.1.** The nonuniformity of the CIE color solid in the *Y* and *z* dimensions. Departures from circles toward ellipses show extent of nonuniformity.

vided the data representing the human observer, and the designers of UCS color systems transformed these data by equations to create scales that:

1. Gave reasonably uniform measures of visual color difference
2. Were readily understandable, usable, and relatable to visually recognized attributes of object color appearance
3. Were adaptable to, or usable with, specific methods of instrumental analysis

Twenty of these scales and their important features are given in Table A.8.

## LIGHTNESS PERCEPTION AND UNIFORM LIGHTNESS SCALES

Before considering the development of the uniform color scales as a whole, we will look at lightness scales as a separate subject. In Chapters 1 and 7 it has been shown that spectrally the visual lightness sense is represented by the so-called luminosity or $\bar{y}$ function. Thus lightness correlates with the $Y$ tristimulus value. In Chapter 4, it was shown that the relationship between $Y$ and perceived lightness intervals is nonlinear. In the general aperture mode of viewing this relationship appears to be logarithmic, following the laws of Weber-Fechner which apply not only to light, but also to other sense stimuli.

In the aperture mode, the observer sees light through a hole in an instrument in darkened surroundings. However, the usual condition for observing the colors of objects in everyday experience is a lighted situation surrounded by other objects. The lighting characteristics and the surroundings may importantly affect the visual lightness intervals that one sees in objects. The most frequently used lightness judgment involves diffusely reflecting surfaces. Here the logarithmic relation of the Weber-Fechner scale applies to the lighter surfaces of high reflectance. As the dark end of the lightness scale is approached, however, light from the surroundings interferes with an observer's ability to discriminate small lightness intervals. Thus, lightness scales for reflecting objects start at zero reflectance and are separated at the dark end of the scale by intervals small in terms of reflectance factor but large logarithmically.

As uniform scales for lightness, both square-root and cube-root intervals of luminous reflectance factor $Y$ have proved to be quite useful. Another nonlinear scale of lightness is the Munsell Value function, which was derived by means of actual experiments involving painted chips and human observers (Munsell, Sloan, and Godlove, 1933). These UCS lightness scales are solely functions of $Y$. However, there has been considerable evidence advanced by Wyszecki and others that both saturation and, to a lesser extent hue, make a contribution to the visual impression of lightness (Sanders and Wyszecki, 1957).

Figure 8.2 compares some of the different functions of *Y* proposed for rendering lightness intervals and shows that the cube-root and Munsell Value functions are very similar in spacing. The rather significant differences in the curves shown come from two sources:

1. Most of the formulas are attempts to use simple equations to approximate visual response (only Munsell Value is a true reproduction of visual data).
2. The reflectance of the background and optical mode of the objects have a major effect on the amount of bending that occurs in the curve.

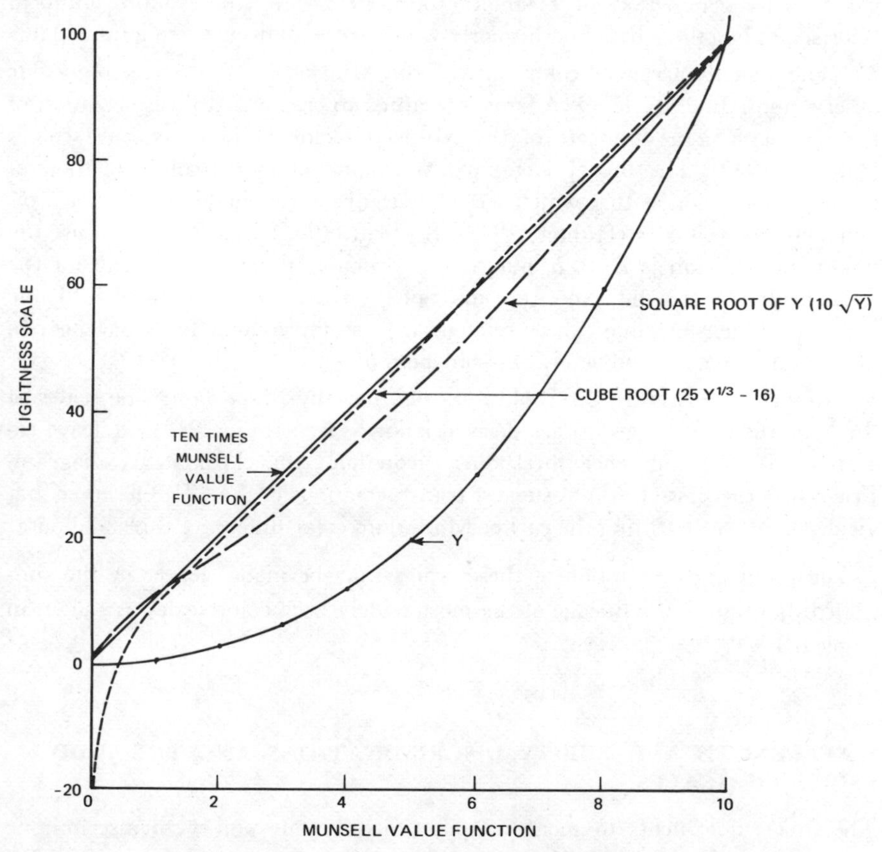


**Figure 8.2.** Comparison of lightness scales with the Munsell Value function.

Thus, no one lightness scale can be said to be "right" unless it compensates for the background. On the other hand, a scale that varies with background would be useless for industrial specifications of color since lightness values could no longer be assigned to the product alone.

## THE DEVELOPMENT OF UNIFORM COLOR SCALES

There were three groups of visual data on color-difference perception that served as bases for the uniform color scales that followed the establishment of the Standard Observer in 1931:

1. Monocular data on wavelength and purity discrimination from the National Bureau of Standards and other laboratories. Judd's pioneer uniform chromaticity scale Maxwell Triangle (Judd, 1935), and the group of uniform color scales by others based on his early work, were all derived from these data.
2. The visually arranged color chips of the Munsell Color System (binocular observation). In 1942 E. Q. Adams attempted to create in a transformation of the CIE system a duplicate of the Munsell Color Solid for surface colors (Adams, 1942). His theory suggested the same opponent-colors method of human color vision as that which served as the basis for the Hunter $L'$, $\alpha'$, $\beta'$ opponent-colors scales (Hunter, 1942). Scofield (1943) was the first to use the now popular symbols *L, a, b,* but Adams probably deserves the credit for the *L, a, b* type color solid. Note that most of the color scales identified in Table A.8 have dimensions that can be related, at least approximately, to lightness *L*, redness-greenness *a*, and yellowness-blueness *b*.
3. Binocular-difference threshold data obtained by MacAdam. The scales of the MacAdam-Friele group are derived from MacAdam's (1942) extensive research on color difference thresholds throughout the chromaticity diagram. Friele was the first (1961) to suggest transformations of the CIE Observer that yielded color scales fitting the earlier MacAdam color-difference threshold data.

The following description of these scales may be made clearer by the simplified diagram of the lineage of the most widely used color scales as shown in Table 8.1.

## WAVELENGTH AND PURITY DISCRIMINATIONS AND THE JUDD SERIES OF SCALES

The first experiments to measure quantitatively our ability to discriminate between colors investigated the human observer's ability to perceive the color differences between adjacent wavelengths of light. Figure 8.3 shows the results

**TABLE 8.1**
FAMILY TREE OF COLOR SCALES

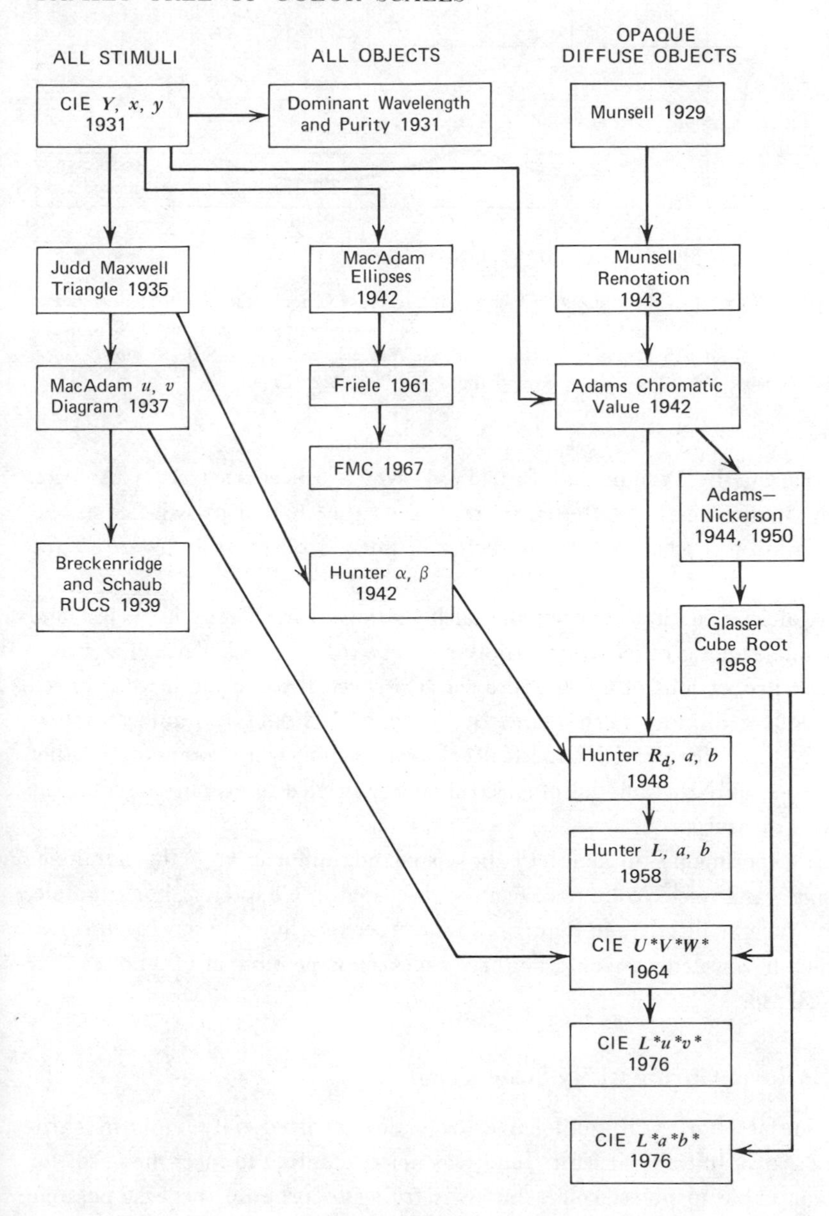

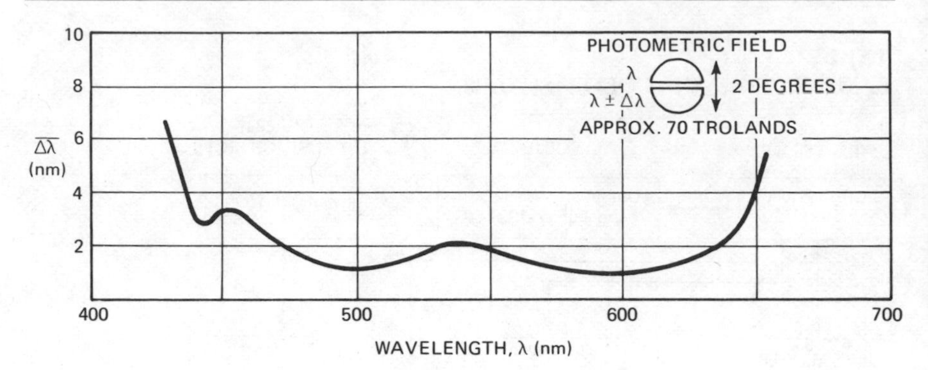


**Figure 8.3.** Wavelength discrimination curve based on observations made by Wright and Pitt (1934). The curve indicates good discrimination at about wavelengths 485 nm and 590 nm, and poor discrimination at 425 nm and 650 nm. From G. Wyszecki and W. S. Stiles, *Color Science,* John Wiley & Sons, 1967. Used with permission of *Proc. Phys. Soc.,* Lond.

of observations by Wright and Pitt (1934) which are characteristic. At each wavelength, the height of the curve represents the change in wavelength (of equal luminosity) that must occur before a normal observer is aware of the change.

Purity discrimination experiments such as those carried out by Priest and Brickwedde, among others, were similar except that instead of varying wavelength the proportions of white and a selected spectral wavelength were varied in an additive mixture (Priest and Brickwedde, 1926). The number of perceivable "steps" between the spectral color and white were counted. In other experiments, only the amount of spectral energy needed to visually discolor the white was measured.

These experiments all point to the visual nonuniformity of the dominant wavelength and purity color specification discussed in Chapter 7. For example, from the Wright-Pitt data in Figure 8.3 we can see that hue differences expressed in dominant wavelength would be three times as perceptible at 600 nm as they are at 450 nm.

## Judd Uniform Chromaticity Scale Triangle

Judd's initial effort with uniform color scales involved only uniform chromaticity scales. Initially at least, Judd was not attempting to meet the needs for scales adaptable to object colors but was trying to propose the best possible scale for the measurement of visual chromaticity differences as represented by threshold discrimination differences in wavelengths and purity. The techniques of achieving linear transformations were the same as the ones used to generate the Standard Observer. In 1935 Judd described a transformation that he called the "Maxwell Triangle" since it proposed three equally important visual responses as suggested by Maxwell in 1860 (Judd, 1935).

Figure 8.4 shows the triangular coordinates of the Judd-Maxwell Triangle. The location and spacing of the spectrum locus and temperature radiators as transformed by this scale are shown within the triangle. In this figure, distance on the transformed chromaticity diagram corresponds to visual chromaticity difference. For example, notice that the widest spacing of wavelengths occurs at two places—near 485 nm and near 590 nm. If we refer to Figure 8.3 we see that these are in fact the spectral regions in which our eyes do achieve the best spectral resolution.

To demonstrate how the Maxwell Triangle predicted color spacing in the *x*,

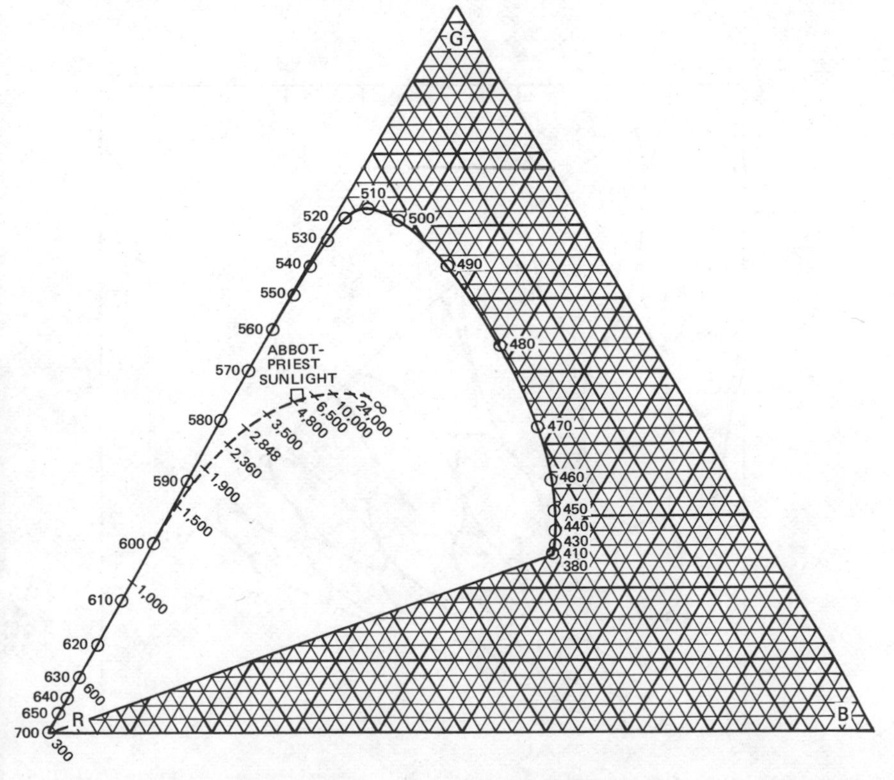


**Figure 8.4.** Uniform chromaticity triangle according to Judd. Distance corresponds to chromaticity difference. The areas which show the widest spacing of wavelengths are near 485 nm and 590 nm. From D. B. Judd, *J. Opt. Soc. Am.*, **25**, 27, January 1935.

*y* diagram, Judd transformed circles of equal size in the Maxwell Triangle back into CIE *x, y* space. These are shown in Figure 8.5 as distorted ellipses, each representing "equivalent visual distances" from their center points as predicted by the transformation. Between any center and any point on the corresponding ellipse there are approximately 100 just-noticeable chromaticity steps. This figure illustrates the nonuniformity of the *x, y* chromaticity diagram as Figure 8.1 showed the even greater nonuniformity of CIE *Y, x, y* color space when the lightness dimension is added.

## MacAdam *u, v* Diagram

In 1937 MacAdam modified the Judd Triangle into a more usable form (MacAdam, 1937). Keeping its essential shape (as can be seen in Figure 8.6) he moved away from an implied trireceptor theory of vision (Young-Helmholtz, Chapter 4) and used rectangular rather than triangular coordinates. His *u, v*

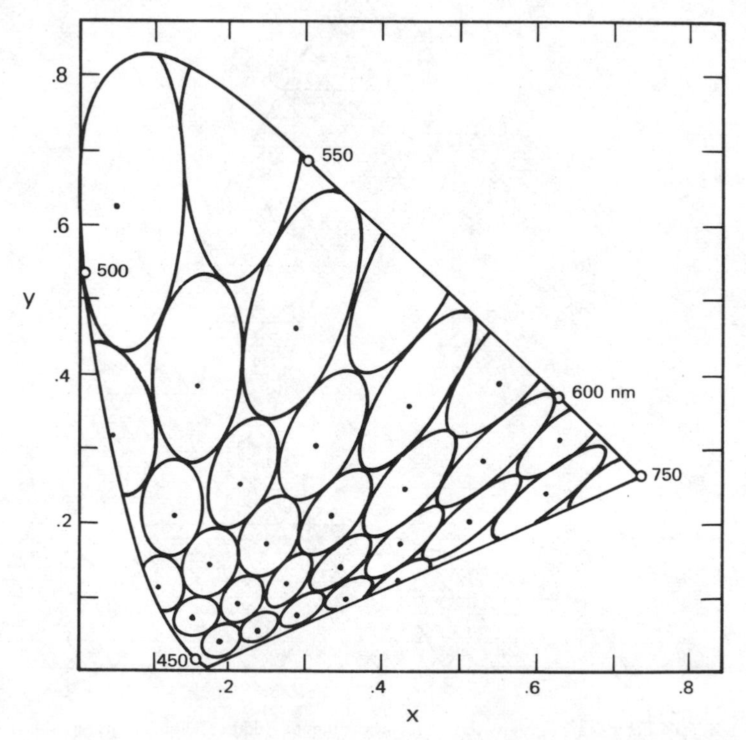


**Figure 8.5.** Uniform chromaticity spacing in the *x, y* diagram. From I. Nimeroff, NBS Monograph 104, 1968.

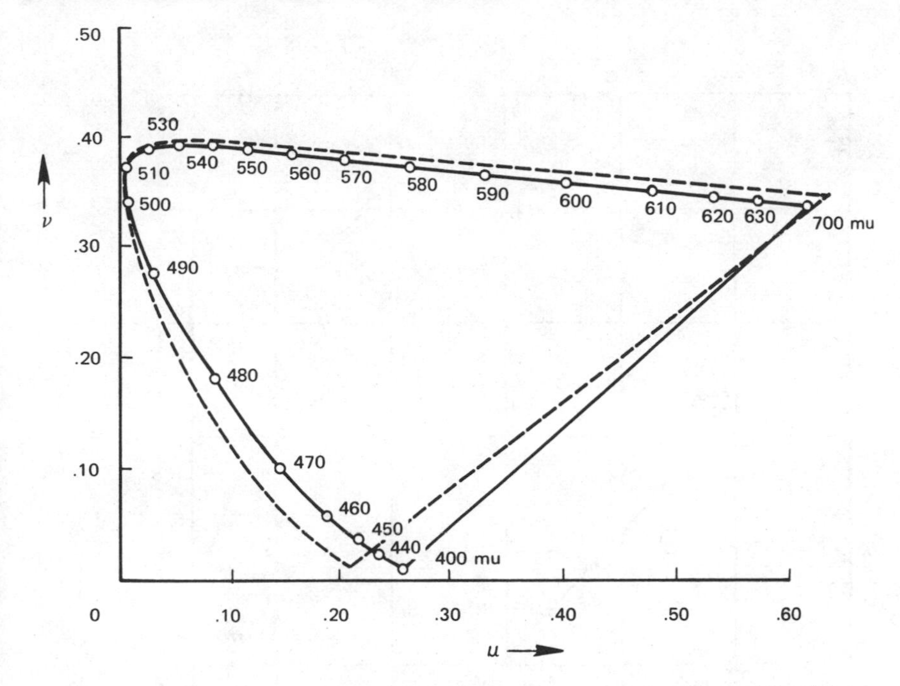


**Figure 8.6.** The loci of spectrum colors corresponding to two transformations of the CIE color mixture—Judd's *r, g, b* Maxwell Triangle system (dotted line) and MacAdam's *u, v* system (solid line). From D. L. MacAdam. *J. Opt. Soc. Am.*, **27,** 1937.

chromaticity triangle was the first to be oriented in red-to-green and yellow-to-blue dimensions. MacAdam's equations are shown in Table A.8. MacAdam's 1937 paper is an excellent source for those wishing to pursue the techniques for achieving linear transformations of *x, y* or *X, Y,* and *Z* into other tristimulus scales.

## Breckenridge and Schaub: RUCS

Breckenridge and Schaub modified the MacAdam *u, v* diagram to achieve another Rectangular Uniform Chromaticity System (Breckenridge and Schaub, 1939). The RUCS system also proposed chromatic scales corresponding to the red-green and yellow-blue opponent colors dimensions of Hering (See Chapter 4). This was accomplished by placing equal energy white at the 0,0 point, and orienting the diagram so that the horizontal axis corresponded to blue-yellow differences while the vertical axis corresponded to red-green differences. Table A.8 gives the RUCS equations, and Figure 8.7 shows the RUCS diagram.

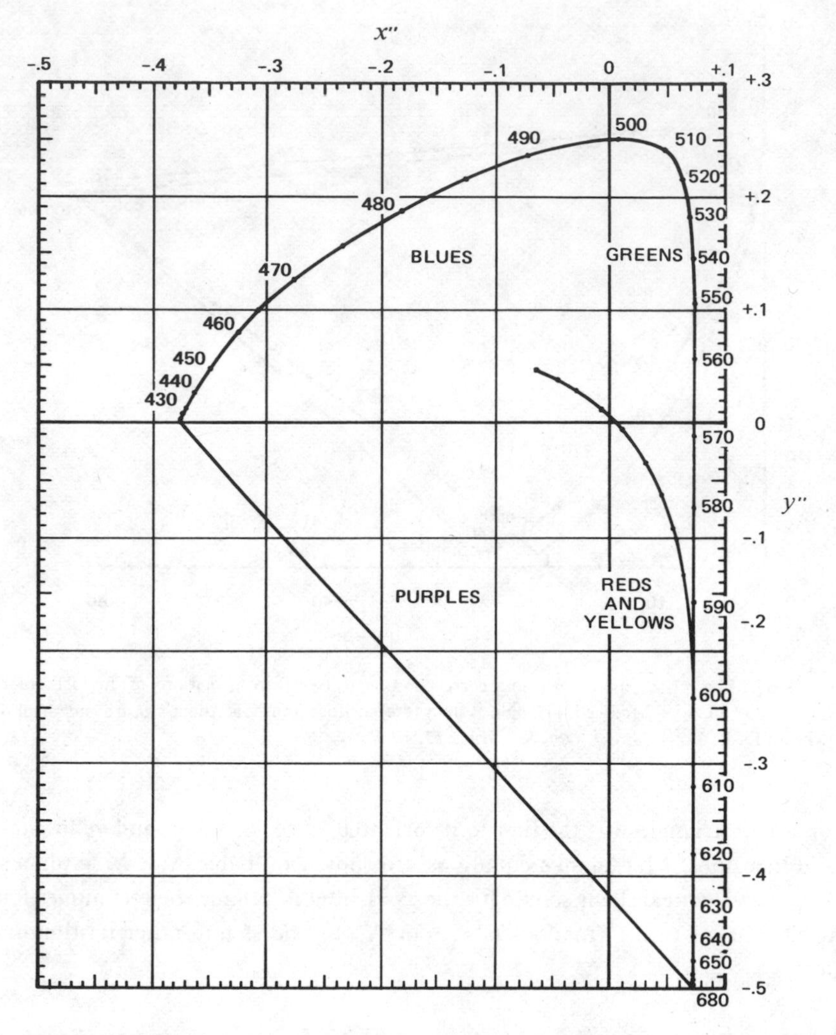


**Figure 8.7.** The RUCS chromaticity diagram proposed by Breckenridge and Schaub in 1939. The equal energy white is placed at the center (0,0) point, with a horizontal yellow-to-blue axis and a vertical red-to-green axis. From F. C. Breckenridge and W. R. Schaub, *J. Opt. Soc. Am.*, **29**, 1939.

To one who thinks in terms of colors of objects, the RUCS diagram emphasizes better than any previous one the necessity for special scales for objects. The RUCS system locates white, in this case equal energy, at the center (0,0 point) of the diagram. But the location of object color moves in the diagram with change of illuminant. There is no provision for the phenomenon of object-color constancy previously discussed (Chapter 4). Breckenridge and Schaub were not interested in object-color identification; their RUCS system was intended for, and has been used primarily for, identification of the colors of lights (signals, identification lights, marker lights).

## Hunter Alpha-Beta Diagram and $L'$, $\alpha'$, $\beta'$ Color Solid

In 1940 Hunter, attempting to develop an instrumental colorimeter, took an important step toward the present uniform object-color scales. Using Judd's Maxwell Triangle as a starting point, but paying particular attention to the chromaticity spacing of a set of ten Munsell colors, he worked to develop a simple transformation that could be employed with the *A, B, G* readings from his tristimulus Multipurpose Reflectometer. (*A* is the amber portion of the CIE $\bar{x}$ function; *B* is the blue, or CIE $\bar{z}$, function; *G* is the same as the CIE green, or $\bar{y}$, function.) Working with Judd, Hunter found that differences between *A* and *G* and between *G* and *B* yielded a relatively good transformation. He also found that $A - G$ and $G - B$ corresponded to the visual opponent-colors attributes of red-greenness and yellow-blueness, respectively. He called these new scales $\alpha$ and $\beta$, where

$$\alpha = \frac{A - G}{A + 2G + B} \qquad \beta = \frac{0.4\,(G - B)}{A + 2G + B}$$

Hunter's work with reflectometers directed his attention to the problems of object color. In describing $\alpha$ and $\beta$ in 1942 he demonstrated both the closeness of his system to Judd's and the spacing of the ten Munsell colors used, as shown in Figure 8.8 (Hunter, 1942). Percent specifications of *A, G,* and *B* were implied so that whites and neutrals remained at the neutral chromaticity point ($\alpha$ and $\beta$ equal to zero.)

The improvement in visual uniformity made by the Hunter diagram over the *x, y* diagram is shown in Figure 8.9. Here the ellipses of Figure 8.5 are plotted in $\alpha$, $\beta$ space, becoming much more uniform in size and more closely approximating circles.

Further, Hunter proposed the addition of the lightness dimension *L*. All previous attempts at uniform transformation had been for chromaticity only.

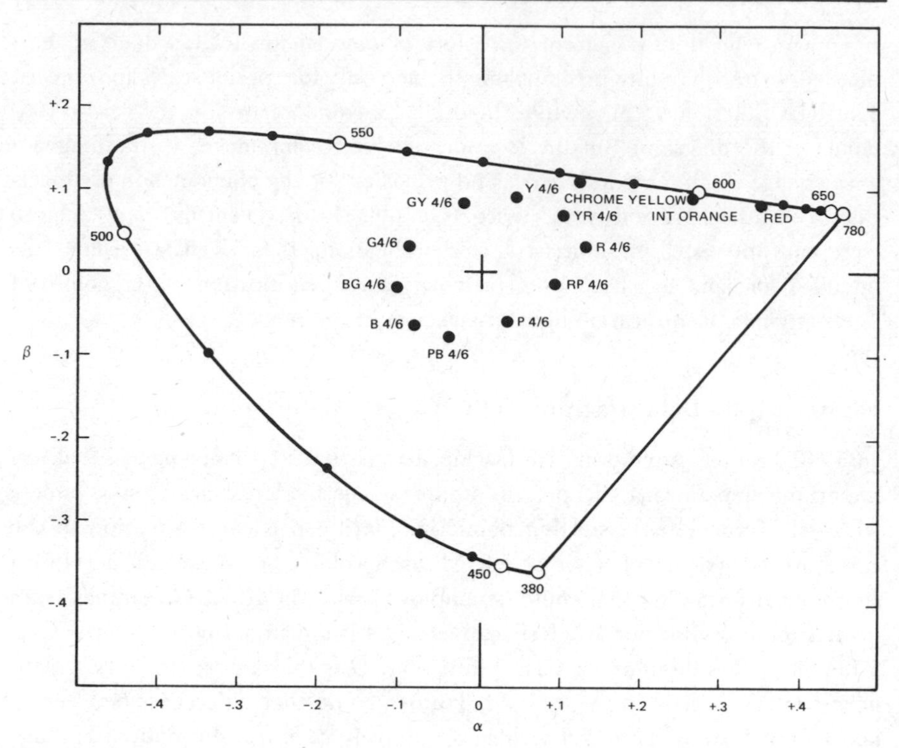


**Figure 8.8.** The Hunter alpha-beta diagram, showing the spectrum locus and points representing ten Munsell colors (Hunter, 1942).

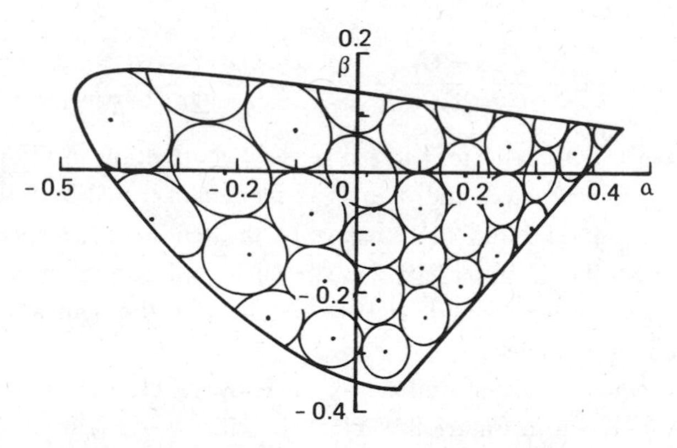


**Figure 8.9.** Tolerance ellipses on alpha-beta diagram. The ellipses of just-noticeable differences on the *x, y* diagram in Figure 8.5 are plotted here on the alpha-beta diagram. Improvement in visual uniformity is shown by the closer approximation of the ellipses to circles in this figure. Reprinted with permission from J. M. Adams, *Optical Measurements in the Printing Industry,* 1965, Pergamon Press, Ltd.

Hunter's $L'$, $\alpha'$, $\beta'$ scales were based on a transformation of CIE space into a three-dimensional, visually uniform, object-color space with the dimensions of opponent-colors space (see Figure 8.10). The visual chromaticity expansion that occurs with changes in lightness (see Figure 8.1) was accounted for by multiplying $\alpha$ and $\beta$ by a function of lightness so that $\alpha' = f(Y)\,\alpha$ and $\beta' = f(Y)\,\beta$. Based on a suggestion by Judd, 700 $Y^{1/4}$ was used for $f(Y)$. The Hunter 1942 equations are shown in Table A.8.

Hunter further investigated the use of $L'$, $\alpha'$, $\beta'$ for color-difference measurement. He recognized at this time, which no other color-difference scale before or since has done, that perceived color difference will depend on the proximity of the specimens intercompared and on their glossiness. Accordingly, the 1942 Hunter color-difference equation includes factors to account for these variables. It is this equation, with selected constants, that is now generally termed the "NBS unit of color difference."

Scofield (1943) proposed a simple version of the Hunter surface-color scales and introduced the symbols $L$ (for lightness), $a$ (for redness-greenness), and $b$ (for yellowness-blueness) now so widely used. His were the first equations to use the square-root of $Y$ in all three terms (see Table A.8).

## The CIE $U^*V^*W^*$ System

In 1964 the CIE adopted a new uniform color system designated $U^*V^*W^*$ (Wyszecki, 1963). The CIE $U^*V^*W^*$ space uses essentially the 1937 MacAdam chromaticity diagram (adopted by the CIE in 1960 and shown in Figure 8.11) and the 1958 Glasser cube-root lightness function. This system may be used for any stimulus mode. When objects are considered, the neutral colors are placed in the center of the chromaticity diagram by measuring chromaticity differences from the illuminant ($u_0$, $v_0$).

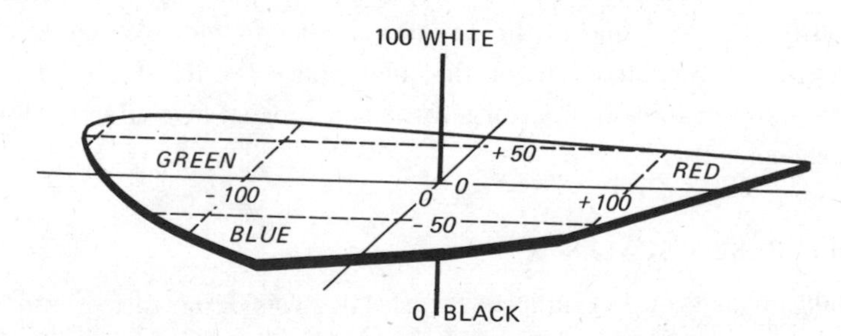


**Figure 8.10.** The $L'$, $\alpha'$, $\beta'$ surface color solid (Hunter, 1942). The $L'$, $\alpha'$, $\beta'$ color solid proposed the addition of the third dimension of lightness to the two-dimensional chromaticity diagrams of previous figures.

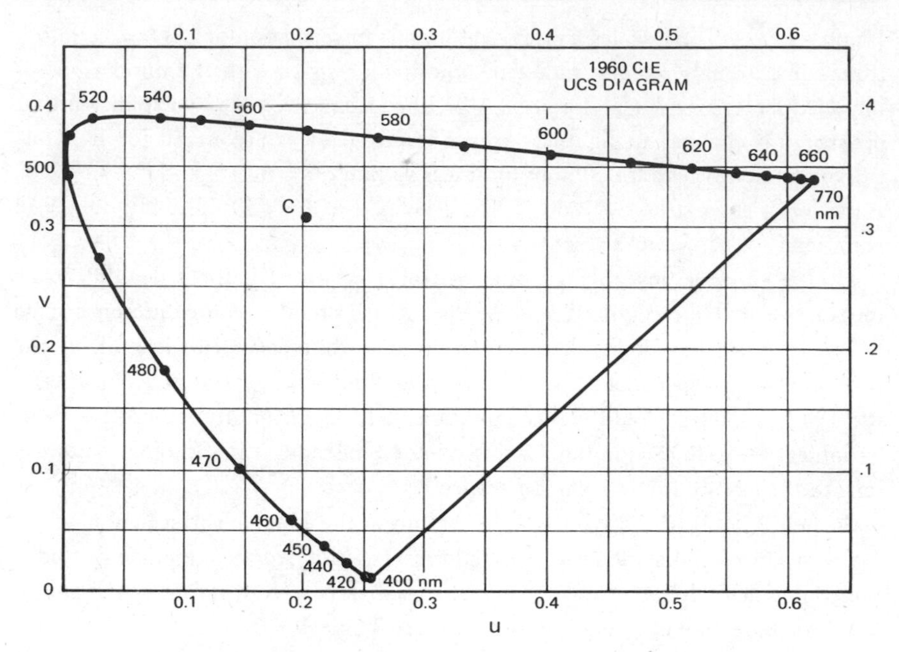


**Figure 8.11.** The 1960 CIE diagram. This diagram is essentially the same as MacAdam's 1937 *u, v* chromaticity diagram. From G. Wyszecki and W. S. Stiles, *Color Science,* John Wiley & Sons, 1967. Used with permission of *Proc. Phys. Soc.,* Lond.

The CIE 1964 $U^*V^*W^*$ space was subjected to extensive field tests involving many visual experiments. As a result the CIE endorsed the system as representative of the limits of visual uniformity that could be achieved, at that time, by linear transformation of the Standard Observer.

In 1974 an improvement over the 1964 $U^*V^*W^*$ space was proposed, particularly with regard to the uniform spacing of object colors with differences of sizes such as those found in the Munsell Book of Color (Wyszecki, 1974). The name under consideration for this new space is the CIE 1976 $L^*u^*v^*$ space—the latest chapter in the story of the star-crossed love affair with the vision of ideal color space.

## MUNSELL BASED SCALES

The second major group of uniform color scales was designed to simulate visually adjusted spacings of surface colors. By far the best known organized spacing of surface colors is the Munsell Color System. Conceived by A. H. Munsell, this system was developed not as an experiment in psychophysics, but to meet his needs as an artist for a means to identify and interrelate colors of surfaces (Munsell, 1905). The system was published in the form of a book, *The Munsell Book of Color,* which contained a collection of colored chips so arranged that differences between adjacent chips were seen to be equal (Munsell, 1929). The coordinates of the Munsell system are value (lightness), chroma (saturation), and hue.

The Munsell system provided the experimental basis for a number of sets of uniform color scales. In the early 1940s the original Munsell color chips were measured spectrophotometrically and specified by $X$, $Y$, and $Z$ values for Illuminant C. Visual spacing was reassessed based on visual experiments with a neutral gray background. In these color-spacing experiments, observers were asked to select chips that were visually between and equally different from two other chips. Then these equal differences were subdivided again in the same way. Results were compared, averaged, and adjusted to smooth out some of the apparent inconsistencies. Extrapolations to colors more saturated than those achievable with real pigments were also made. The derived relationships between instrumentally computed Munsell colors and visual estimates are called the Munsell Renotation System (Newhall et al., 1943).

Figure 8.12 shows the locations of some Munsell renotation colors in the $x$, $y$ diagram. Note the indication that the diagram is visually nonuniform. For example, we see from the figure that, if the spacing between Munsell chips is visually uniform, the eye is far more sensitive to changes in $x$ and $y$ in the lower (violet) portion of the diagram than it is in the upper (green) portion.

The Munsell renotation effort carried out in the early 1940s gave workers the first good instrumental data on the visually uniform spacing of diffuse object colors under daylight. The result was a number of Munsell based color scales.

## Adams: Chromatic Value and Chromatic Valence

The Adams surface-color scales were proposed initially to support a theory of color vision based on the opponent-colors theory of Hering (Adams, 1923). Adams decided to test his theories with the then newly obtained data on the Munsell System. It was through his efforts that the validity of taking tristimulus response differences (as Hunter had done with $\alpha$ and $\beta$) became generally appreciated. Adams published his papers relating $X - Y$ and $Z - Y$ to Munsell color spacings in 1942 and 1943. He proposed two sets of color measurement systems, chromatic value and chromatic valence (Adams, 1942, 1943).

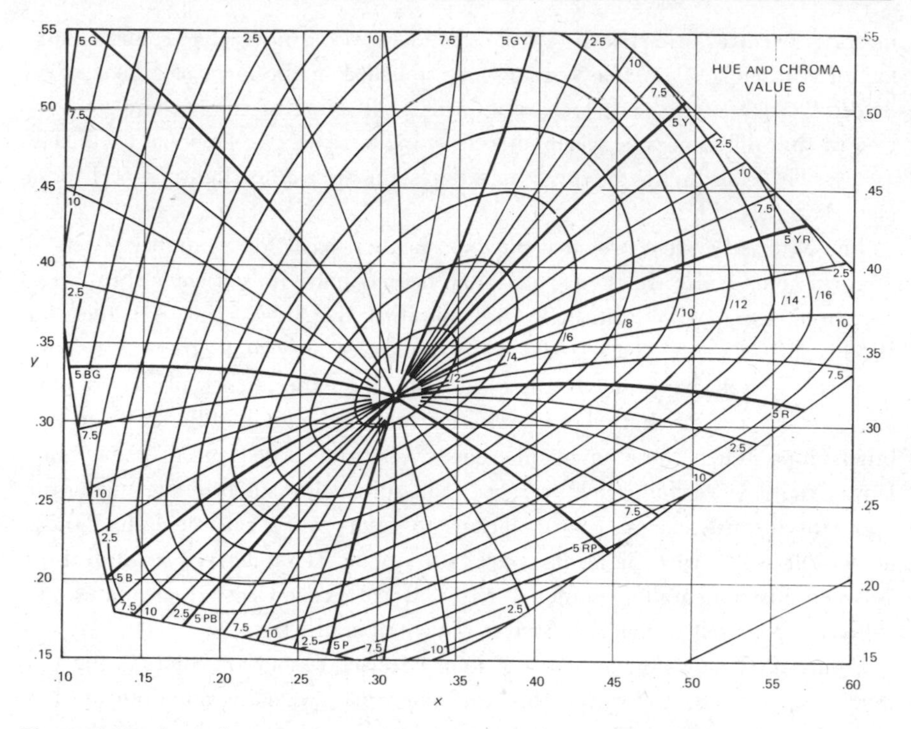


**Figure 8.12.** Loci of constant hue and constant chroma in the *x, y* diagram at value 6. The nonuniformity of the *x, y* diagram is shown in this figure by the contracted spacing between Munsell colors in the lower left of the diagram. From S. M. Newhall, D. Nickerson, and D. B. Judd, *J. Opt. Soc. Am.*, **33,** 1943.

In both of Adams' systems the three receptors in the eye are assumed to actually have $\bar{x}$, $\bar{y}$, and $\bar{z}$ spectral responses. The nerve signals relate to $Y$ for lightness response and to differences of $X - Y$ and $Z - Y$ for chromaticity, as shown in Figure 8.13. In producing his value and valence scales, Adams utilized the preliminary Munsell spacing data. In the first Chromatic Value Scale he apparently reasoned that the nonlinear response to stimuli so characteristic of all sensory functions would apply to all three visual receptors alike. Since lightness is simply the response from the $Y$ receptor in Adams' model, it further followed that the formula for visual nonlinearity must be the Munsell Value Scale. He accordingly applied the value function of the Munsell system to the $X$ and $Z$ responses as well, before taking the differences $V_x - V_Y$ and $V_z - V_Y$.

Adams further found that the $Z - Y$ difference had to be contracted to 0.4 times that of the $X - Y$ difference to properly space the Munsell colors. Thus, the value system became

$$L_{val} = V_Y \qquad a_{val} = V_X - V_Y \qquad b_{val} = 0.4\,(V_Z - V_Y)$$

Adams then graphed the data on Munsell colors in his new color space, with astonishing results. The application of the value function in the above equations generated good hue circles and almost perfectly compensated for the lightness expansion needed with changes in lightness level. Figure 8.14 shows these chromaticity graphs for light, medium, and dark Munsell colors. Note how much nearer to circles are the ellipses of Figure 8.14 than those of Figure 8.12. Note also how the circle sizes are much the same at different lightness levels.

In fact, the Adams Chromatic Value Scale spaced the Munsell colors so well that infrequent inconsistencies found were later used as a basis for altering some of the Munsell designations in the Munsell Renotation System.

Adams' Chromatic Valence System (1943) followed essentially his same visual model. In the valence scales, however, $X - Y$ and $Z - Y$ differences were computed prior to applying the lightness function compensation. In this respect the formulas were similar to those of the Hunter $L$, $a$, $b$ system. Adams' 1942 paper originated the practice of using $X_c$, $Y_c$, and $Z_c$ to represent percent tristimulus values of $X$, $Y$, and $Z$ under Illuminant C (herein designated as $X_{\%}$, $Y_{\%}$, and $Z_{\%}$). The Adams Chromatic Value and Chromatic Valence Scales are shown in Table A.8.

## Adams-Nickerson and Modified Adams-Nickerson Scales

Starting with Judd's introduction in 1939 of the NBS unit of color difference (Chapter 9) and Hunter's proposal of the $L'$, $\alpha'$, $\beta'$ color scales, there developed a trend toward surface color scales with exactly or approximately 100 units between black and white. The Adams Chromatic Value and Chro-

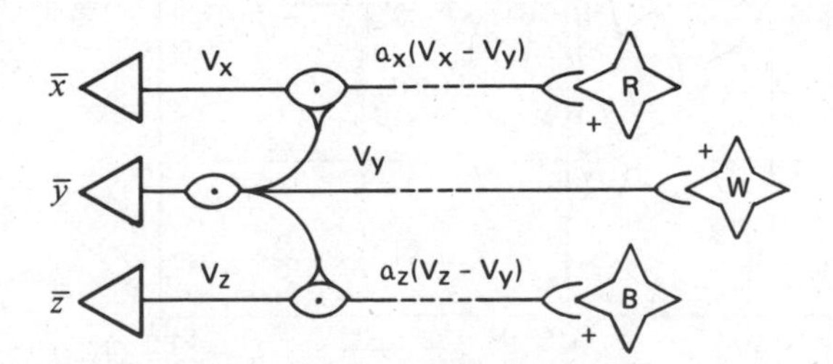


**Figure 8.13.** Diagram representing Adam's theory of color vision. After conversion to value function, the $V_Y$ function serves not only as a lightness function, but also is compared with the $V_X$ function and with the $V_Z$ function; $V_X - V_Y$ gives red-green chromaticity, and $V_Z - V_Y$ gives yellow-blue. From E. Q. Adams, *J. Opt. Soc. Am.*, **32,** 1942.

matic Valence Scales have only ten units between black and white. Miss Nickerson (1950) proposed rounded constants to approximate, in scales intervals now called Adams-Nickerson, the Judd NBS unit of color difference. Glasser and Troy (1952) selected what they considered the ideal (unrounded) constants for the same purpose and used them in the scales now called Modified Adams-Nickerson. Adams originally described the yellowness-blueness dimension as being the result of the $(V_Z - V_Y)$ difference. In present day use of Adams-based scales the difference is considered to be $(V_Y - V_Z)$, resulting in a plus $b$ value indicating yellow and a minus $b$ value for blue. The difference between these two Adams-Nickerson scales is slight, as shown:

| Adams-Nickerson | Modified Adams-Nickerson |
|---|---|
| $L_A = 9.2V_Y$ (92 units black to white) | $L_M = 10\ V_Y$ |
| $a_A = 40\ (V_X - V_Y)$ | $a_M = 41.86\ (V_X - V_Y)$ |
| $b_A = 16\ (V_Y - V_Z)$ | $b_M = 16.74\ (V_Y - V_Z)$ |

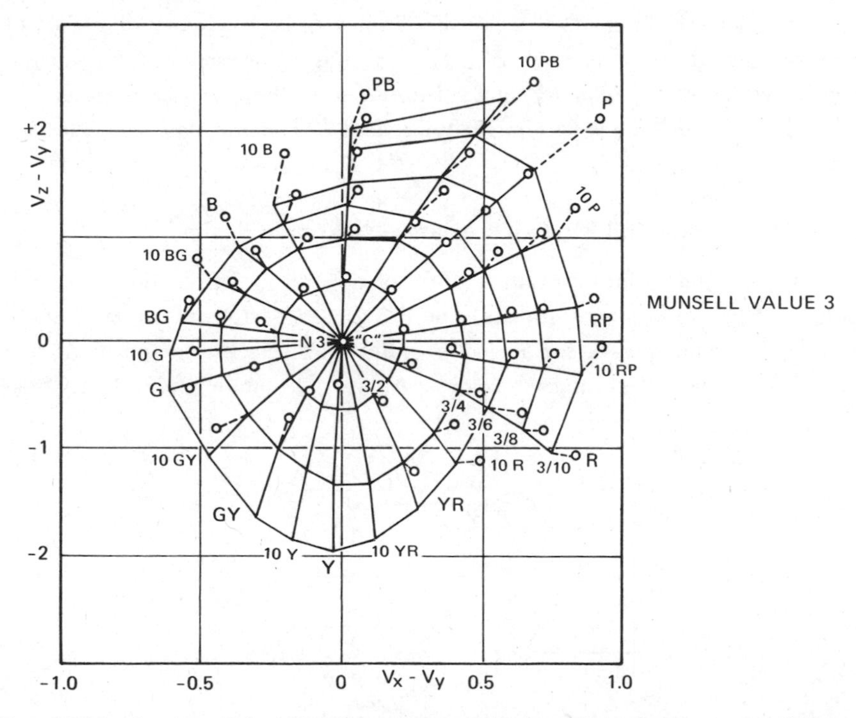


**Figure 8.14.** Spacing of Munsell Colors in Adams' Chromatic Value System, at three lightness levels. The improvement in uniformity is shown by the closeness of the ellipses to circles, and the fact that the ellipses do not become radically larger at lower lightness levels. From E. Q. Adams, *J. Opt. Soc. Am.*, **32,** 1942.

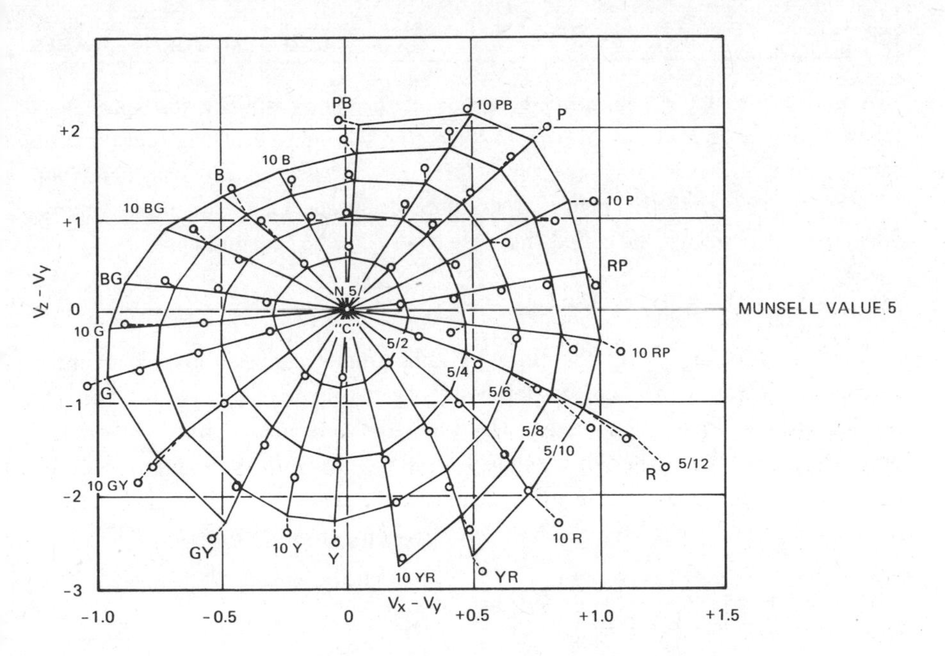


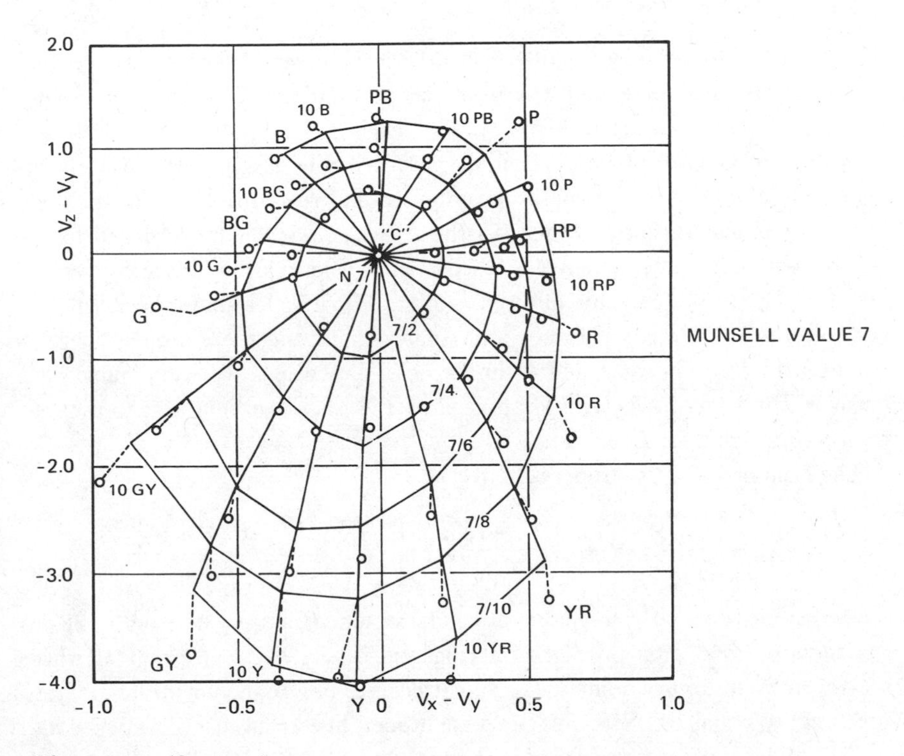


**Figure 8.14** (Continued)

A new formula for a simplified version of the Adams-Nickerson space was taken under consideration by the Colorimetry Committee of the CIE, in the same proposal in which the 1976 CIE $L^*$ $u^*$ $v^*$ space was described (Wyszecki, 1974). If the proposal is adopted by the CIE, the new Adams-Nickerson formula will be called the CIE 1976 $L^*$ $a^*$ $b^*$ formula.

## Hunter $R_d$, *a*, *b* and *L*, *a*, *b* Scales

In 1948, continuing along the lines he and Adams had established, Hunter developed the $R_d$, *a*, *b* Color Scales, which were intended for use in a direct-reading photoelectric colorimeter. His objective was not to improve on the previous scales, but rather to develop formulas that could be computed automatically by analog devices in an instrument. He was able to partially accomplish this by developing an algebraic lightness-compensation function $f(Y)$.

Where $f(Y) = \dfrac{0.51\,(21 + 0.2Y)}{1 + 0.2Y}$ in the $R_d$, *a*, *b* scales

$$L_{Rd} = Yf(Y) \qquad a_{Rd} = 1.75f(Y)(X_{\%} - Y) \qquad b_{Rd} = 0.70f(Y)(Y - Z_{\%})$$

The earliest Hunter Color Difference Meter (Hunter, 1948) computed and read $a_{Rd}$ and $b_{Rd}$ directly, but the instrumental readings of lightness $R_d$ do not correspond to $L_{Rd}$ but are rather equal to $Y$ itself.

In the period 1950 to 1958 Hunter achieved his desired objective of uniform opponent-colors scales that could be computed and read directly from the dials of an instrument (Hunter, 1958). In the $L_L$, $a_L$, $b_L$ scales he returned to the square-root lightness function and used Adams' chromatic valence concept of taking CIE tristimulus value differences first. These scales show the 2.5 to 1.0 relationship used by both Hunter and Adams to expand the *a* dimension relative to *b*. In addition the *a* and *b* chromaticity dimensions were expanded by a factor of 7 to bring intervals into approximate visual correspondence to those of the 100-unit lightness $L_L$ scale.

The Hunter $L_L$, $a_L$, $b_L$ functions are

$$L_L = 10\sqrt{Y} \qquad a_L = \frac{17.5\,(X_{\%} - Y)}{\sqrt{Y}} \qquad b_L = \frac{7.0\,(Y - Z_{\%})}{\sqrt{Y}}$$

These scales produced an opponent-colors system (Figure 8.15) with the dimensions originally proposed by Hering and others. The units on all three scales represent approximately equivalent degrees of visual color difference and are roughly equal to NBS units in magnitude. The relationship between $L_L$, $a_L$, $b_L$ values and perceived color is shown when CIE specifications (in percent

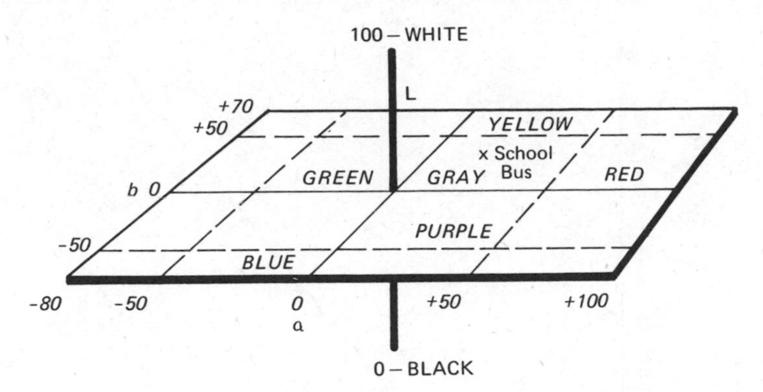


**Figure 8.15.** The Hunter *L, a, b* diagram. The location of the yellow school bus is shown. This system is considerably more uniform than is the CIE System. (See also Figure 1.6.)

form) of the yellow school-bus color (see Table A.5) are converted into $L_L$ = 63.5, $a_L = +11.9$, and $b_L = +35.4$, a light, saturated reddish-yellow.

### Glasser Cube Root

Glasser, Reilly, and others developed scales that are designed to provide an approximation to the Adams Chromatic Value Scales (Glasser et al., 1958) and which, like Hunter's *L, a, b* scales, can be computed by automatic analog devices. The formulas apply a lightness-response function to *X, Y,* and *Z* separately before taking differences. Then the differences are multiplied by constants to achieve approximate correspondence to NBS units of color difference. The cube-root scales are so called because they employ a cube-root function to approximate the Munsell value function. This simplifies computations without introducing significant changes in spacing of the Adams system. Figure 8.16 shows how closely the cube-root scale matches the Munsell Value Scale. The Glasser equations are given in Table A.8.

## MacADAM-BASED SCALES—THE MacADAM-FRIELE GROUP

In 1942 MacAdam published results of experiments in which a single observer made 25,000 observations of color differences from 25 selected points in the *x, y* chromaticity diagram. As in the previous wavelength and purity experiments carried out by others, the observer viewed a split screen and was asked to determine the point at which one half of the screen became visually different from the other. The experiment was designed to "measure" in the *x, y* diagram the

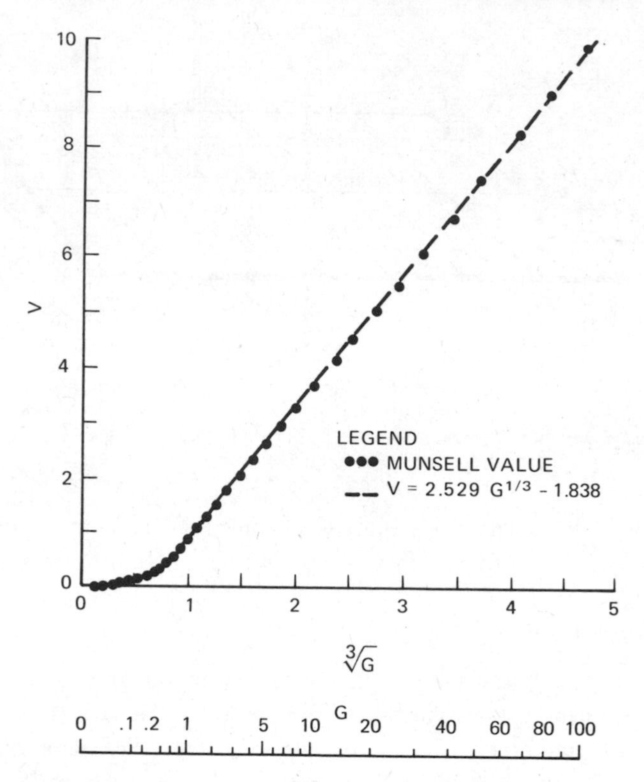


**Figure 8.16.** Comparison of cube-root and Munsell value functions. Note: $G = Y$ (see also Figure 8.2). From L. G. Glasser, A. H. McKinney, C. D. Reilly, and P. D. Schnelle, *J. Opt. Soc. Am.*, **48,** 1958.

distance in each direction that one would have to shift chromaticity to produce a just-noticeable, or threshold, color difference. Because of the dependence of such experiments on the level of illumination and adaptation of the eye, MacAdam surrounded his test field with a neutral gray field to simulate the object mode.

The results of his experimentation are summarized in Figure 8.17. The ellipses shown are for equiluminous colors only, and each represents the way MacAdam's observer "spaced" just perceptible differences from the 25 standard colors in the center of the ellipses. The ellipses are magnified ten times; that is, every radius of each ellipse is ten times the amount of chromaticity difference that one would have to shift in order to produce a just-perceptible change in color. Note that, as with the Munsell spacing in Figure 8.12, the eye can discriminate between small changes in $x$, $y$ in the violet region, while the eye is least sensitive to chromaticity changes in the green region of the diagram.

For many years the MacAdam data were used only for color difference analysis, not for color measurement. Only recently have the MacAdam small-difference measurements been associated with continuous color scales. Friele (1961, 1965, 1971) used the Mueller (1930) theory of color vision to provide direction for the opponent-colors scales he was seeking. He used the MacAdam data on equally perceptible intervals to guide the selection of tristimulus response functions.

Friele followed the Mueller theory in proposing that the lightness-response function $L$ is the result of summing red-receptor and green-receptor signals $(R + G)$. The redness-greenness $a$ response function, on the other hand, results

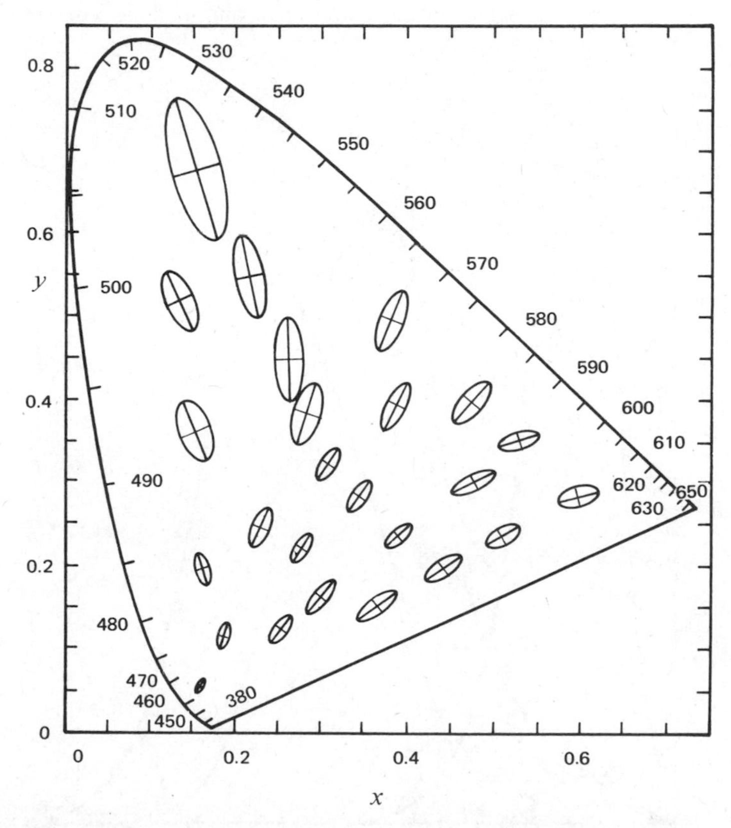


**Figure 8.17.** MacAdam ellipses of equal chromaticity difference. The observer spaced just perceptible differences from the 25 standard colors in the center of the ellipses. The ellipses are magnified ten times. The contraction of the ellipses in the lower left corner, indicates that the eye is more sensitive to changes in $x$ and $y$ in the violet region than in the green region. From *Color in Business, Science, and Industry* by D. B. Judd and G. Wyszecki, John Wiley & Sons, 1963. Used with permission of *J. Opt. Soc. Am.*

from taking the difference between the same two signals ($R - G$). The yellowness-blueness $b$ response function corresponds to subtracting the blue-receptor signal from the red-plus-green average, thus;

$$\left(\frac{R+G}{2}-B\right).$$

## MacAdam (*P, Q, S*) and the FMC Metric

In 1965 MacAdam modified the formulas of Friele to improve correlation with his own data, and generated the *P, Q, S* color specification. Chickering (1967) optimized the Friele-MacAdam equations generating the set of response primaries shown in Figure 8.18. These equations contained no corrections for

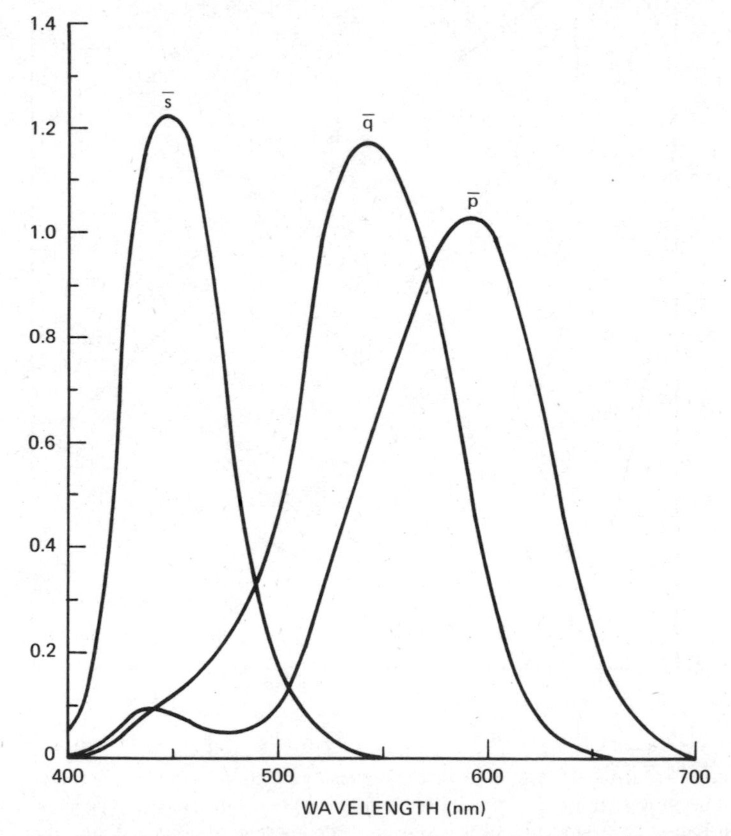


**Figure 8.18.** Chickering's $\bar{p}$, $\bar{q}$, $\bar{s}$ primary functions. From K. D. Chickering, *J. Opt. Soc. Am.*, **57**, 1967.

variation in lightness and chromaticity as lightness varied. Correction terms for this purpose were later suggested by Chickering and recommended by the CIE (Wyszecki, 1968; Billmeyer, 1968). These equations have become known as the FMC metric because of the contributions of Friele, MacAdam, and Chickering.

## OTHER UNIFORM COLOR SCALES

There were other scales proposed during these years of experimentation with uniform color space. Quite close approximation to the uniformity of the Munsell system was achieved, but often at the sacrifice of simplicity in the mathematical calculations involved. In 1943 Moon and Spencer defined a metric that would take into account the adaptation of the eye to the test stimuli. The resulting system has been recommended for studying scientific color harmony (Moon and Spencer, 1943).

In 1946 Saunderson and Milner proposed a modification of Adams' Chromatic Value Scales that afforded close approximation to Munsell spacing but involved complicated calculations (Saunderson and Milner, 1946).

In 1954 an automatic photoelectric colorimeter for direct reading of Munsell coordinates was proposed by G. L. Gibson and D. A. Neubrech. The Gibson-Neubrech Cylindrical Coordinates Scale converted $X$, $Y$, $Z$ values into approximate correlates of lightness, hue, and saturation (Gibson and Neubrech, 1954). Later Taguti and Sato suggested the use of three kinds of exponential functions of the Adams chromatic value diagram (Taguti and Sato, 1962).

## UNIFORM OBJECT COLOR SCALES FOR ILLUMINANTS OTHER THAN DAYLIGHT AND FOR OBJECTS OTHER THAN DIFFUSELY REFLECTING SURFACES

The uniform color scales so far described were designed primarily to evaluate the colors of diffusely reflecting surfaces under daylight illuminants. The $U^*V^*W^*$ scales are adaptable to illuminants other than daylight, but their use for these specialized illuminants has been rather small. Uniform color scales for transparent films and liquids, and for shiny metallics have barely been proposed. With some of these specular materials it is quite apparent that the lightness scale needs to be more highly curved than the cube-root and square-root scales. That is, the chromaticity dimensions of the darker colors need to be expanded by comparison with the standard $a$ and $b$ dimensions.

Two specific proposals have been made by Hunter for expanding the range of UCS scales.

1. $L'$, $a'$, $b'$ scales for specular colors and for very dark colors viewed in dark surroundings. In 1967 Hunter proposed a special $L'$, $a'$, $b'$ scale which expands the $L$, $a$, $b$ system in the regions near black. This expansion has resulted in greatly improved correlation with visual estimates of differences in dark colors seen in dark surroundings. The equations are shown in Table A.8.

This same kind of expansion appears to be required to achieve uniform spacing for metallic and transparent objects. The high intensity of the specular component from both metallic and transparent products permits the eye to discriminate between small differences in chromaticity even when they are dark.

2. Natick $Q$ scales for surface colors under illuminants other than Illuminant C. A further optimization of $L$, $a$, $b$ relates to its use under other illuminants. Factors that alter the $a$ and $b$ expansion coefficients for chromaticity when the illuminant is changed appear to be necessary. In the present proposal these expansion factors are varied with the absolute magnitudes of $X$ and $Z$ of the illuminant. Initial tests have produced good visual uniformity under Illuminant A (Hunter, 1966). See Table A.8 for the equations.

## INTERCOMPARISON OF COLOR SCALES

The many color scales in use differ from each other substantially, but they are all improvements over the $X$, $Y$, $Z$ and $x$, $y$ specifications in terms of visual uniformity. The wavelength data, the Munsell data, and the MacAdam data are all in agreement, at least to a first approximation, as to the general direction in which the $x$, $y$ diagram must be altered to improve its uniformity. This is made apparent by a comparison of Figure 8.5 (Judd's perceptual ellipses in the $x$, $y$ diagram), Figure 8.12 (loci of constant hue and constant chroma in $x$, $y$ diagram), and Figure 8.17 (MacAdam ellipses of equal chromaticity difference in the $x$, $y$ diagram). In all three cases, the specimens are contracted in the lower left and expanded at the top. In addition, the orientation of the ellipses in both Figures 8.5 and 8.17 is generally the same—roughly toward an area slightly to the lower left of the diagram.

In the Family Tree of Color Scales, Table 8.1, the development is traced of some of the tristimulus psychophysical scales available for the measurement of color. Note that the three columns represent (from left to right) increasing specialization in the use for which the scale is intended. A major factor responsible for variety in the scales is the large assortment of visual information on perceptual color intervals available. Different sets of color-interval data have led to the generation of different groups of uniform color scales. Another factor causing differences among scales is that some of them have been designed specifically to fit theories of color vision. For example, the theories of Hering have influenced the selection of dimensions of all the opponent-colors (*L, a, b*) scales, whereas Adams' specific proposals had an important bearing on the formation of the scales that now bear his name. More recently the somewhat more elaborate opponent-colors theories of Mueller have influenced the design of the series of scales starting with Friele's work.

To compare some of the color scales for uniformity of visual spacing, we have used the Munsell system as a base in preparing Figure 8.19. This obviously favors those scales based on it, but it will permit us to see the effect of the different approaches to scale construction, and to assess the magnitudes of nonuniformity among scales. Figure 8.19 shows the chromaticity spacing achieved by seven color scales. Five Munsell renotation colors at value 5 and chroma 10 were converted to the seven scales and then graphed. Visual uniformity is evaluated by the degree to which the graphs approximate circles.

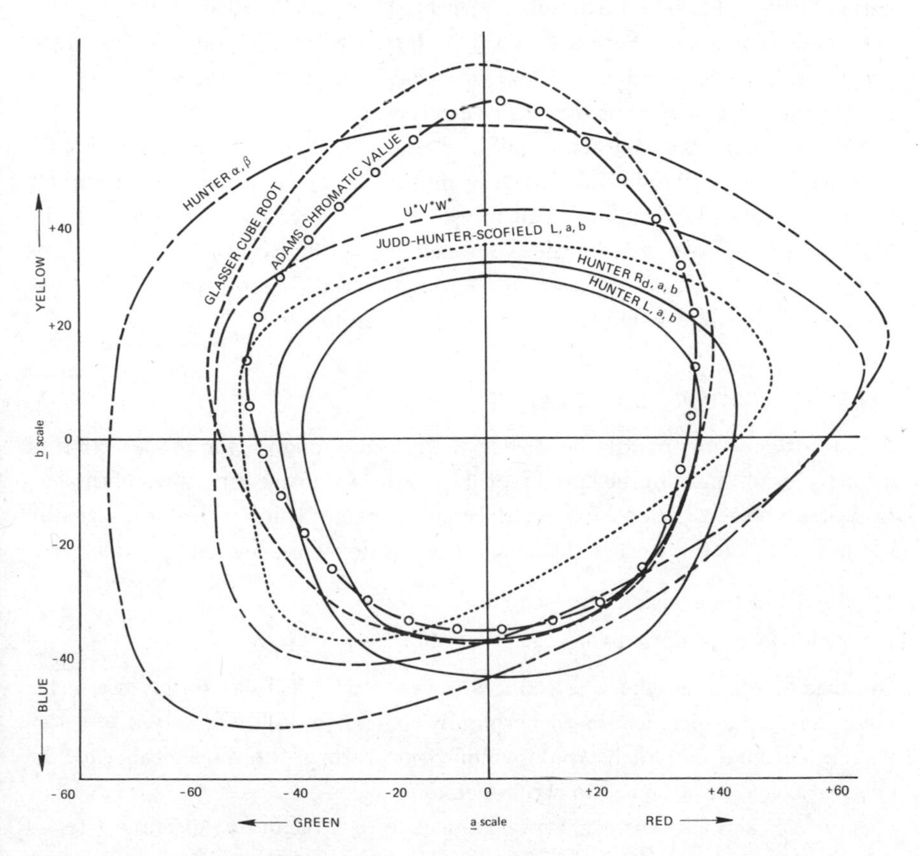


**Figure 8.19.** Munsell 5/10 hue circles in seven scales.

## INTERCONVERSION AMONG COLOR SCALES

All of the foregoing uniform color scales are relatable by equations to the tristimulus scales of the CIE observer, and therefore they may properly be called tristimulus scales for the measurement of color. Through the relationship of each set of scales with the CIE observer, it is possible to convert color values from any one scale to those of any other scale. There have been many mechanisms devised to facilitate transforming each set of color scales to the CIE coordinates scales and thence to any other set of color scales. Often it is expedient to prepare special conversion formulas or charts to speed computation, and many of the basic formulas can be simplified. For example, Table A.9 gives the mathematical relationships between a number of color scales for Illuminant C in general use today in instrumental measurement.

Graphic techniques for conversion have been and can be developed, and are particularly useful when a number of conversions are needed within a small region of color space. Recently, with the increased availability of computers, programs have been generated and are readily available that will carry out most of the transformations that may be needed.

Note in Table A.8 that about half of the opponent-color scales listed call for subtraction of tristimulus values before multiplication by a nonlinear lightness function, while the remainder call for subtraction after nonlinearization. The Adams Chromatic Valence Scale is an example of the former, and the Adams Chromatic Value Scale illustrates the latter. It is difficult to say which practice yields the better scales, but the differences are significant.

## SELECTION OF COLOR SCALES

Let us attempt to provide the reader with some guidance for the selection of color scales to apply to the specific problems that confront him. Two of the factors cited above as responsible for differences among color scales are normally not important to the worker choosing a color scale for his own application:

1. Model of color vision represented
2. Color-interval data used

In other words, one who is selecting a scale for practical day-to-day use in his laboratory, or a specification, is not vitally concerned with which type of color-perception data, or which type of color-vision theory, the numerical scales he chooses have been designed to accommodate.

A person should consider, however, as the first factor in selecting a set of color scales, the class, or mode, of specimens for which the method is intended.

It can be noted in the Family Tree, Table 8.1, that these methods vary from those applicable to all modes (the CIE *Y, x, y* scales) to methods intended only for diffusely reflecting surfaces observed in daylight (Illuminant C). The Hunter $L_L$, $a_L$, $b_L$ scales are an example of scales restricted to this limited but most important group of specimens. In general the color scales with the greatest applicability for different specimen modes will be least intelligible or meaningful when applied to only one specific mode. So first, the consideration of scales should be limited to those appropriate for the specimen mode.

The second factor to consider in selecting color scales is the relatability of the dimensions of the color solid selected to the attributes of color appearance that the users of the color-measurement system will understand. It is primarily because workers in the field of colored materials find it easy to appreciate and interpret the *L, a, b* opponent-colors dimensions that these enjoy such widespread use.

The third factor to consider is the uniformity of the color-scale intervals as they relate to visually perceived intervals of color. In the past many workers have considered this matter of perceptual uniformity of scales to be the fundamental goal of uniform color-scales selection. Now it is known that perceptual intervals of color are susceptible to change in relative magnitudes with changes in observer adaptation, surroundings, and level (also spectral distribution) of illumination. Interference by gloss, texture, and other geometric attributes also can change a perception of color. Whether the visual intervals are measured on an estimate-of-magnitude or threshold basis can also make a difference. It is apparent from what is now known that visually ideal, uniform color scales must each apply only to strictly limited types of specimens and observing situations. Visually uniform general-purpose color scales are not attainable. This third factor of scale uniformity is thus probably less important than thought in the past.

A fourth factor that should receive attention in selecting color scales is the size of the unit of color measurement. Here the trend is toward the use of color scales in which the unit of color measurement, for surface colors at least, is based on 100 units between black and white. This unit is quite similar in magnitude to the NBS unit of color difference (Judd 1939). In 1939, 1 unit was designed to represent the break point between commercially acceptable and unacceptable color matches. In practice, maximum commercially acceptable color differences vary all the way from 0.2 to 4 or 5 NBS units depending on product, use, and discrimination of the purchaser.

A fifth factor affecting choice of color scales is the adaptability of the scales to efficient instrumental measurement. Scales like the Hunter $L_L$, $a_L$, $b_L$ were designed not necessarily for best perceptual uniformity, but rather for direct precise measurement with a relatively inexpensive analog color-difference meter. At the time these scales were first developed, 20 years ago, the advantage of measurements with the simple analog instruments was somewhat greater than it is now with the widespread existence of digital computers. Nevertheless, the direct-reading *L, a, b* instruments have not only an advantage in simplicity and usability of readout but, as Billmeyer (1969) has shown, also tend to have the highest precision available for small color-difference measurement. This is because tristimulus differences in $X_{\%} - Y$ and $Y - Z_{\%}$ are taken directly to the readout device. Uncertainties of only one signal, not uncertainties in separate $X$ and $Y$, or $Y$ and $Z$ signals, affect the repeatability of *a* and *b* readouts, respectively.

## SUMMARY

Visual discrimination experiments have permitted Standard Observer data to be converted into scales based on visual units of color perception and that are easy to interpret and relate to perceived colors and color differences. Differences among the scales relate primarily to the visual data and color theory on which they were based. All of the uniform color scales are a significant improvement over the *x, y* spacing of colors. There is no best scale, but some are better than others for specific uses. The best is that one most closely duplicating the visual situation encountered. An ideal system is impossible and would be useless for industrial specification because it would change with changes in light intensity, background conditions, gloss, and perhaps other variables.

All the various surface color scales for objects under Illuminant C are interrelated through the CIE Standard Observer data, on which they are based. Therefore color values in one scale can be converted to those of another scale. Conversion among scales has been simplified by the use of special formulas, tables, and graphs.

# CHAPTER NINE

# SCALES FOR THE MEASUREMENT OF COLOR DIFFERENCE

A measured value of small color difference is intended to be a quantitative representation of the ease of visual differentiation between the colors of two specimens. Precise analyses of small color differences are used in research and technology to

1. Compare industrial production specimens with standards in appraisals of conformity to color-tolerance specifications. This is probably the most frequently used type of measurement in all appearance technology.
2. Determine the nature and magnitude of ingredient or process adjustment required to convert a product with unacceptable color to a product having acceptable color.
3. Measure color changes of products resulting from their exposure to weather, light, laundry, or other real or simulated usage.
4. Measure the metamerism index between pairs of similarly colored specimens where each has different spectral curves. Such procedures are still in the process of development.

There are a variety of numerical scales used to measure color difference, many of which involve uniform color scales described in the previous chapter. There are also a variety of criteria and techniques involved in the psychological-visual evaluation of color differences. Thus neither the numerical measurements of color difference nor the relationships of these measurements to visually observed color differences can be considered to be well standardized or reproducible from one experimental setup to another. Only when the same scales and instrument conditions are used can numerical results of instrumental color-difference measurement be reproduced. Only when identical observing conditions are employed and observers of similar training make the observations can visual estimates of color difference normally be repeated.

## CONCEPT OF COLOR DIFFERENCE

Most of the scales used in industry for color-difference measurement are approximations traceable to one of three following psychophysical units:

1. The Nickerson Index, based on Munsell scales of hue, value, and chroma. The maximum permissible fade of a commercially lightfast fabric from 20 hr Standard Fademeter exposure. This is called a "just noticeable fade" (Nickerson, 1936).
2. The NBS Unit of Color Difference originally represented the average maximum difference acceptable in a series of dye-house commercial matches in 1939 (Judd, 1939). The NBS Unit generally referred to today is the Hunter rectangular-coordinate version of Judd's equation (Hunter, 1942). ASTM, in Method E 284, defines a similar NBS Unit in Munsell terms.
3. The MacAdam Unit represents the minimum perceptible (threshold) color difference determined by MacAdam's experiments reported in 1942 and 1943 (MacAdam, 1942, 1943).

For most colors, 1 NBS Unit is equal to 2 or 3 MacAdam Units of color difference. The MacAdam Unit is the standard deviation of 25,000 threshold color-difference settings. "The interpretation of 'just perceptible' is restricted to the conditions of viewing in the MacAdam experiment. It is not the minimum difference that can be seen under ideal conditions. There are, in fact, some industrial tolerances that are well under one MacAdam Unit [Friele, private correspondence, 1971]."

All but the earliest color-difference scales are based on the concept that $Y$, $x$, $y$ space can be transformed so that magnitudes of color difference can be measured by distances between points representing colors in the transformed space, and that these differences will have a relationship in size to the magnitude of the visual estimation of color difference. Figure 9.1 shows how distance is measured in a color space having rectangular coordinates and then in a color space having cylindrical coordinates. Many of the scales described in Chapter 8 simply measure color differences in the units of the uniform color scales used.

The magnitude of the total color difference is frequently represented by a single number such as $\Delta E$ or $I'$ in Figure 9.1. Since a single number only tells the size of the difference rather than the direction or nature of the difference from standard, the trend in recent years has been to examine the character of each color difference by recording its three components separately. Sometimes only two components of the total color difference, lightness difference and chromaticity difference, are used. For most analyses, however, all three aspects of the difference should be recorded.

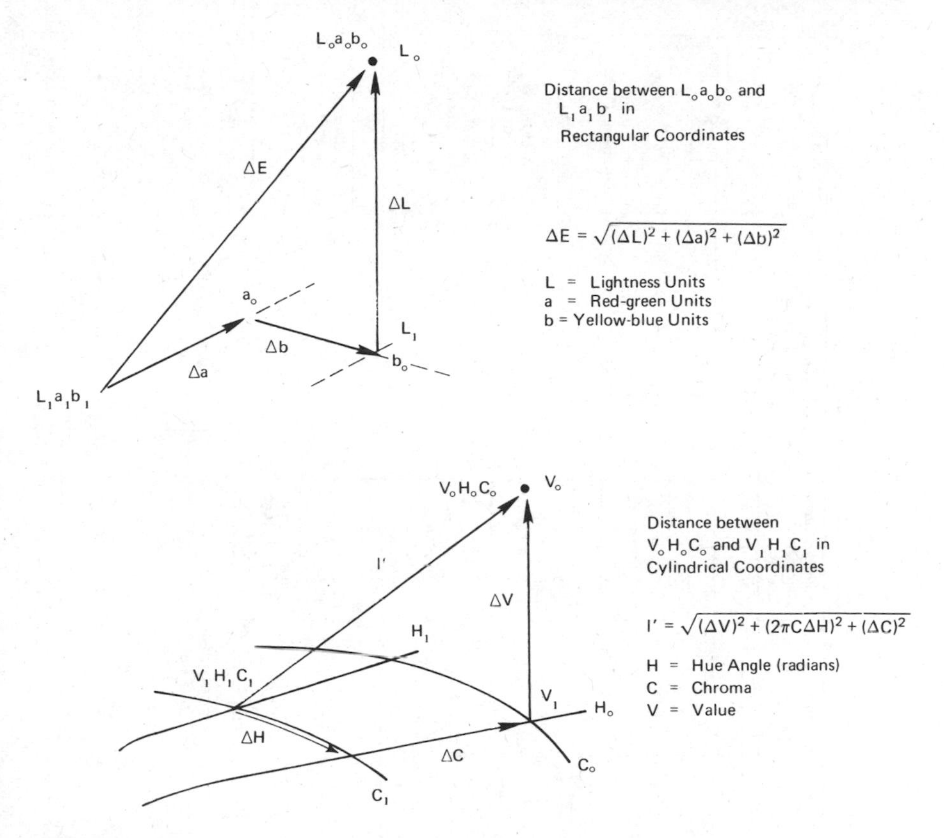


**Figure 9.1.** Distance between two object colors in rectangular-coordinate space and in cylindrical-coordinate space.

## THE HISTORICAL DEVELOPMENT OF METHODS OF COLOR-DIFFERENCE MEASUREMENT

As with the uniform color scales to which many of them are related, the number and variety of scales used for the quantitative measurements of color difference are large. From the point of view of historic development these measurement scales can be divided into four groups:

1. Munsell System Color Difference Scales.
2. An Adams-Nickerson series, based on the opponent-colors concept and its similarity in spacing to the Munsell color system.
3. A series starting with the application of Judd's uniform-chromaticity Maxwell Triangle. This series has tended to merge with the Adams-Nickerson series.

**TABLE 9.1**

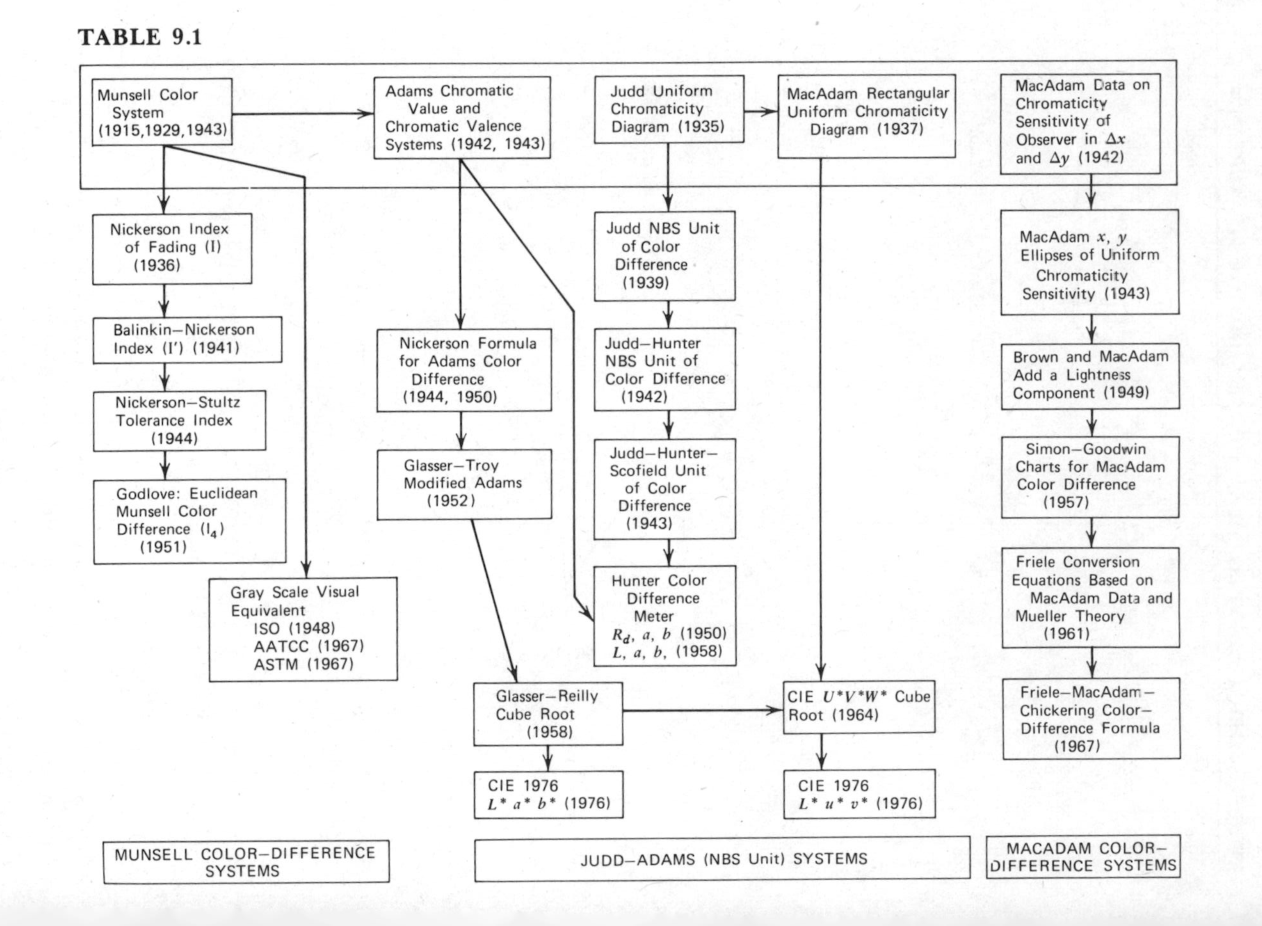

Munsell Color System (1915,1929,1943)
Adams Chromatic Value and Chromatic Valence Systems (1942, 1943)
Judd Uniform Chromaticity Diagram (1935)
MacAdam Rectangular Uniform Chromaticity Diagram (1937)
MacAdam Data on Chromaticity Sensitivity of Observer in $\Delta x$ and $\Delta y$ (1942)
Nickerson Index of Fading (I) (1936)
Balinkin–Nickerson Index (I′) (1941)
Nickerson–Stultz Tolerance Index (1944)
Godlove: Euclidean Munsell Color Difference ($I_4$) (1951)
Gray Scale Visual Equivalent ISO (1948) AATCC (1967) ASTM (1967)
Nickerson Formula for Adams Color Difference (1944, 1950)
Glasser–Troy Modified Adams (1952)
Glasser–Reilly Cube Root (1958)
CIE 1976 $L^* a^* b^*$ (1976)
Judd NBS Unit of Color Difference (1939)
Judd–Hunter NBS Unit of Color Difference (1942)
Judd–Hunter–Scofield Unit of Color Difference (1943)
Hunter Color Difference Meter $R_d$, $a$, $b$ (1950) $L$, $a$, $b$, (1958)
CIE $U^*V^*W^*$ Cube Root (1964)
CIE 1976 $L^* u^* v^*$ (1976)
MacAdam $x$, $y$ Ellipses of Uniform Chromaticity Sensitivity (1943)
Brown and MacAdam Add a Lightness Component (1949)
Simon–Goodwin Charts for MacAdam Color Difference (1957)
Friele Conversion Equations Based on MacAdam Data and Mueller Theory (1961)
Friele–MacAdam–Chickering Color-Difference Formula (1967)
MUNSELL COLOR–DIFFERENCE SYSTEMS
JUDD–ADAMS (NBS Unit) SYSTEMS
MACADAM COLOR–DIFFERENCE SYSTEMS

4. Systems developed from MacAdam's chromaticity-difference threshold series of experiments.

The first two groups relate measurements to the Munsell coordinates for color. The third group started by using Judd's unique triangular coordinates but was converted as the methodology grew to use of several opponent-colors *L, a, b* coordinate systems. MacAdam color differences originally were related to differences in CIE *x, y,* and *Y* color space, but in their newer forms are measured also in opponent-colors coordinates.

The following descriptions of these scales may be made clearer by following their historical development and interconnections as diagrammed in Table 9.1.

## MUNSELL SYSTEM COLOR-DIFFERENCE SCALES

This series of color-difference scales has no parallel in Chapter 8 and includes scales based solely on the Munsell Color System and its coordinates. The Nickerson Index of Fading (I) was intended for use with textiles and was the first measured index of color difference (Nickerson, 1936). To derive this index Miss Nickerson obtained Munsell designations for each of several pairs of specimens, computed differences in each attribute (hue, value, and chroma), weighted the component differences on the basis of observers' estimates of degree of fading, and then added them together. The observers were members of the Silk Subcommittee of ASTM Committee D-13 on Textiles, and the specimens were dyed cotton, woolen, and silk textile materials. Her equation was purely an empirical creation without consideration of any color-space analogy. In terms of Munsell units of hue, value, and chroma, Miss Nickerson's Index of Fading was

$$I = \frac{C}{5}(2\,\Delta H) + 6\,\Delta V + 3\,\Delta C$$

Balinkin altered the Nickerson equation to make it measure distance between points in the Munsell cylindrical-coordinate system instead of simply using the sum of the three coordinate components of this distance. This required that an increase of weight be given to the chroma (radius) dimension (Balinkin, 1941).

Nickerson and Stultz developed a color-difference equation to fit ratings of a number of differences in camouflage-color swatches (Nickerson and Stultz, 1944). They intercompared the then available color-difference equations and proposed a unit that was later the basis of the Adams-Nickerson AN40 unit.

Bellamy and Newhall were looking for least-perceptible differences, not estimates of difference magnitudes. They gave much greater weight to the value component than did Nickerson (Bellamy and Newhall, 1942). Godlove, working with textile dyings, found that he needed to use much the same relative weightings as did Bellamy and Newhall, but his units were much larger than the Nickerson Index, the NBS Unit, or the MacAdam Unit (Godlove, 1951).

One of the earliest methods of making visual assessments of color differences goes under the name of the Gray Scales because intervals in the Munsell Value Gray Scale serve as references for rating visually equivalent differences in color. This method came into use both here and in England in the 1930s and 1940s. In 1948 the International Organization for Standardization (ISO) adopted the Gray Scale method, and it has gained international acceptance. It is used primarily in the textile industry as a means of visually assessing color change and staining in textiles. It has been promoted by the American Association of Textile Chemists and Colorists (AATCC), whose yearbook describes 37 different standardized exposure procedures for testing color fastness, all of which involve the use of the Gray Scales. Its use has also been promoted by the Society of Dyers and Colourists in England and by the American Society for Testing and Materials (ASTM).

## ADAMS-NICKERSON SERIES

Although they are quite similar to the later Hunter scales for color difference, the Adams-Nickerson series are here treated separately because they are intermediate between the purely Munsell series of scales and the Hunter series in the Judd-Hunter group.

As was noted in Chapter 8, the Adams series of color scales was not initially designed for the measurement of color as such. In designing his Chromatic Value and Chromatic Valence Scales for color, Adams was attempting to justify an opponent-colors theory of color vision that he was proposing. Miss Nickerson promoted the use of these Adams parameters for the measurement of both color and color difference. She suggested in 1944 an Adams color-difference equation in which the Munsell value step was the unit of color difference (Nickerson, 1944). This is a large interval, about 10 to 30 times the size of the units in current use. Later Miss Nickerson proposed instead the Adams-Nickerson unit of color difference equal to 1/92 of the whole value scale (Nickerson, 1950). Miss Nickerson was deliberately attempting to define a unit midway in magnitude between her original Index of Fading (1936) and the Judd NBS unit, being close to both. Her 1950 proposal is now frequently identified as ANLab 40, or AN40, since it describes *L, a, b* type opponent-color scales with the constant 40 applied to $V_X - V_Y$.

The modified Adams-Nickerson color-difference formula differs from the 1950 Adams-Nickerson formula only in the multiplying constants it uses. (See Chapter 8.) It was suggested by Glasser and Troy as a scale to be used in conjunction with their photoelectric filter colorimeter (Glasser and Troy, 1952). They used the optimum constants, rather than the rounded constants suggested by Nickerson, to derive NBS units of color difference. There is only a 4 to 5% difference between the regular and modified Adams-Nickerson formulas.

Glasser and Reilly discovered that the cube root of the tristimulus value of reflectance could be substituted for the Munsell value function without producing generally significant changes in measured magnitudes of color difference. This is the basis for the cube-root version of the Adams-Nickerson series of color-difference scales (Glasser et al., 1958). Most recently, Reilly and his co-workers have used the amber tristimulus reflectance in place of the *X*-function. This gives a *b* axis in their chromaticity diagram, which is rotated in a clockwise direction from the *b* axis in the Adams and Hunter *L, a, b* chromaticity diagrams.

Recently the CIE Colorimetry Committee TC-1.3, with Wyszecki as Chairman, undertook the consideration of a proposal for an Adams-Nickerson cube-root formula to be applicable mostly to small color differences such as those found in the textile industry. This formula was tentatively named the CIE 1976 $L^*a^*b^*$ Color Difference Formula, reflecting its anticipated formal adoption by the CIE in 1976. Table A.8 gives the $L^*a^*b^*$ formula (Wyszecki, 1974).

## JUDD-HUNTER SYSTEMS

This series of scales for color difference started with Judd's uniform chromaticity scales triangle (Judd, 1935). Judd made an instrumental study of a number of woolen swatches rated by dyers for their acceptability as commercial color matches to standard. The NBS Unit of Color Difference was designed by Judd to be the maximum difference commercially acceptable to these dyers at the time the study was carried out (Judd, 1939). The NBS Unit generally referred to today is not Judd's 1939 unit, but the Hunter 1942 version of the Judd unit with 100 units falling between black and white.

Since 1942 this unit, or others approximating it, has been used with a number of the opponent-colors scales. If a color scale is approximately uniform in the relationship of visual intervals to numerical differences in lightness and chromaticity, and has about 100 units between black and white as well as rectangular coordinates for chromaticity, color difference in approximate NBS units can be specified in units of that system. The color difference can be estimated in terms of distances between pairs of points representing pairs of colors in each of the different color spaces. With rectangular coordinates the formula is

$$\Delta E = [(\Delta L)^2 + (\Delta a)^2 + (\Delta b)^2]^{1/2}$$

Similarly if only the chromaticity component of difference is desired, the formula is

$$\Delta C = [(\Delta a)^2 + (\Delta b)^2]^{1/2}$$

Following Judd's proposed unit of color difference, Hunter proposed the measurement of much the same quantity by a photoelectric tristimulus method in 1942, employing the $L'$, $\alpha'$, $\beta'$ scales identified in Table A.8. In the next year Scofield proposed a quantity that was quite a bit easier to compute (Scofield, 1943). These Scofield scales are identified as $L_s$, $a_s$, $b_s$ in Table A.8. As can be seen, Scofield used the square root of reflectance factor instead of the fourth root as the multiplier for adjusting chromaticity (*a* and *b*) values for uniform lightness intervals. This is the only difference between the Scofield and the Judd-Hunter formulas for color difference. For dark colors, however, this difference is quite significant.

In 1948 Hunter started to develop a tristimulus instrument that would read chromaticity dimensions of opponent colors directly. He was seeking to improve on the precision and usefulness of measured color differences. The $R_d$ scales developed in the period 1948 and 1950 did not have a uniform lightness readout but did have direct reading, fairly uniform *a* and *b* scales. The Hunter *L* scales, in which there is approximate perceptual uniformity in all three dimensions, were created in the period 1950 to 1952, but were not described in a formal publication until 1958. These two sets of Hunter Color Difference Meter Scales (identified as $L_R$, $a_R$, $b_R$ and $L_L$, $a_L$, $b_L$ in Table A.8), enjoy wide use because they can be read directly from a tristimulus instrument with high precision and offer instrumental computation of color difference by the $\Delta E$ formula given above.

The CIE $U^*V^*W^*$ color scales, adopted by the CIE in 1964, have elements of similarity to the Judd NBS unit scales and the Glasser cube root scales (Wyszecki, 1963). The MacAdam (1937) *u, v* chromaticity triangle is used to measure the chromatic intervals, and lightness intervals are measured by the cube root of luminous reflectance factor. As the multiplier to adjust chromaticity intervals for lightness level, the $W^*$ cube root of reflectance factor is used. (See Table A.8.)

Recently, when the CIE Colorimetry Committee TC-1.3 took under consideration the formal adoption of the $L^*a^*b^*$ formula for the measurement of small color differences, it also considered a formula mostly applicable to large color differences, such as those found in the color-reproduction industry. This formula, identified as $L^*u^*v^*$ in Table A.8, is a modification of the CIE 1964 $U^*V^*W^*$ formula. It was named tentatively the CIE 1976 $L^*u^*v^*$ color difference formula pending its formal adoption by the CIE in 1976.

## MacADAM SYSTEMS

This series is based on MacAdam's original chromaticity-discrimination experiments (MacAdam, 1942 and 1943). When MacAdam identified his ellipses in 1943 he showed three $x$, $y$ diagrams from whose contours one could determine, for any point in the diagram, the ellipse having a radius in the $x$, $y$ diagram of just one perceptible chromaticity difference. These MacAdam ellipses are shown in Figure 8.17. One of the ellipse-defining charts is shown in Figure 9.2. The $g_{11}$ factor relates to the variability of these threshold ellipses in $x$ and $y$.

Brown and MacAdam (1949) and Brown (1951) supplemented the original MacAdam chromaticity scales with lightness-interval data. Their lightness dimension is a logarithmic function. Davidson and Friede (1953) found that 2½ MacAdam units of chromaticity difference made a good limit of commercial acceptability in dyings.

Davidson and Hanlon (1955), Simon and Goodwin (1957), and Foster (1966) have prepared graphic techniques for quick computation of color difference in MacAdam units from CIE data. The Simon-Goodwin graphs, of which examples are shown in Figure 9.3, each represent a small area of the chromaticity diagram for which a single linear transformation is a good approximation of all points. Lightness compensation is applied as a separate operation. These early MacAdam measures of color difference did not identify direction, only magnitude.

The Friele, MacAdam, and Chickering developments discussed in Chapter 8 culminated in the FMC metric (Chickering, 1967), for which a computer program has been published (Billmeyer and Smith, 1967). A later development by Chickering (1969) presented the values of color difference obtained in this system in terms of opponent-colors dimensions: redness-greenness, yellowness-

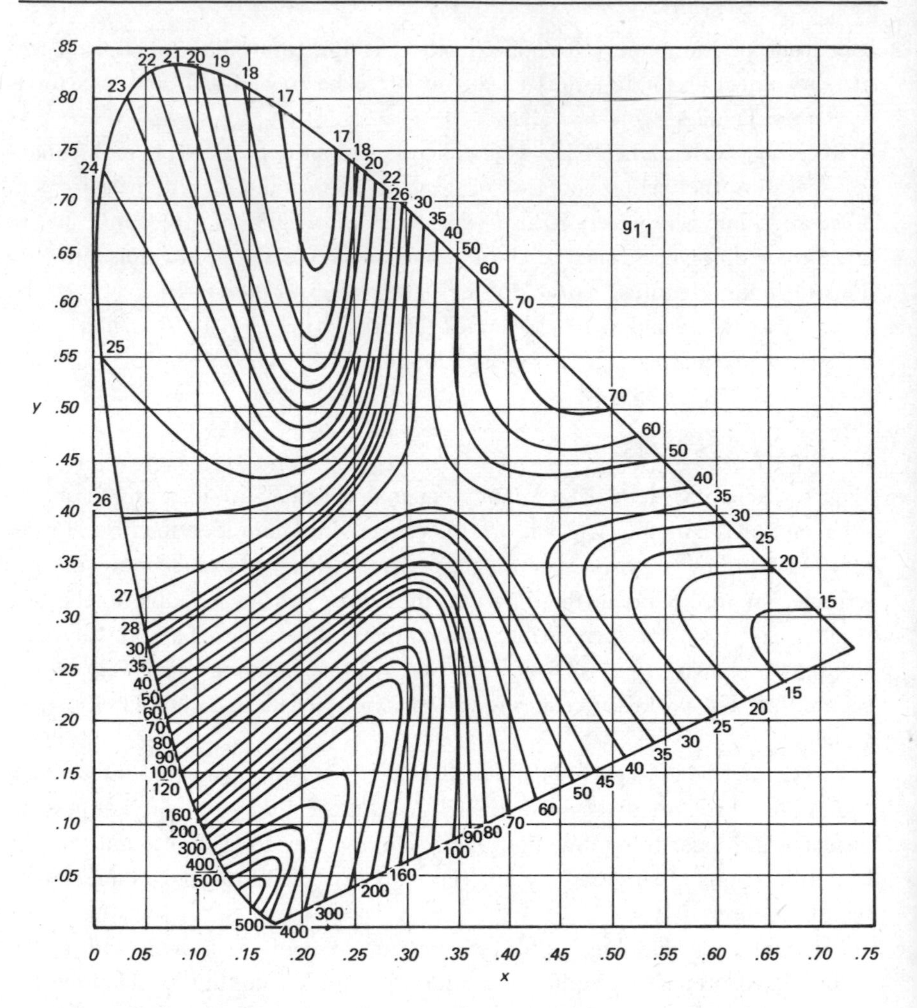


**Figure 9.2.** The 1931 CIE *x, y* diagram showing contour lines of constant metric coefficient $g_{11}$ based on MacAdam ellipses. From D. L. MacAdam, *J. Opt. Soc. Am.*, **33,** 1943.

blueness, and lightness. These three-dimensional measurements indicate both magnitude and direction of color difference (Billmeyer, 1969).

## INTERRELATIONSHIPS OF COLOR DIFFERENCE SCALES

Relationships between the different units of color difference are decidedly nonlinear. Thus even though they all are purported to measure more or less the same thing, the numbers are different for the different scales. In the past it has been suggested that conversion factors are feasible to convert color difference units in one system to those in another system. For instance, some have said that the number of NBS (commercially acceptable) color difference units multiplied by 3 will approximate the number of MacAdam (just-perceptible) units that would define the same color difference. Actually, the multiplication factor of 3 between an NBS unit and a MacAdam measure of the same color difference may apply at one point and direction in color space, and not hold true at some other point or direction.

Figure 9.4 is a specific example of the way factors relating chromaticity differences vary with direction in color space. It shows a series of near ellipses of varying size and shape representing the magnitudes of difference measured by

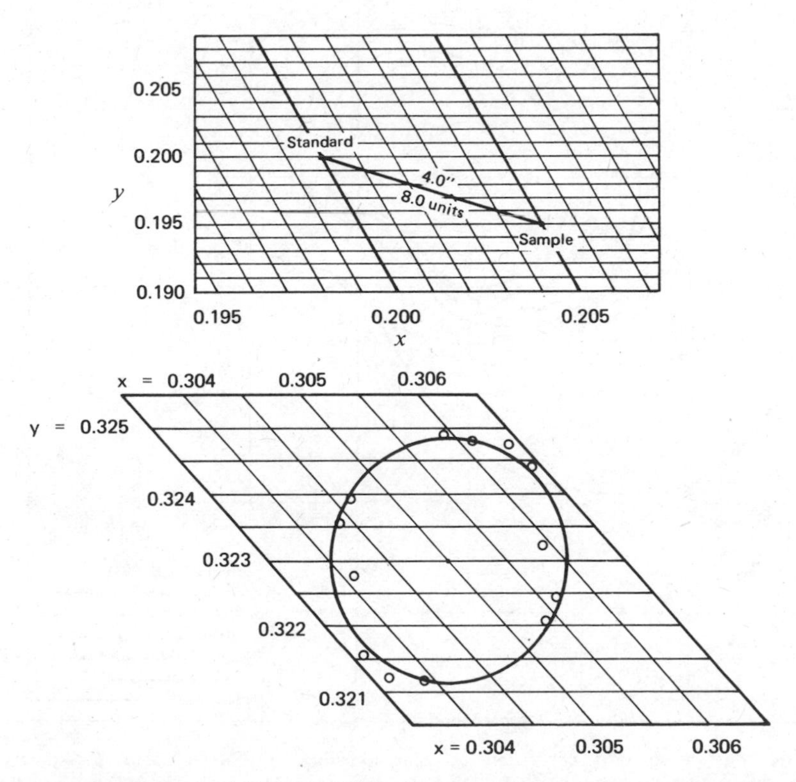


**Figure 9.3.** Simon-Goodwin Charts for two small areas of the $x$, $y$ diagram. In effect these charts change the ellipses of Figure 8.17 to circles by distortion. From F. T. Simon and W. J. Goodwin, *Rapid Graphical Computation of Small Color Differences,* Bakelite Company, 1957. Used with permission of Union Carbide Corp.

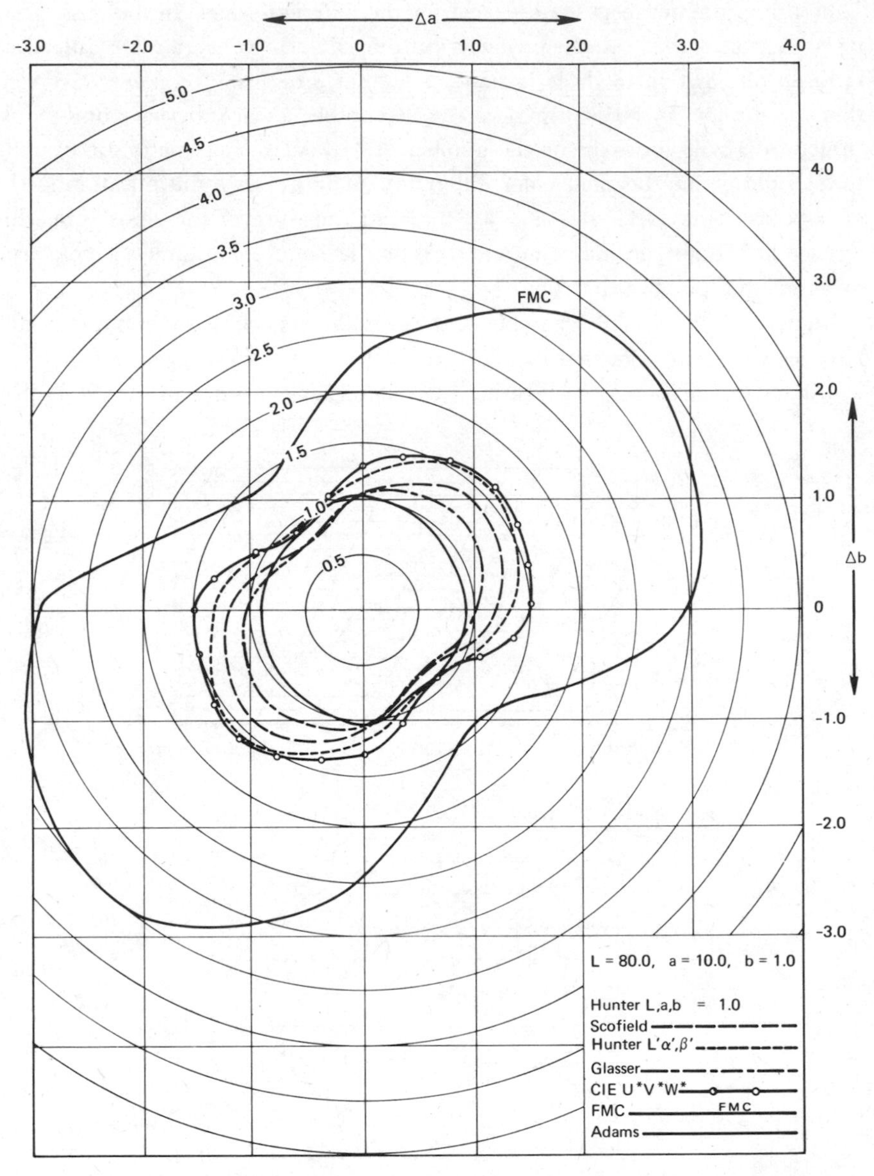


**Figure 9.4.** Magnitudes of 1 unit of color difference in Hunter *L, a, b* scale $\left(\Delta C = \sqrt{(\Delta a)^2 + (\Delta b)^2}\right)$ when converted to six other color difference scales.

different scales. For this figure, a light pink central color was selected. Four neighboring colors were selected so that small color differences in four directions could be computed. These color differences were computed in seven different color-difference scale units, and plotted on the Hunter $\Delta a$, $\Delta b$ graph. The circles define color difference units in the Hunter *L, a, b* system. The comparison of the magnitudes of the color difference units of various scales is made to a color difference of 1 unit in the Hunter *L, a, b* system as shown by the circle of radius 1.0. Similar studies done of colors at lower lightness levels show a significant increase in the sizes of the ellipses relative to the Hunter *a, b* circle.

## WHICH COLOR DIFFERENCE SCALE IS BEST?

With the amount of disparity existing among color difference scales, the question is frequently asked, "Which of the different scales is best?" By best, one usually means the one which most accurately represents visual judgments of color difference made by experts in the technology of the product involved. In order to determine this, many studies have been and are still being made (Jaeckel, 1973; Strocka, 1971; McLaren, 1969; Schultze, 1969; Kuehni, 1971; Morton, 1969; among others). So far, they have shown agreement primarily on one point alone—that there is no universally "best" color difference formula that gives good correlation between visual judgments and calculated color difference of all types of products for all regions of color space. Many of these studies, such as those reported by Jaeckel and Strocka, show correlation coefficients that seldom go over 77 for any of the samples or color difference formulas used. This falls well below the mid-90s range that is normally expected in successful psychophysical experiments. Each of these studies shows some particular system for color difference measurement to be superior to others in its ability to correlate with observers' estimates of specific color differences. However the same system is not favored by all of the studies. One color difference system may perform favorably in one area of research, while in another situation, an entirely different system will be preferred.

This lack of unanimity in the results of researchers' studies is easy to understand when one considers all the physiological, psychological, and psychophysical connotations involved in visual and instrumental evaluations of color difference, and in attempts at correlation between them. One must remember that the scales for tristimulus measurement of color are intended only to identify, by identical numbers, color stimuli that match visually to the normal observer under prescribed standard conditions. Questions of magnitude of difference did not enter into the CIE 1931 formulation of trichromatic scales for color. With color difference measurements, however, comparisons are made of psychophysical quantities with psychological estimates that can be importantly affected by physiological variables in the observing situation.

## FACTORS AFFECTING VISUAL COLOR DIFFERENCE EVALUATIONS

L. F. C. Friele has suggested three factors that explain why color differences as measured by the tristimulus psychophysical methods may fail to correlate reliably with visual estimates of color difference. First, Friele, among others, suggests that distance in "uniform" tristimulus color space is not a proper measurement of the magnitude of observer response. It is not reasonable to expect that there is any sensor or device in the human brain taking the square root of the sums of the squares of three differences, as is done for distance in rectangular space. Rather, the magnitude of color difference is more properly related to the component of the particular difference that is greatest. Other things being equal, it is reasonable to expect that the signal component which shows the greatest incremental difference will be the one which will be signaling most strongly for attention and recognition by the observer.

Second, the threshold criterion for color difference is quite another matter from estimates of easily seen differences. This is because threshold discrimination differences correspond to minimum detectable signal-to-noise ratios in the human signaling system. On the other hand, estimates of easily seen differences correspond much more probably to the actual signal increments.

Third, Friele suggests that numerical tristimulus values, even though they are proper integrations of receptor spectral responses, may nevertheless fail to be valid measurements of neural signals.

"The handling of retinal responses by combining them in opponent or antagonistic channels enables strongly enhanced differential sensitivity as compared to nonopponent signal transfer. A further complication is the decrease in sensitivity to incremental differences with increasing signal level which may be described as a saturation effect [Friele, private correspondence, 1974]."

This explanation would readily show why, when one is dealing with saturated colors, sensitivity to small differences in hue always appears to be so much greater than sensitivity to small differences in saturation. The hue difference corresponds to a change of balance in the opponent channels, whereas the saturation difference is involved mainly with one channel only.

Friele expressed his views in general terms as follows: "The colour sense functions in that way that maximum information is obtained from the visual field by using the best differential sensitivity which is available from the two opponent components [Friele, 1971]."

## VARIABLES AFFECTING VISUAL AND INSTRUMENTAL ESTIMATES OF COLOR DIFFERENCE

The questions raised by Friele concerning the validity of using the tristimulus system to represent visual color difference evaluations suggest some reasons for lack of good correlation between visual and instrumental measurements. There are, in addition to these, certain variables that may also contribute to this poor correlation:

1. Variables in the relation of any color scale to actual perceived color differences.

   a. Uniform color scales differ markedly from each other in dimensions.

   b. The mode of the observation (aperture, illuminant, or object). In the object mode, the character of the object (metallic, transparent, liquid, or turbid) introduces variability in judgment. It also may necessitate changes in the color difference formula used. Equation factors relating lightness and chromaticity for metallic colors, clear transparent liquids, and translucent objects must be quite different from those used in color difference formulas for diffuse surface colors.

   c. The magnitude of differences seen may not be the same when comparing readily discernible differences as that seen when the measurement involves just-perceptible, threshold differences.

2. Variables in the observing situation affecting visual estimates of color difference.

   a. Level of illumination.

   b. Color of illumination.

   c. Diffusion and geometric uniformity of illuminant.

   d. Proximity of specimens to each other. This variable was recognized by Hunter in his 1942 report, in which he suggested that lightness differences vary in their contribution to color difference, depending on the proximity of specimens being compared, by a factor of from 0.2 to 1.2. The normal factor in most of the present equations is 1.0. For the very high 1.2 factor, the specimens were compared in almost perfect proximity to each other.

   e. Size and flatness of specimens.

   f. Presence of gloss on specimens. Again, in his 1942 work, Hunter incorporated a gloss factor into his color difference equation. Dark glossy specimens were said to have visually smaller color differences than dark matte specimens with the same reflectance factor, since the high gloss tended to mask small color differences.

g. Presence of texture distracts attention from perception of color difference and lessens the magnitude of the perceived color difference.

h. Surroundings and backgrounds for the specimens. If these contrast strongly with specimens, they will diminish the observer's ability to discriminate and identify small color differences. Backgrounds on the gray side of the color of the specimens being compared are best for visual sensitivity to small differences. The single best background for use with objects of all different colors is a medium gray.

3. Variables in the observer that affect his estimates of the magnitude of color difference.

a. Prior experience of observer and the difficulty he foresees in correcting the specimen to match the standard target color.

b. Prior adaptation to observing situation.

c. Color difference perceptibility and commercial acceptability do not always correlate.

1. Hue differences are less accepted than lightness and chroma differences.

2. Mismatches on the yellow side of standard are more often rejected.

3. Acceptance practices vary with industry, with company, within industry, and with economic practices.

d. There is variation among observers even about a judgment of "perciptibility." Figure 9.5 shows where 12 different observers see the loci of just-perceptible difference from a standard color.

## SELECTION OF METHODS OF COLOR DIFFERENCE MEASUREMENT

Procedures for the selection of methods of color difference measurements to apply to specific problems are not standardized. There is still much experimentation in the preparation of specimens, in the selection of scales for measurement, and in techniques of measurement. Within an industry concerned with a particular product, such as textiles, however, certain practices tend to be followed, particularly where color difference measurements are being made as a means of communication within that industry. Industry-wide practices tend to develop in quality control, and in customer acceptance on the basis of published tolerances.

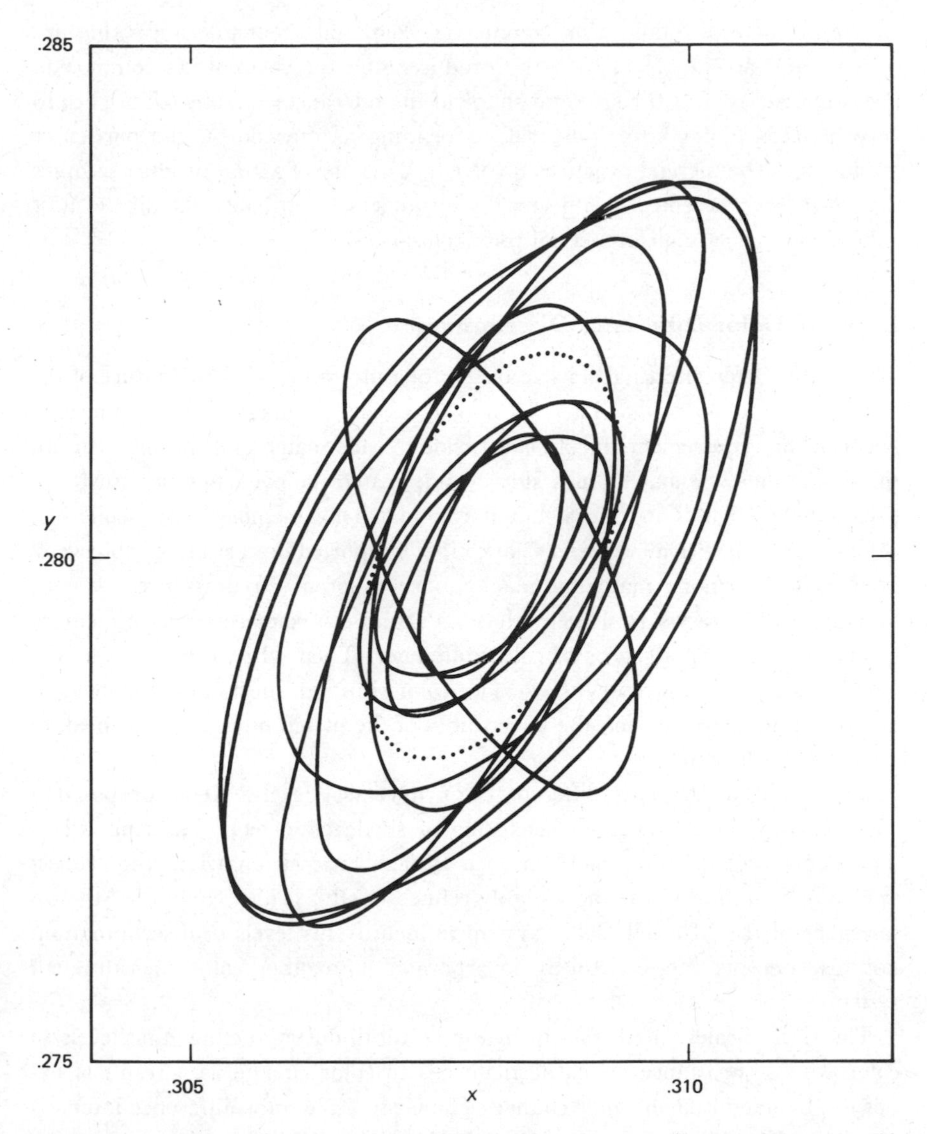


**Figure 9.5.** One example of how observers differ in color perception. The loci of just-perceptible difference from a standard color, as seen by 12 observers. From W. R. J. Brown, *J. Opt. Soc. Am.*, **47,** 1957.

The problem of establishing commercial color difference tolerances that are realistic and acceptable to both the producer and the customer is common to many industries. But the size and shape of the tolerances established are apt to vary between and within industries, depending as they do on the particular product and commercial practices involved. A history of actual product samples that have been accepted or rejected by customers is probably the single most valuable aid in setting commercial color tolerances.

## Levels of Color Difference Discrimination

The major factor affecting the size of the tolerance allowed is the nature of the color match to be made. There is a whole order of magnitude of difference between the requirements for an exact, side-by-side match and those for an approximate memory match. In a side-by-side match the eye can distinguish approximately 10 million colors, but it can match by memory only about 300 colors. Newhall, Burnham, and Clark (1957) reported on a study of observers' abilities for memory matching made under different circumstances. It was found that an observer could reproduce an actual test color retained in memory to only about 7 AN40 units of color difference. Then when the observer was asked to select from memory the surface color chip that most closely matched a familiar, but absent, object, the uncertainty of the match more than doubled, to about 18 AN40 units.

As an aid in defining different color tolerance levels, Kelly proposed a "Universal Color Language" consisting of six levels of color discrimination. The number of discrete colors in each system varies from 13 in the crudest system to 5 million in the most highly refined (Kelly, 1965). Kelly used the dimensions of the Munsell Color System to identify his levels of discrimination, but his concepts can be readily incorporated into other color measurement systems.

The Gray Scales, used mainly in the textile industry, define nine levels of colorfastness (resistance of textile materials to color change as a result of exposure) by magnitude of color change. The Gray Scale color difference intervals are defined in AN40 units, the 1950 Nickerson Unit which attempted to approximate the magnitude of the NBS Unit of color difference. The International Organization for Standardization has adopted the use of the Gray Scale (ISO, 1964, 1974), and ASTM Method D2616 describes evaluating a change in color with a Gray Scale. The American Association of Textile Chemists and Colorists (AATCC) Evaluation Procedure 1, Gray Scale for Color Change, identifies the nine levels used to assess lightfastness as follows:

| Fastness Rating | Terms for Fastness to Light | Color Difference in AN40 Units |
|---|---|---|
| 5 | good | 0.0 |
| 4–5 | | 0.8 |
| 4 | fairly good | 1.5 |
| 3–4 | | 2.1 |
| 3 | fair | 3.0 |
| 2–3 | | 4.2 |
| 2 | poor | 6.0 |
| 1–2 | | 8.5 |
| 1 | very poor | 12.0 |

The AATCC yearly technical manual describes 37 standardized exposure procedures for testing fastness, and in all of these Gray Scale estimates are used to establish the amount of color change.

Another proposal for relating quality of color match to an established color difference unit (in this case, the Nickerson Index of Fading) was made by the Munsell Research Foundation (Judd and Wyszecki, 1963; Hale, 1970). These Munsell Color tolerances and their relations to NBS Units of color difference are as follows:

| Munsell Foundation Designation of the Closeness of Color Match | Nickerson Index of Fading | Approximate NBS (Judd, Hunter) Units of Color Difference | Industry Term for Quality of Color Match |
|---|---|---|---|
| AAAA | 0.4 | 0.3 | Excellent |
| AAA | 1.4 | 1.0 | Fair |
| AA | 2.4 | 2.0 | Poor |
| A | 3.7 | 3.0 | Unsatisfactory |

## Nonsymmetry of Color Tolerances

Color difference tolerances are designed as boundaries in color space within which acceptable colors of the product must fall. The boundaries do not necessarily correlate with perceptibility of difference but rather with the limits of acceptability. The standard color, furthermore, may not be in the center of the bounded region but may be displaced to one side. For example, where subsequent yellowing may occur, the tolerance for the yellow-blue dimension might be +0.1, −0.8 units. For defining such color difference tolerances, a three-dimensional description of the color difference allowed can be considerably more helpful than a one-number tolerance such as $\Delta E$ (total color difference). A color difference expressed in terms of $\Delta L$, $\Delta a$, $\Delta b$ provides a better guide to the needed formulation correction. Such three-term tolerances are readily reducible to graphs for showing acceptability.

### Necessity for High Precision

High precision is almost essential for useful side-by-side match color difference measurements. Only with precise instruments is it possible to measure the commercially important color differences that are about as small as those the eye can see. It is often unappreciated that a very small change in a tristimulus value, particularly in the red-green dimension, as indicated by the $X$ value, can represent a fairly substantial color difference. Figure 9.6 shows the change in Hunter Color Difference Units caused by a 0.1% change in $X$ at various lightness levels. The changes are even larger when measured in MacAdam just-perceptible color difference units. Chickering, after making a study of this subject, had this to say: "For many colors of practical interest a 0.02 variation (in $X$) causes a 1.0 MacAdam unit change; a fact which should concern instrument manufacturer, instrument user, and tolerance setter alike [Chickering, 1969]."

## METAMERISM INDEXES

Metamerism represents a unique problem in color difference that is in the process of being solved. Metamerism relates to the change in color difference as two objects of nearly the same color, but different spectrally, are moved from one light source to another. There are two types of metamerism indexes, general and special:

1. A general index of metamerism is, according to Wright (1969), derived by some formula in which the difference in shapes of two metameric spectral distribution curves can be evaluated numerically *without* regard to the illuminant incident on the object. Wright indicated that no formula defining a general index of metamerism had been adopted, either nationally or internationally, and this is still the case in 1975. Nimeroff and Yurow (1965) studied the cor-

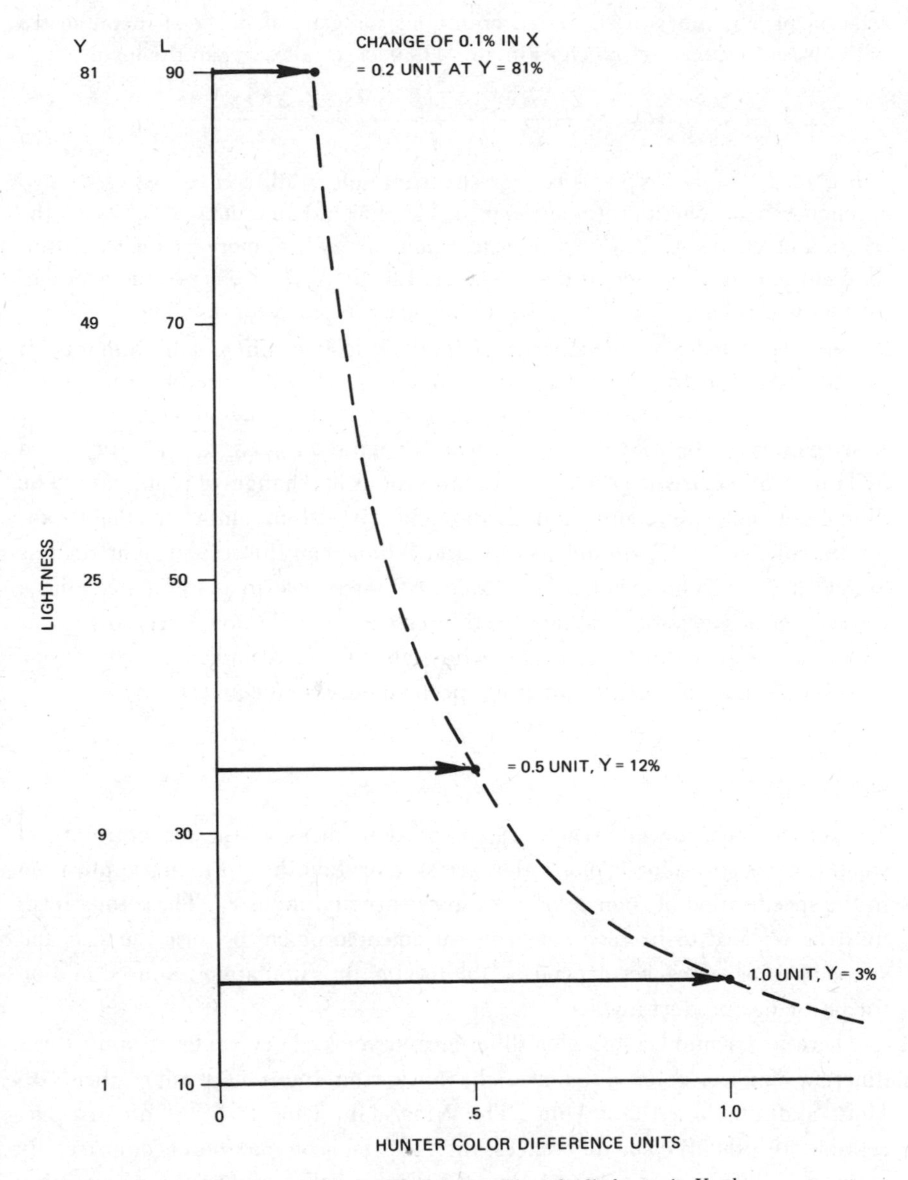


**Figure 9.6.** Change in color difference units caused by a 0.1% change in $X$ value.

relation of a number of suggested equations for general index of metamerism with observer spread of match estimate. Their best index was in the form

$$MI_{\text{gen}} = \frac{\Sigma(\Delta R\bar{x})^2}{X^2} + \frac{\Sigma(\Delta R\bar{y})^2}{Y^2} + \frac{\Sigma(\Delta R\bar{z})^2}{Z^2}$$

where $\Delta R$ is the reflectance (or transmittance) difference between two specimens as a function of wavelength. The spectral functions $\bar{x}$, $\bar{y}$, $\bar{z}$ and the tristimulus values $X$, $Y$, $Z$ can refer to equal energy but, more frequently, standard Illuminant C is used as the reference. Ideally $X$, $Y$, $Z$ refer to the averages for the two specimens, but in practice they are computed for only one.

2. A special index of metamerism is specific to two different illuminants. It can be arrived at by measuring the change in color difference beween a standard color and the test color with specified change of illuminant. Both the MacAdam series of color-difference formulas and the $L_Q$, $a_Q$, $b_Q$ scales proposed by Hunter and Christie (1966) contain provisions for changes of illuminant. The illuminants most frequently used are daylight (CIE Illuminant C or D6500), incandescent light (CIE Illuminant A), and commercial fluorescent light such as Standard Cool White. Ultraviolet excited fluorescence in materials can be a cause of serious metamerism since the fluorescence contribution is very low under Illuminant A, but can be high under daylight. There is no general consensus, however, on how properly to compute special indexes of metamerism.

## SUMMARY

Magnitudes of color difference are intended to measure the perceptibility of small contrasts in color. Color difference systems find their primary application in the specification of color tolerances in science and industry. These tolerances must be selected to fit each color measurement problem because the size and shape of color tolerances depend on the level of discrimination required and on the particular product involved.

There are a number of color difference systems. They center around three different units of color difference, the Nickerson Index of Fading, the NBS Unit, and the MacAdam Unit. The MacAdam Unit is based on just-perceptible (threshold) color differences, the NBS Unit on maximum commercially acceptable differences, and the Index of Fading on the amount of fade resulting from a specified exposure. The units bear no constant relation one to the other because, as shown in Chapter 8, the scales are not linearly related. However within a limited area of the color solid usable approximate relationships may obtain.

# CHAPTER TEN

# SPECIAL SCALES FOR WHITE COLORS

White is the color of purity, freshness, and cleanness. The consumer judges the whitest shirt to be the freshest and cleanest and will automatically select it from several on display. In some products whiteness is not only psychologically associated with purity but also is actually an indicator of freedom from contaminants, and as such it can be a measure of the quality of the product. Wool and cotton fibers that are dirty, that is, not white, will tend to lose strength and deteriorate rapidly. Good white base materials are necessary where products are to be dyed, printed, or otherwise colored. For example, to achieve the full range of colors available from the inks applied in color printing, the base paper must have excellent whiteness.

Visually, whiteness is something more than mere lightness measured by $Y$ or $L = 10\sqrt{Y}$. Special scales have been developed to measure it. In this chapter the source and perception of white colors are discussed, and a number of scales used specifically in the measurement of whites are described.

## THE OPTICS OF WHITE MATERIALS

In physical terms, a white surface is one that reflects strongly (more than 50%) throughout the visible spectrum. The higher and more uniform this spectral reflectance, the whiter the surface usually appears. As examples of white surfaces Figure 10.1 shows spectral reflectance curves for a porcelain panel, a sheet of newsprint, a swatch of white cotton broadcloth, and a panel of semi-gloss wall paint.

From the point of view of geometry, a white surface is one that reflects diffusely in all directions. Mirrors, because they reflect light in only one direction, are not called white although a good mirror reflects strongly throughout the visible spectrum and would thus be judged white by its spectral reflectance curve. The difference is, of course, that white materials have high scattering coefficients as well as low absorption coefficients (see Chapter 3).

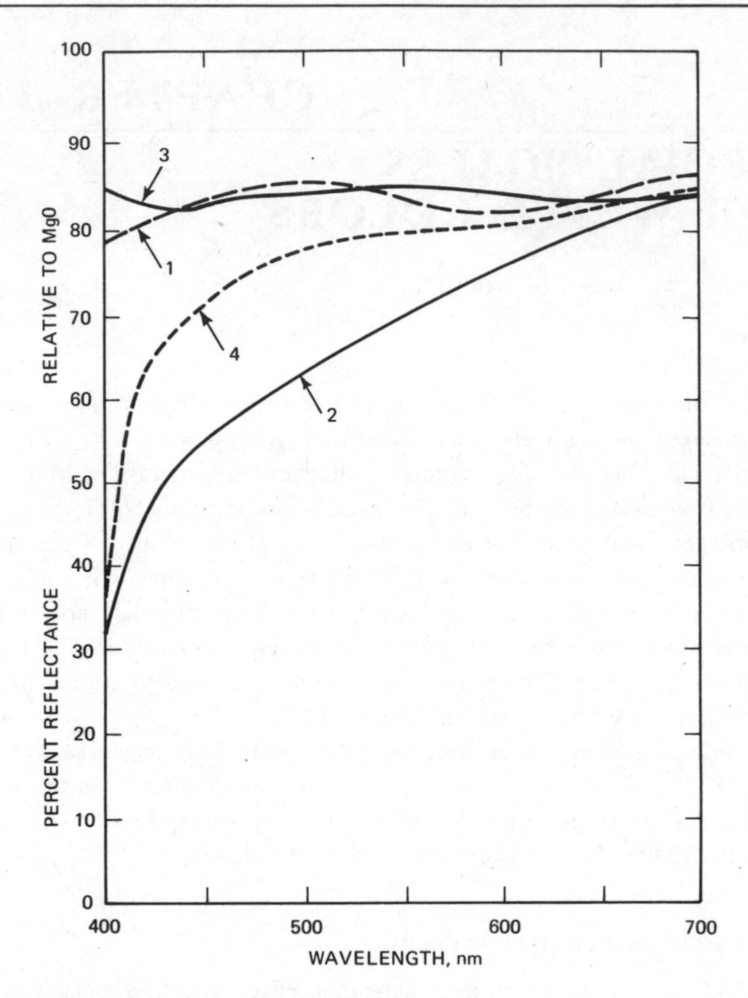


**Figure 10.1.** Spectrophotometric curves for four white surfaces: (1) cotton broadcloth; (2) newsprint paper; (3) porcelain enamel; (4) interior paint.

In three-dimensional color systems, whites occupy an area near the top of the color solid as shown by Figure 10.2. Depending upon the conditions of examination, there are about 5000 distinguishable colors called "white" and 30,000 more called "-ish whites"—that is, "bluish white," "yellowish white," and so on (Kelly and Judd, 1955).

White materials, because of their high reflectances throughout the visible spectrum, are always high in all three tristimulus values, $X$, $Y$, and $Z$. However, the tristimulus value that is most important and receives the greatest attention is $Z$ reflectance factor for the blue end of the spectrum. Reflectance for blue light is important for three reasons. First, the $\bar{z}$ (blue) tristimulus function stands by itself, separate from the $\bar{x}$ and $\bar{y}$ functions as shown in Figure 7.4. As a consequence of this separation, any tipping of a specimen's spectrophotometric curve from flatness is more likely to change the value of $Z$ relative to $Y$ than to change the value of $X$ relative to $Y$. It is these values of $Z$ and $X$ relative to $Y$ that determine chromaticity. This tendency of $Z$ to differ from the other two tristimulus reflectance factors for whites has its major visual impact on yellowness (or its opposite, blueness). When there is an excess of blue reflectance, the product is bluish; when blue light is absorbed—which happens far more often—the product is yellowish.

The second reason for concern with blue reflectance is the tendency of the

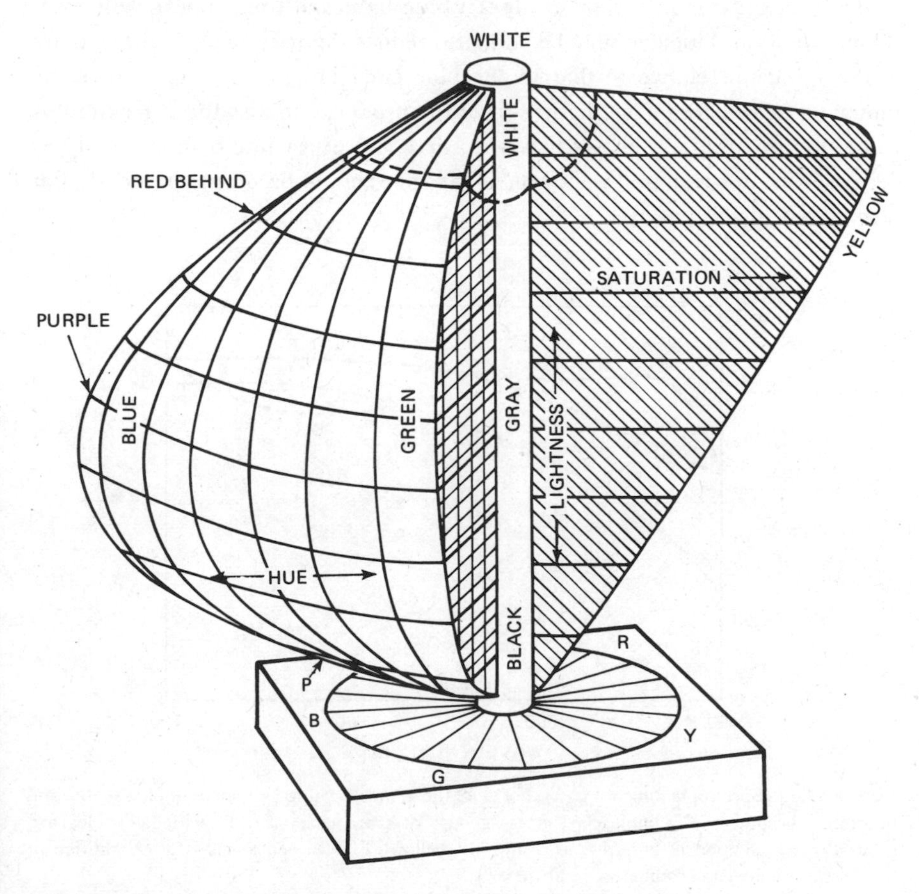


**Figure 10.2.** The color solid with region of whites shown at the top.

majority of naturally white materials to absorb most strongly at the blue end of the spectrum. Frequently this absorption is magnified by the presence of organic impurities and contaminants that have a tendency to absorb blue.

Third, it is reflectance of blue light that should be monitored as one bleaches or otherwise corrects for yellowness. The primary function of bleaching is to remove blue-absorbing contaminants. In Figure 10.3 we see that as one bleaches a textile, reflectance is increased throughout the spectrum. However, this increase is greatest and is most important in the blue. The same is true in bleaching wood fibers for paper, clay for use in paper and paint, and many other materials. To make an optical evaluation of any bleaching process, one pays primary attention to reflectance at the blue end of the spectrum.

After bleaching, there is normally a need to counteract the remaining tendency of organic materials to absorb blue light and thus appear yellowish. As one solution, blue dye may be added to reduce the reflectance in other parts of the spectrum relative to that in the blue end. This, of course, reduces the luminous reflectance factor (lightness) of the product and also the $X$ tristimulus value. Figure 10.3 also shows how the curve (D) of a white base material (B) decreases in the green and red parts of the spectrum to become more nearly flat when a blue dye is added.

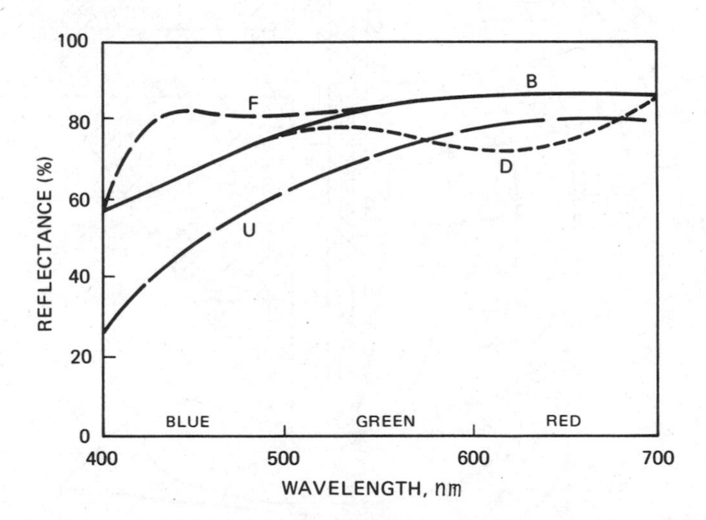


**Figure 10.3.** Spectrophotometric curves for a textile after bleaching and addition of blue dye and fluorescent brightener. U: unbleached greige goods; B: same goods after bleaching; D: bleached goods B after addition of blue dye to counteract yellow; F: bleached goods B after addition of fluorescent brightener to counteract yellowness.

A newer technique to compensate for the low natural-blue reflectance of a yellowish material is to add a fluorescent brightener. A fluorescent brightener converts ultraviolet light, which is invisible to the eye, to blue light and makes the product look bluer than it otherwise would. The result is high apparent-blue reflectance. With brighteners it is possible to make up for deficiency in blue light and gain whiteness without the corresponding loss of luminous reflectance that accompanies the use of blue dyes. Figure 10.3 shows, by curve F, this change in spectral reflectance with the addition of a brightener.

## MEASUREMENT SCALES FOR WHITES

As with any other color, three numbers are necessary for the complete identification of any white. The CIE color-coordinate system $Y$, $x$, $y$ may be used for this purpose. Sometimes the dominant wavelength and purity specification (Chapter 7) is used in place of $x$ and $y$. However, more widely used and more easily understood are the uniform color scale systems described in Chapter 8. The one enjoying the widest use for whites is the Hunter $L$, $a$, $b$ system.

When using $L$, $a$, $b$ it is almost always shown that the $b$ dimension measuring yellowness is the most critical of the three. In fact, the use of bluing, which decreases $L$ but increases blueness, produces a visually "whiter" product. In studies $b$ is typically found to be three or four times as important as $L$, which is the next most important. The $a$ dimension is the least important in normal practice.

Since the $L$ and $b$ dimensions are generally those of major concern, white colors are very frequently represented on graphs having the two dimensions of Figure 10.4. Here the blueness-yellowness dimension is horizontal, increasing in yellowness toward the right. Lightness increases vertically toward the top. In this figure the processing of a white material has been "mapped" showing how its lightness and yellowness change with cleaning, bleaching, and bluing or brightening.

As has been shown in the three foregoing chapters, three numbers are necessary to completely identify either a color or a color difference. However, as shown in Figure 10.4, two dimensions, $L$ and $b$, are sometimes adequate for the identification of a white color. Frequently one who is dealing with white materials finds that for a specific problem, a one-number color index identifies the properties of the material of interest to him. Following are descriptions of four different types of one-number color indices used to evaluate white materials. (See also Table A.10.)

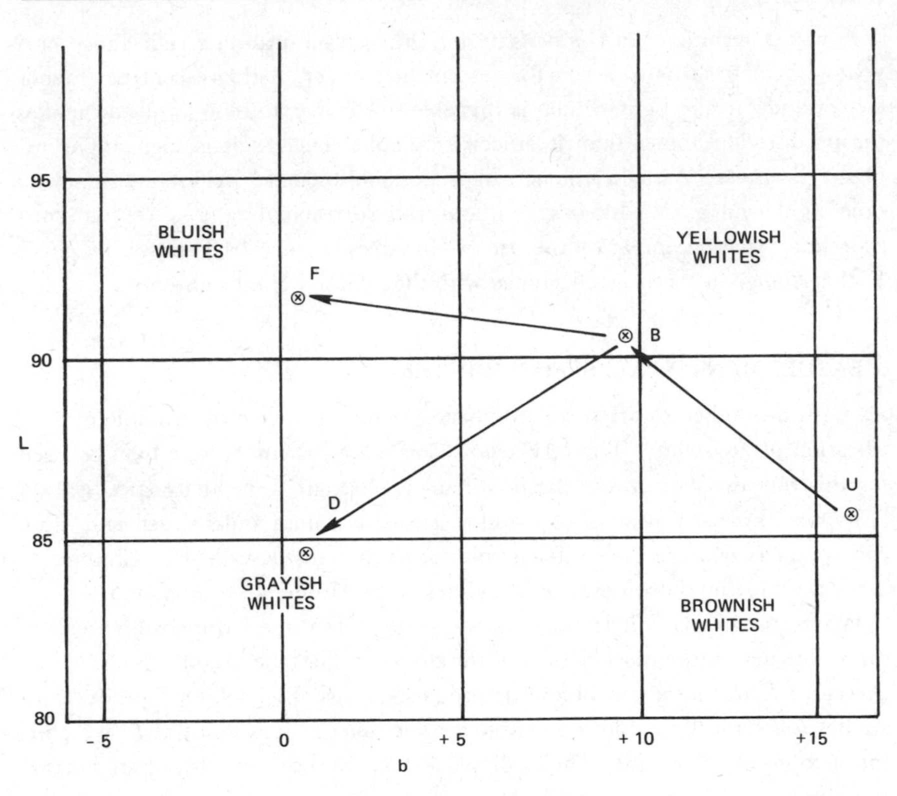


**Figure 10.4.** Position of a white material on a lightness-yellowness diagram. U; unbleached; B: bleached; D: with blue dye added; F: with fluorescent brightener added.

## Luminous Reflectance *Y* or Lightness *L*

Lightness and percent reflectance of light are widely used to evaluate the efficiency of surfaces as reflectors of light. As such these measurements are used first for measuring the overall optical quality of white materials prior to bleaching. As an example, raw cotton is graded primarily for lightness because yellowness can be eliminated by bleaching. If raw cotton is dark to start with, it is usually of inferior quality.

A second use of reflectance factor *Y* measurements involves determinations of the efficiency of reflectors used in lighting systems. Thus, the reflectors used in outdoor luminaries must possess a minimum acceptable luminous-reflectance factor.

A third use of luminous-reflectance-factor measurements is for testing textile materials for amount of dirt removed. Here the luminous-reflectance-factor measurements are made before soiling, after soiling, and then finally after washing. The increase in reflectance factor with washing is said to measure the percentage of dirt removed from the soiled cloth. This is used as a measure of cleaning efficiency. In recent times the same types of measurements have been used to evaluate redeposition, or the tendency of dirt to settle on unsoiled cloth during the washing process.

### Blue Reflectance

Reflectance for the blue end of the spectrum is used to follow the progress of bleaching. Since 1937 the paper industry has been using blue reflectance as its primary determinant of optical quality of pulp after bleaching (Davis, 1934). The textile industry has established within the past few years a similar test procedure used to measure bleaching progress in cotton cloth (AATCC Method 110-1968).

In measurements of white materials that contain either fluorescent brighteners or $TiO_2$ pigment, the spectral character of the ultraviolet side of the blue reflectance function must be closely controlled. White products containing either of these materials are characterized by spectral reflectance curves that drop very steeply at the violet end of the visible spectrum. Typical curves are shown in Figure 10.5. Note that all four curves drop sharply in going from the visible into the ultraviolet. Note also that curve B representing a fabric with fluorescence has apparent reflectance factors above 100. These can be measured properly only by spectrophotometers that disperse light by wavelength after it is reflected from the sample.

The problem of obtaining a suitable blue function for the accurate measurement of these whites is not necessarily one of duplicating the well-known Standard Observer $\bar{z}$ function. In 1948 Jacobsen argued that the $\bar{z}$ function should be more violet than it is. He showed that it was necessary to use a tristimulus function more violet than the Standard CIE function in order to convert spectral data for titanium whites to color values that agreed with visual observation.

It was partly in view of this need that the CIE in 1960 standardized the second set of tristimulus functions designed to match an average observer working with a field 10° in diameter instead of the 2° field of the 1931 Observer. The 10° $\bar{z}$ function is about 4 nm to the violet side of that for the 2° field. (See Figure 7.13.)

Recently additional work by White and Jacobsen (1965) has shown that the 10° $\bar{z}$ still may not be violet enough for titanium white assessments, and they

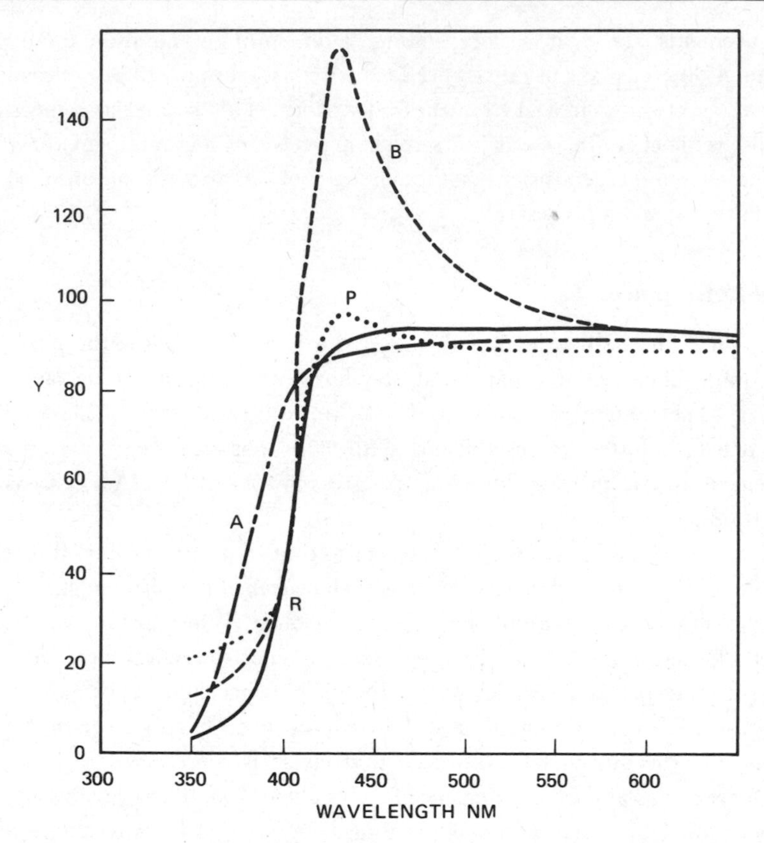


**Figure 10.5.** Spectrophotometric curves (3000 K incident energy) showing slopes of $TiO_2$ pigments and fluorescent brighteners. B: Cotton, with high concentration of fluorescent brightener; P: Cotton, with low concentration of fluorescent brightener; A: Anatase $TiO_2$; R: Rutile $TiO_2$.

recommended a much more violet function first proposed by Judd in 1949. (See Figure 10.6.)

To further complicate measurements of blue reflectance, the paper industry uses still another blue response function $B$ to make readings of "TAPPI brightness." This response is shifted to the green side of the 2° $\bar{z}$ function. While it is in extensive use and is adequate for monitoring processes such as bleaching, it is too far toward the green to produce readings that properly evaluate the normal color appearance of whites, especially those with optical brighteners. The four blue-reflectance wavelength-weighting functions are shown in Figure 10.6.

## Yellowness Index *YI*

In terms of visual appearance, absorption in the blue part of the spectrum causes yellowness. Visually we associate yellowness with scorching, soiling, and general product degradation by light, by chemical exposure, and by processing. Yellowness indices are used chiefly to measure these types of degradation. The most widely used equation was proposed by Hunter in 1942: $YI = 100(A - B)/G$. Since Adams' work, colorimetric evidence has suggested adequacy of an equation involving only $G$ and $B$. Some of the numerical index scales used to measure yellowness are

$YI_{\text{(MacAdam 1934)}}$ = Pe (excitation purity)

$YI_{\text{C429 (ASTM D 1925, Hunter 1942)}} = 100(A - B)/G = 100(1.28X - 1.06Z)/Y$

$= 125(X_{\%} - Z_{\%})/Y$

$YI_{\text{CDML (Hunter 1958)}} = b_L = 7.0(G - B)/\sqrt{G} = 7.0(Y - Z_{\%})/\sqrt{Y}$

$YI_{\text{diff (Hunter 1960)}} = G - B = Y - 0.847Z$

$YI_{\text{(ASTM E-313)}} = 100(G - B)/G = 100(Y - Z_{\%})/Y$

## Whiteness Index *WI*

"Whiteness" is the attribute by which an object is judged to approach the preferred white. It is important because in many cases observer ratings of whiteness correspond to consumer preferences. The earliest studies merely used average reflectance through the visible spectrum as the scale for whiteness. This

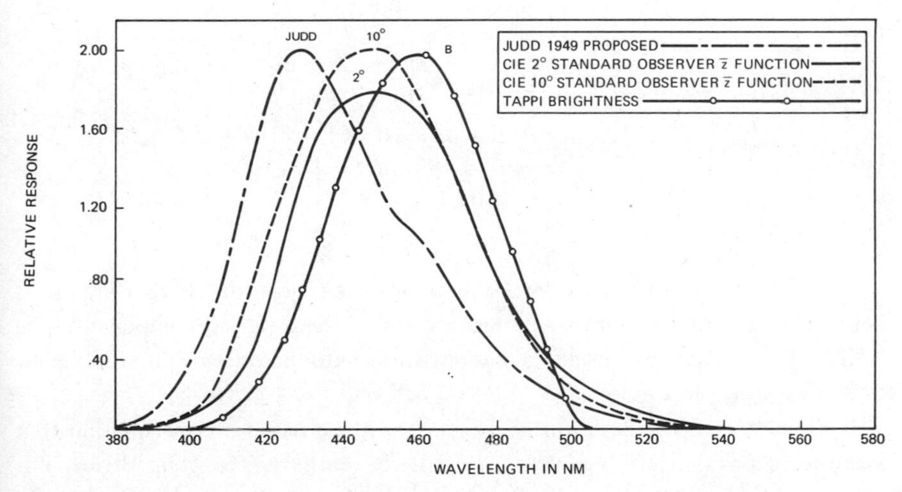


**Figure 10.6.** Four blue spectral response functions.

was hardly satisfactory, since whitening by addition of blue dye lowers average reflectance but increases visual whiteness. MacAdam (1934) was the first to use a combination of luminous reflectance and yellowness as a measurement of whiteness. He plotted contours of equal whiteness in terms of color departure from ideal white toward yellow and gray.

We know now that individual preference for whites makes optical criteria of whiteness variable from one observer to another. Judd in his study of paper-whiteness ratings found two ideal whites. One was MgO, the standard white color, and the other was "natural paper white," the best white available in paper without dyeing. In general, color of actual white can depart from color of ideal white in two directions—toward yellow and toward gray. For normal observers, we know that yellow is more serious because one of the techniques of improving whiteness is to use blue dye or pigment that diminishes yellowness by increasing grayness.

The earliest equations for whiteness index assigned a neutral white standard such as MgO the highest *WI* rating of 100. Whites varying in any manner from this white neutral color (toward blue, as well as toward gray or yellow) were penalized. Now most of the formulas in use rate bluish whites higher than neutrals, in accordance with the judgment of most observers. This means that whites rated at considerably higher than $WI = 100$ are possible.

Some of the equations used for the measurement of whiteness using tristimulus values are

$$\mathrm{WI}_{(\text{Judd 1936})} = \sqrt{Y - 6700(\Delta S)^2}$$

$$\mathrm{WI}_{(\text{Coppock 1965})} = 10\sqrt{Y - 2Pe^2}$$

$$\mathrm{WI}_{\text{C429 (ASTM D 1925, Hunter 1942)}} = 100 - \left\{\left[\frac{220(G-B)}{G+0.242B}\right]^2 + \left[\frac{100-G}{2}\right]^2\right\}^{1/2}$$

$$\mathrm{WI}_{(\text{CDML Hunter 1960})} = L - 3b = 10\sqrt{Y - 21(Y - Z_{\%})/\sqrt{Y}}$$

$$\mathrm{WI}_{(\text{ASTM E-313})} = G - 4(G-B) = 4B - 3G$$
$$= 0.01\,(L - 5.7b)$$

$$\mathrm{WI}_{(\text{Stensby 1967})} = L - 3b + 3a$$

The last three equations for whiteness index give the blue component between three and four times as much weight as the lightness component. The ASTM E-313 equation produces straight-line contours on a graph of $G$ versus $G - B$ as shown in Figure 10.7.

It is likely that the best whiteness index will be based on an equation that locates the position of an *ideal white* (not the perfectly reflecting perfect diffuser, but an imaginary blue-white) and measures whiteness as the distance in color space from the sample to that ideal.

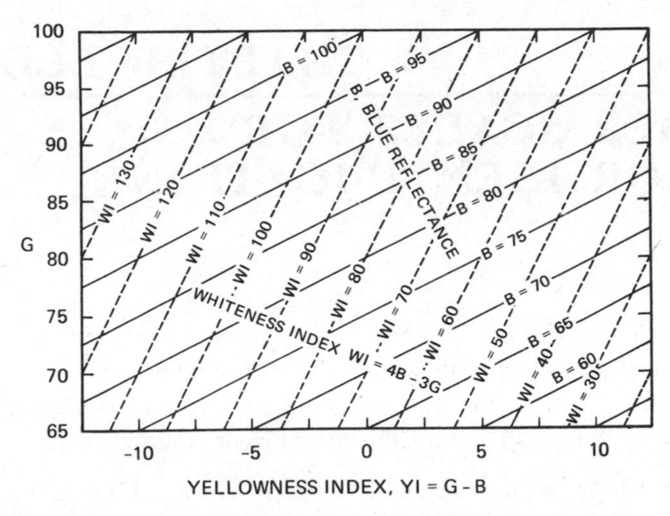


**Figure 10.7.** Yellowness versus luminous-reflectance factor. $G = Y$; $B = Z_{\%}$.

## GRAPHIC TECHNIQUES

There are several graphic techniques useful in working with whiteness specification. The $Y$ versus $Y - Z$ chart has been discussed as has the $L$ versus $b$ graph. These are probably the best means of showing the relations of white materials graphically in two dimensions. For quick computation of whiteness, $4B - 3G$, and yellowness, $Y - Z$, and also for colorimetric purity, nomographic techniques have been developed.

## SUMMARY

White materials are measured to assess their quality, to control bleaching or other processes, to determine yellowness, and to evaluate their visual whiteness.

For many problems involving whites, the red-green color dimension can be ignored. On the other hand, the blue-yellow dimension is critical—three to four times more so than the lightness dimension.

Whiteness indices, used to provide correlation with visual whiteness rankings, evaluate blue whites higher than neutral or yellow whites of the same lightness.

Table A.10 outlines the attributes of white surfaces of interest to particular industries.

# CHAPTER ELEVEN

# OTHER SCALES FOR COLOR IDENTIFICATION

To complete the study of methods for color identification, it is necessary to list a number of color scales, some of which are not traceable back to the trireceptor Standard Observer. These are divided into four groups:

1. Single-number color scales, each designed for the color quality evaluation of a specific product such as lubricating oil or tomato juice.
2. Analyses based on subtractive color mixture, such as Lovibond designation and color densitometry.
3. Techniques using additive color mixture, such as disk-mixture analysis.
4. Identifications by systematic collections of colored chips, such as the Munsell Color System.

## SINGLE-NUMBER COLOR SCALES FOR SPECIFIC PRODUCTS

The single-number, or single-locus, color scales fall generally into two groups. The first group consists of about 20 scales that are used to measure the yellowness (or, more importantly, the freedom from yellowness) of amber-colored oils, resins, liquids, and chemicals. The second group of single-number color scales is comprised of those used to measure the color quality of specific food products, although the techniques used to derive them could be applied to the development of color-quality scales for other products.

### Single-Number Scales for Yellowness of Amber-Colored Liquids and Resins

Over the years, a number of color-index scales have been developed for yellowness of amber-colored liquids and resins. Initially each was based on some visual "comparison of a solution of unknown concentration with a series of suitably prepared standard solutions. . . . In these comparisons the colors of the unknown exhibit a one-dimensional change with concentration [Judd and

Wyszecki, 1963]." All of these scales are designed to measure yellowness and/or brownness, but for some the saturation range of yellowness is the important attribute, while others are designed to measure a change in hue, as from greenish to reddish brown.

The Inter-Society Color Council (ISCC) Subcommittee for Problem 14 completed an interim report in 1962 on single-number transparent standards for yellowness (Johnston, 1971). This report, and Judd's description of one-dimensional color scales (Judd and Wyszecki, 1963) are recommended to those needing further information on this subject. Figure 11.1 shows the path in the *x, y* diagram of the Gardner (ASTM D1544) one-dimensional scale for yellowness of resins and oils used in the paint industry. Table A.11 identifies a number of these scales for transparent liquids and resins.

The Hazen Standards (APHA) determining the color of light-colored liquids and also the Barrett Color Standards vary primarily in saturation (along a straight line from the colorless illuminant point). On the other hand, the U.S. Rosin Standards and the FAC darker standards vary primarily in hue from greenish to reddish brown.

## Scales Used to Measure Color Quality of Specific Food Products

Several single-number scales have been developed for rating the color quality of specific foods. *X, Y, Z,* or *L, a, b* color specifications are converted to numbers

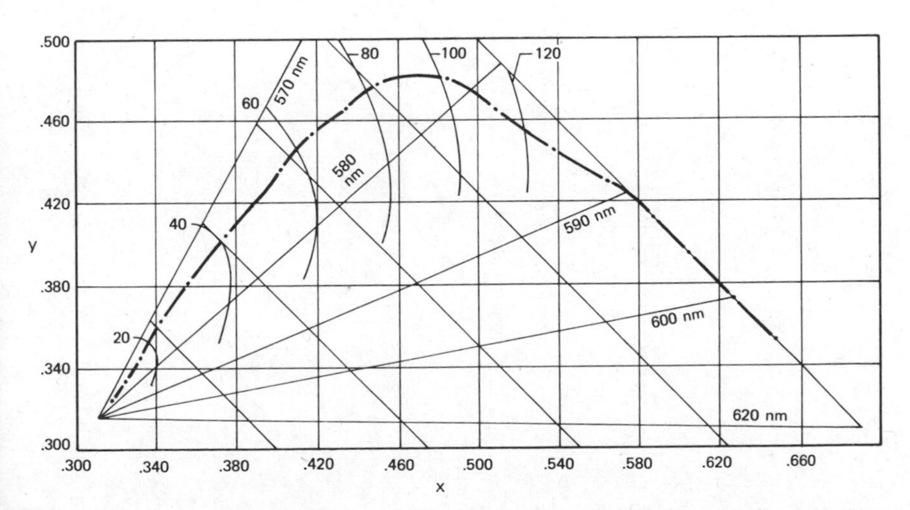


**Figure 11.1.** The path in the *x, y* diagram of the Gardner (ASTM D1544) scale for yellowness of resins and oils. Reprinted, with permission, from the February 1971 Journal of Paint Technology, a publication of the Federation of Societies for Paint Technology.

indicating grade or quality on the desired scale. Such a scale was developed by Yeatman and others to measure the color quality of raw tomatoes (Yeatman, Sidwell, and Norris, 1960). Hunter then devised an analog measurement circuit and designed a direct-reading Tomato Colorimeter (Hunter and Yeatman, 1961). The equation for the Tomato Color Index is

$$TC = \frac{216}{Y^{1/2}} - \frac{30}{Y^{1/2}} \left( \frac{Y - Z_{\%}}{X_{\%} - Y} \right)$$

or

$$TC = \frac{216}{0.1L} - \frac{75b}{0.1La}$$

Another such scale evolved out of the needs of the citrus industry. Orange juices occupy the part of color space shown in Figure 11.2. The need for an instrumental measure of the colors of orange juices led to the development of the Citrus Colorimeter. This instrument has two scales:

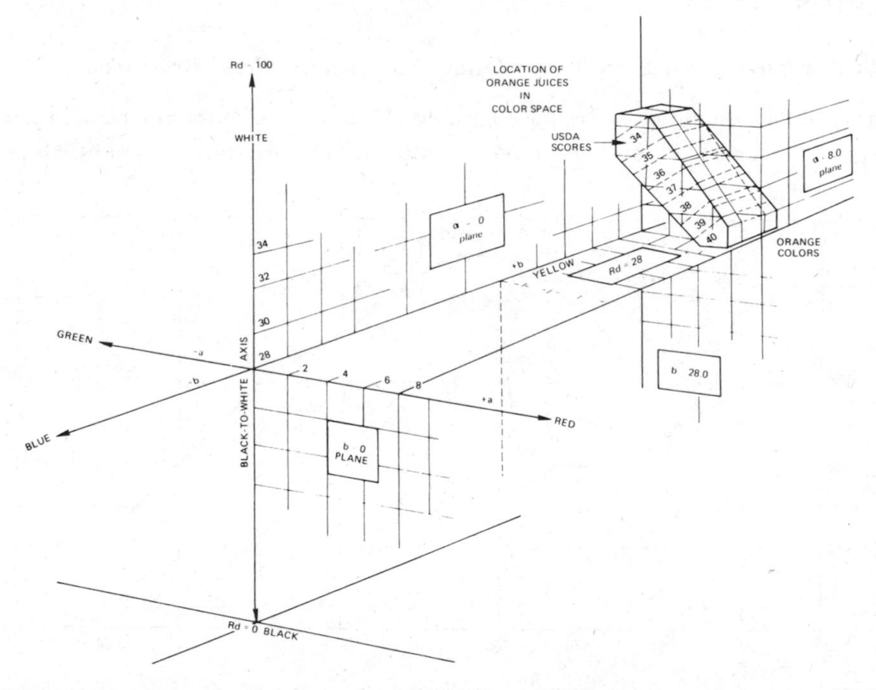


**Figure 11.2.** How the colors of orange juices of different USDA color scores fit into the yellow-orange region of Hunter *Rd, a, b* color space. Note that *Rd* = *Y*.

1. Citrus Redness $CR$ scale for the measurement of orange juice (Hunter, 1967). The formula for this scale is:

$$CR = 200 \frac{A}{Y-1} \text{ or } CR = 200 \frac{1.2777X - 0.213Z}{Y-1}$$

2. Citrus Yellowness $CY$, a subsidiary scale was also developed

$$CY = 100 \frac{1 - 0.847Z}{Y}$$

The Equivalent Orange Juice Color Score $ECS$ is derived from $CR$ and $CY$ values as follows

$$ECS = 22.51 + 0.165CR + 0.111CY$$

Figure 11.3 shows graphically how Equivalent Color Scores relate to grade score of reconstituted frozen orange juice.

Other food rating scales are represented by the nomographs shown in Figures 11.4 and 11.5. These have been used to obtain grade scores for color of applesauce and processed tomato juice from $L$, $a$, $b$, measurements.

## COLOR IDENTIFICATION BY SUBTRACTIVE COLOR MIXTURE

There are two techniques of color identification based on subtractive color mixture.

1. Identification by amounts of subtractive primaries mixed to match the unknown in either color or chromaticity. An example is the Lovibond System.
2. Color densitometry. Optical densities for three regions of the spectrum correspond roughly to amounts of three subtractive primary colorants used to generate colors in color printing and color photography.

### Lovibond Color System and Tintometer

A visual colorimeter that uses a comparison field adjusted by introducing calibrated amounts of three subtractive primary colorants into the field is called a subtractive colorimeter. The Lovibond Color System is used for color identification of beer, vegetable oil, some opaque objects, and many transparent materials. The instrument associated with it is a visual subtractive colorimeter called the Lovibond Tintometer. The chromaticity of each sample measured is visually matched to a light whose chromaticity has been adjusted by passing it through superimposed colored glasses selected from series of red, yellow, and

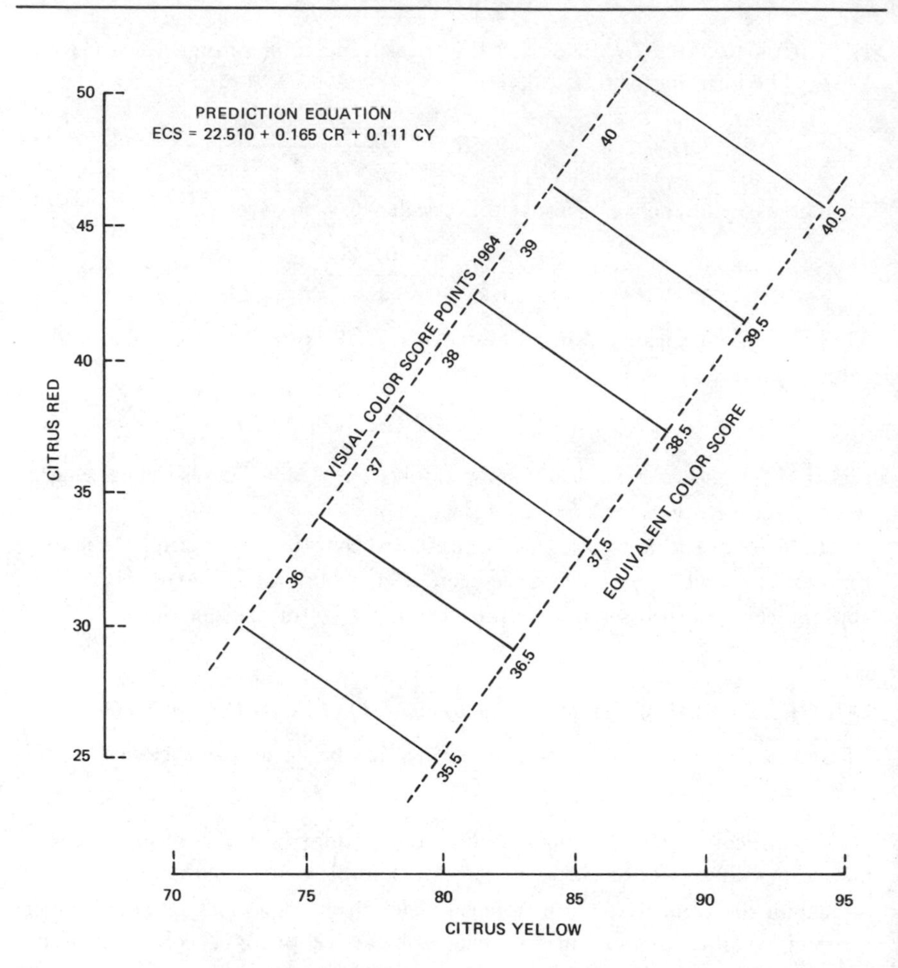


**Figure 11.3.** Nomograph for estimating USDA color score of reconstituted frozen concentrated orange juice in terms of Hunterlab citrus colorimeter values. Citrus yellow *CY* and Citrus Red *CR* readings of the Citrus Colorimeter are combined in the formula shown for Equivalent Color Score *ECS*. The *ECS* score is then related to the visual color score by means of the nomograph.

blue glasses. These are available in different densities and are selected by the operator of the Lovibond Tintometer to make a visual match to the chromaticity of the specimen. The units of the Lovibond scales are arbitrary, but the glasses in each set of standards are graded and marked with values. The color of the matched specimen is specified by adding together the values of the component glasses that were used to make the subtractive light mixture that matched the sample. The units of the three scales are also adjusted so that a

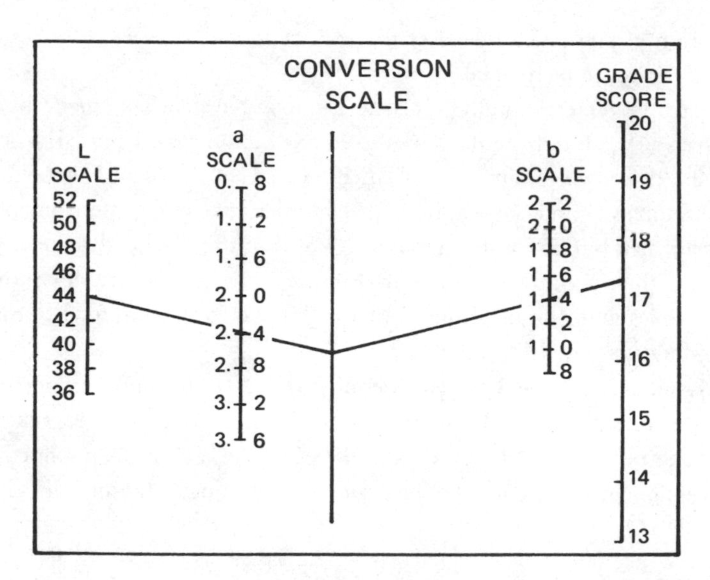


**Figure 11.4.** Nomograph for calculating U.S. grade for color of canned applesauce in terms of Hunter *L, a,* and *b* values. From *Quality Control for the Food Industry,* 3rd. ed., Vol. 1, by A. Kramer and B. A. Twigg, published by Avi Publishing Co., Inc., Westport, Conn., 1970.

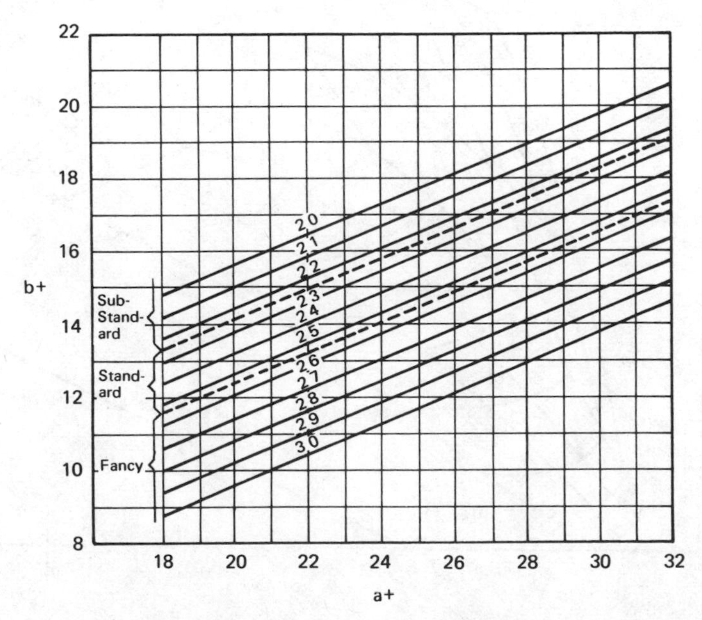


**Figure 11.5.** Nomograph for calculating U.S. grade for color of tomato juice in terms of Hunter *a* and *b* values. From *Quality Control for the Food Industry,* 3rd ed., Vol. 1, by A. Kramer and B. A. Twigg, published by the Avi Publishing Co., Inc., Westport, Conn., 1970.

combination of one unit of each of the red, yellow, and blue scales results in a filter that is seen to be neutral in color.

A spectrophotometric analysis of the Lovibond system was made by Gibson and Harris (1927) and the data was used by Tintometer, Limited, to compute the chromaticity coordinates *x, y* of all the colors of the ideal Lovibond system for CIE Sources B and C (Schofield, 1939). Similar computations for Source A were performed by Haupt and Douglas (1947) at the National Bureau of Standards, and the results are shown in Figure 11.6. A more recent publication presents tables and graphs of the ideal Lovibond color system for Illuminants A and C (Haupt, et al., 1972).

The Lovibond-Schofield system (Schofield, 1939) uses an auxiliary device to control the luminance of the matching field. Luminance can be read from a scale on the instrument. The CIE specification *Y, x, y* can be obtained quickly by conversion graphs from the instrumental readings obtained by the Lovi-

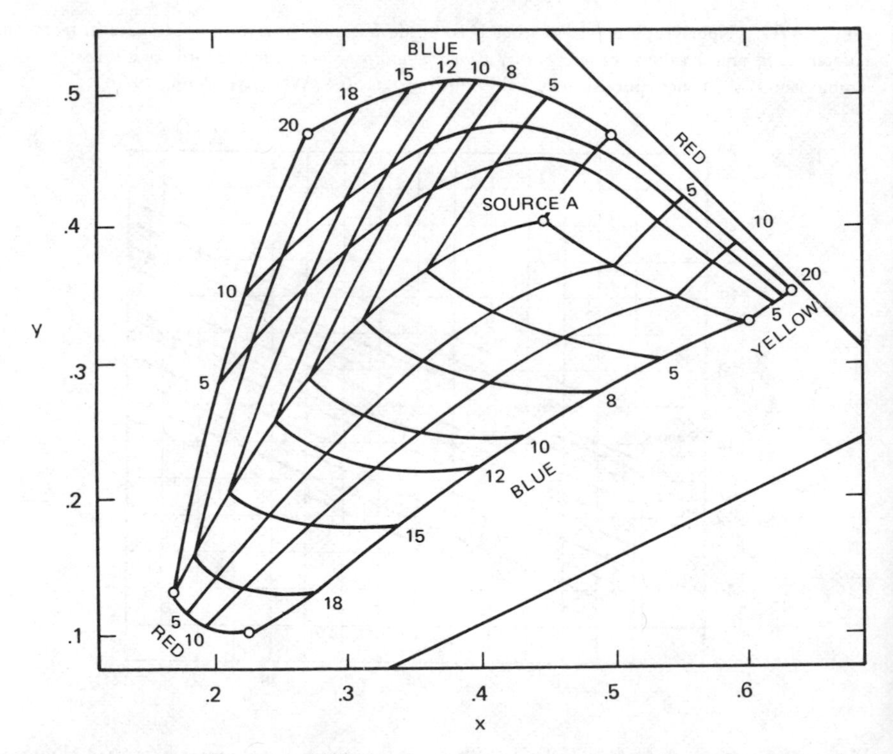


**Figure 11.6.** Chromaticities of two-part combinations of Lovibond glasses for Source A. After I. Nimeroff, NBS Monograph 104, 1968.

bond-Schofield system. These graphs are similar to Figure 11.6, but they are larger and are applicable to Illuminant C.

## Densitometry and the Subtractive-Mixture Color Solid

Densitometry is used primarily in photography and printing where subtractive mixtures of dyes or inks are used to produce desired color appearance. The colorants used are designed to mix subtractively; that is, one is designed to absorb blue light (yellow colorant), one to absorb green light (magenta colorant), and one to absorb red light (cyan colorant). Density is measured through three filters, each transmitting only the spectral region of maximum absorption of one of the colorants.

The function of densitometry is to identify the colors reproduced. Densitometric measurements are values of optical density for the spectral regions of absorption of each of the specific dyes or inks involved. Reflected or transmitted light is measured through one of a number (usually three or four) filters and converted to density

$$D = -\log 10\ (T \text{ or } R)$$

With ideal subtractive primary dyes or inks, each value of density is a measure of the amount of the one dye or ink that absorbs in the same spectral region. With real dyes and inks, each value of density is affected somewhat by all three or four of the dyes or inks. Nevertheless, densitometry is widely used to assess amounts of each of the colorants present at the point of measurement.

The Foss Color Order System is based on a cubic color solid that provides a visual relationship between printed color and the concept of a color solid, and the measurements of printed color by density (Foss, 1973). Figure 11.7 shows the cube standing on its black corner with white at the top. The cube corners for cyan, yellow, and magenta are on the arms connected to the top, while the darker subtractive primary colors of green, blue, and red are at a lower level on the arms connected to black. Foss claims reasonable visual uniformity for his solid and suggests increments in density required to make the arms fairly uniform in visual spacing. Of course, the actual location of colors in this Foss color solid will change whenever the specific subtractive primary dyes or inks are changed, or when the white base on which they are printed is altered. A printer might envision the cube with black at the top and white at its base.

The relationship of densitometric measurements to tristimulus values of color is not well developed. To understand it, let us identify first the gray component. According to Yule, every color can be said to have a gray component and a

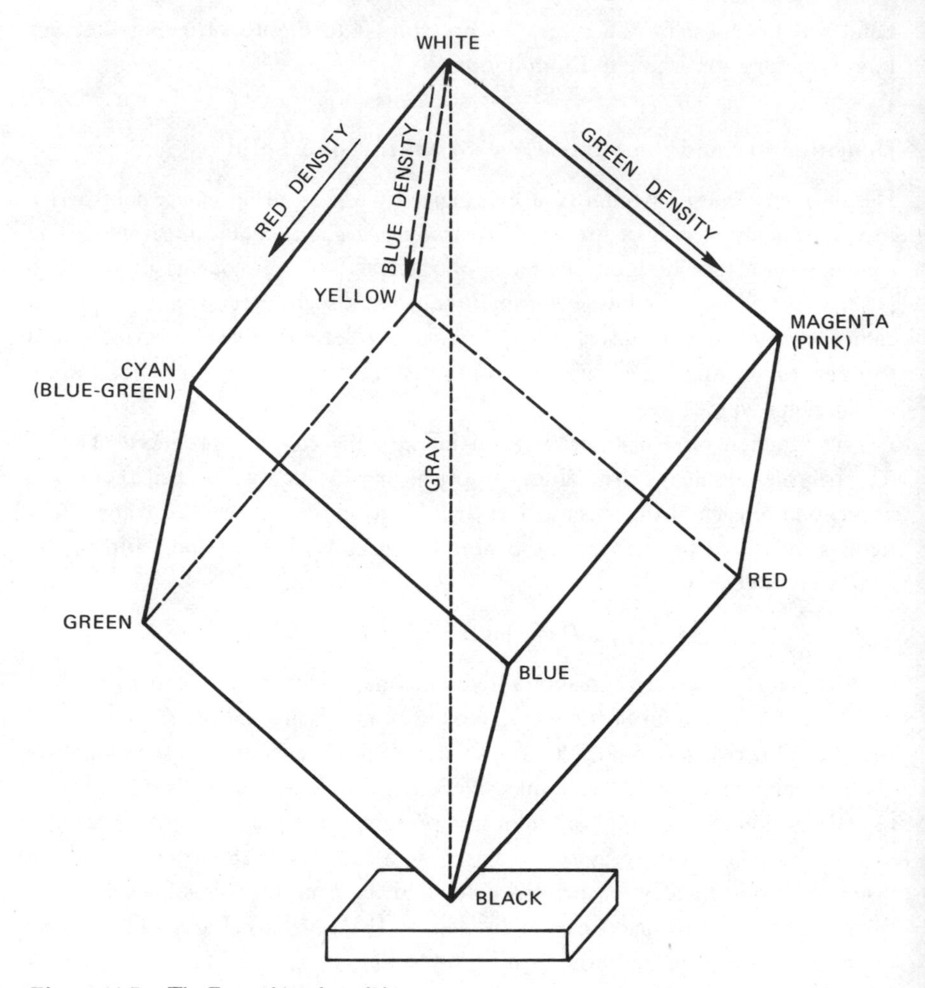


**Figure 11.7.** The Foss cubic color solid.

chromatic component (Yule, 1967). The gray component is established primarily by black-ink density in printing, where no more than two chromatic inks with black ink need ever be used at one location. In the three-dye mixtures used in photography the gray component is set by the level at which all three dyes are present equally. This is because the amounts of the three dyes present at equal density levels contribute to gray component, not chromaticity.

The chromatic component relates to density in the manner shown in Figure 11.8, by the densities of the two chromatic inks present in any printed area. For three-dye mixtures as in photography, the chromatic component would be established by density differences of the two dyes present in greater densities than the third dye. Figure 11.8 is intended to be qualitative not quantitative. Its grid will change with every change of colored inks or dyes. It will also be different for printing (halftone ink dots) and photography (dyed film). Note the similarity of chromatic scales in Lovibond (Figure 11.6) and densitometry (Figure 11.8). However, the identifying names are in contrasting colors. This is because blue-light density (absorption) in Figure 11.8 corresponds to Lovibond

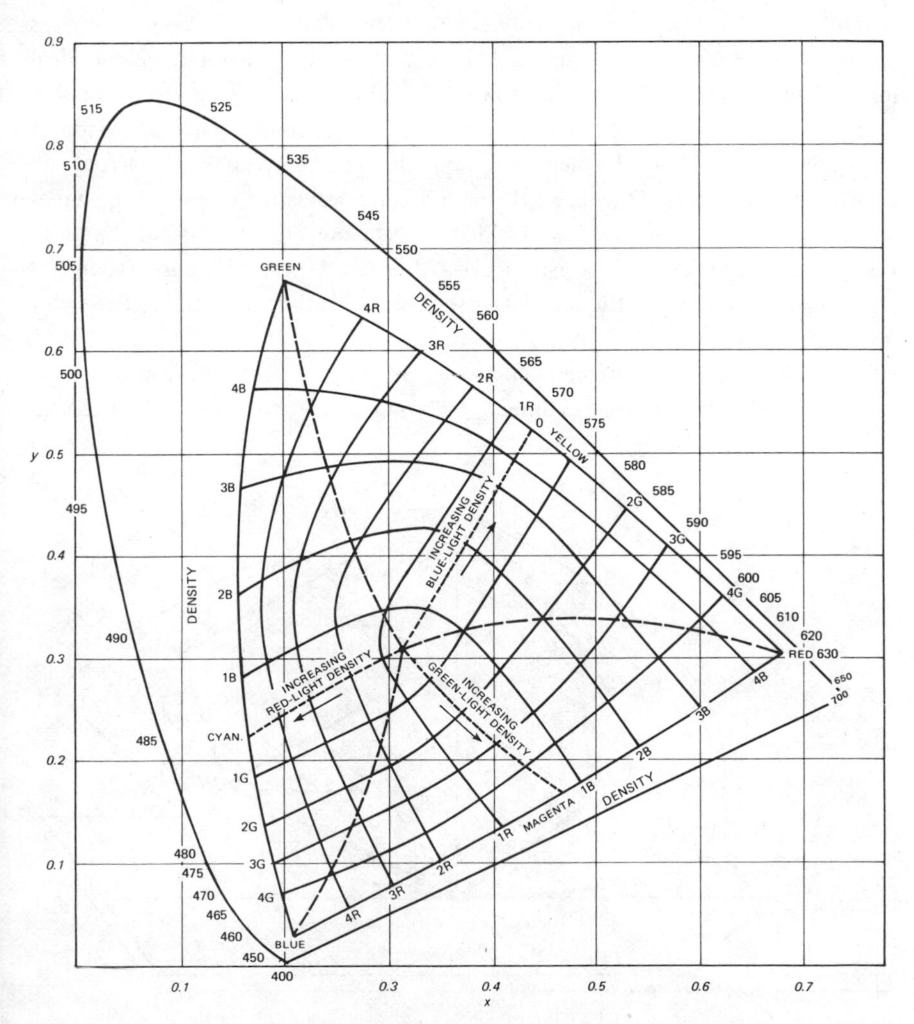


**Figure 11.8.** A qualitative representation of chromaticities in the $x$, $y$ diagram of mixtures of a typical set of subtractive primaries. Densities are measured from the gray-level density.

yellow-glass content of the color match. Similarly, red-light density corresponds to Lovibond blue-glass content, and green-light density to Lovibond red-glass content.

Only with ideal subtractive primary inks or dyes would chromaticities in $x$ and $y$ remain fixed with changes in the gray component. Ideal colorants, if they existed, would have absorption curves changing from zero to maximum absorption at single wavelengths. Such colorants are never found in practice.

Pearson and others at the Rochester Institute of Technology have used the Hunter *L, a, b* coordinate system to study the relationships between colors available in actual printing inks, and those that would be available if the ideal inks existed (Pearson, 1968). Pearson used a Hunter *L, a, b* space except that he turned it upside down, putting white at the bottom and black at the top as shown in Figure 11.9. He then compared the gamuts available in actual ink colors with the ideals. Of course, the actual color solids were of varying shapes and none matched the uniform six-sided cube of the Foss color solid shown in Figure 11.7. The Pearson work suggests that the Hunter *L, a, b* system has significant usefulness in the evaluation of color printing and subtractive color mixture.

A densitometric measurement of a color is suitable for identification of the color only with the same instrument and colorant system. Methods of densi-

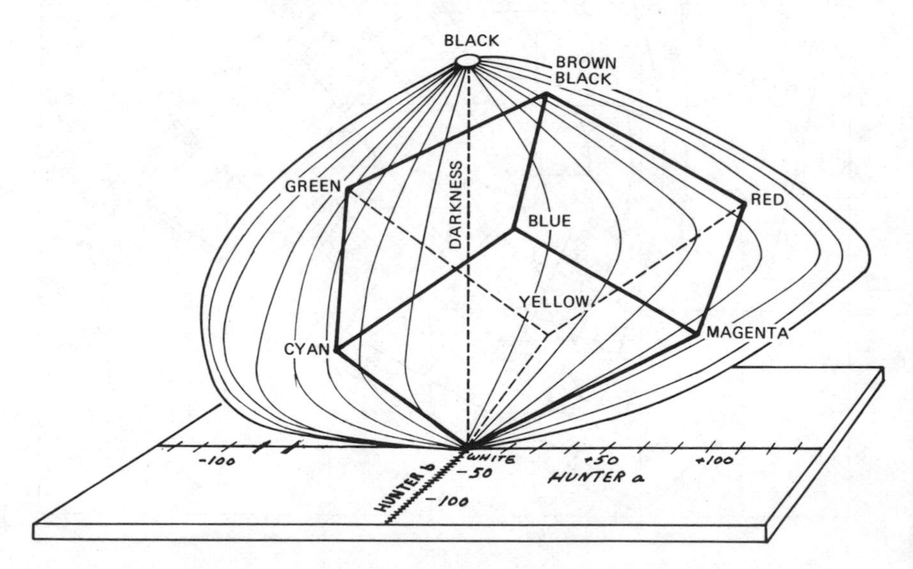


**Figure 11.9.** Pearson-Yule use of inverted Hunter *L, a, b* color solid to represent ink colors achievable in process inks.

tometry can be used for accurate color identification only within each single system of dyes and base materials. Thus densitometry does not have the broad ability of tristimulus colorimetry to identify color appearance regardless of materials used. On the other hand, densitometric measurements within a single system can provide precise separation of small color differences.

## COLOR IDENTIFICATION BY ADDITIVE COLOR MIXTURE

There are two techniques of additive color mixture worth noting. Visual additive colorimeters are used primarily for research relating to color vision of observers, although they can also be used for object-color and illuminant-color identification. Disk-mixture color matching in conjunction with the Munsell Color System is used for foodstuff quality evaluation and for identification of colors in designs.

### Visual Additive Colorimeters

This type of instrument is used to derive color-mixture functions of observers, such as those that served as the basis for the CIE Color System. Figure 11.10 shows the Donaldson Colorimeter, in which an observer matches light from an unknown specimen by adjusting an additive mixture of six primary lights. The use of six primary lights lessens the metamerism between the matched colors of the unknown light and the test light. The results can be specified in CIE terms.

Another colorimeter for visual research, the MacAdam Binocular Colorimeter, was utilized by Wright and Wyszecki (1960) in their field trials of the proposed 10° color-mixing functions. This instrument consisted of two colorimeters of the Donaldson type, each supplying one half of the bipartite field. The viewing area is large and can be viewed with both eyes. The matches made using this instrument were all nonmetameric, since the same three filters that were used to generate the test light were used to duplicate the color in the other half of the field.

### Disk Colorimetry

Disk colorimetry involves rotating several colored papers on a disk. The spinning disk creates a true additive color mixture by high-speed fusion of the light from each of the component colors. The method was originally suggested by Clerk Maxwell, whose name is still attached to the disks. The disks are cut radially so that several can be slipped together, showing portions of each. By adjusting the proportions of each color, an unknown sample's color can be

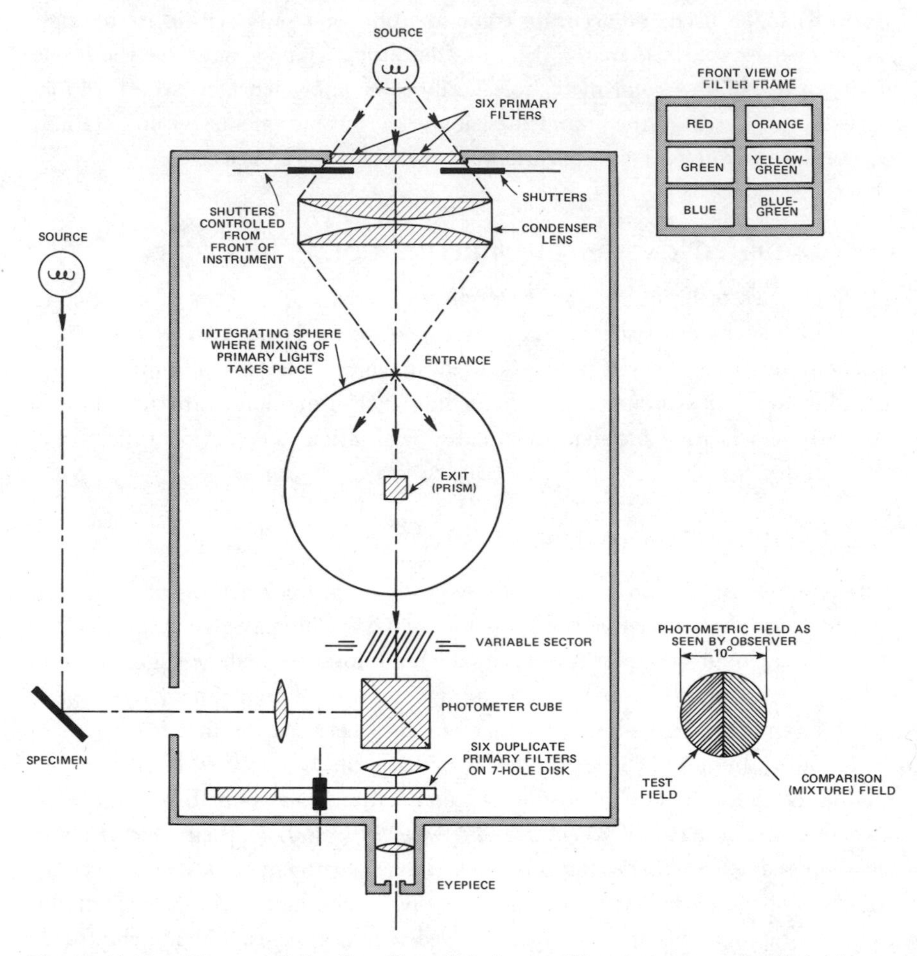


**Figure 11.10.** Diagram of the Donaldson Colorimeter. The observer matches light from a specimen by adjusting six primary lights. From *Color in Business, Science, and Industry* by D. B. Judd and G. Wyszecki, John Wiley & Sons, 1963. Used with permission of *J. Opt. Soc. Am.*

matched by the additive color mixture of the spinning disks. (See Figure 11.11.) The method was used for color measurement of agricultural and other products by Nickerson at the Department of Agriculture (Nickerson, 1946). She used colored papers from the Munsell System, but any disks for which the CIE values are known can be used.

Food-color identifications in Munsell terms can be made more precisely by using disk colorimetry than by referring only to the Munsell chips nearest in color to the sample.

## COLLECTIONS OF COLORED CHIPS FOR IDENTIFICATION OF COLORS OF SURFACES

The last group of color identification systems represents identifiable colors by charts or collections of colored chips. A number of different systems are used for the arrangement and identification of the chips in these collections. It is by the method of chip arrangement that the various systems are known.

### Munsell Color System

The Munsell system was conceived by A. H. Munsell as a means of showing the perceptual arrangement of colors (Chapter 1) and assigning Munsell designations to them. The *Munsell Book of Color* contains actual chips of the Munsell colors arranged by value and chroma on pages, each of which shows a specific hue. There are also charts, each of a specific value, arranged in polar layouts of hue and chroma (Munsell, 1929).

In the Munsell system, there are five primary hues (red, yellow, green, blue, and purple) and five intermediate hues. These, with a further subdivision of each of these ten hues, are shown in Figure 11.12. Value (the Munsell designation for lightness) and chroma (representing saturation) form the other two dimensions of the system. The value scale goes from 0 for perfect black, to 10 for perfect white. The chroma scale extends to the limits established by stable colorants, with numbers as high as 12 or 14 for actual colored surfaces. The actual Munsell chips do not achieve all of the theoretically conceivable colors, although they do cover a wide range. Figure 11.13 shows the arrangement of the page in the Book of Color for yellow-red yellow. The darker colors are at the bottom, and the more saturated colors are to the right. A complete designation of color in Munsell terms is given as hue-value/chroma. The closest visual

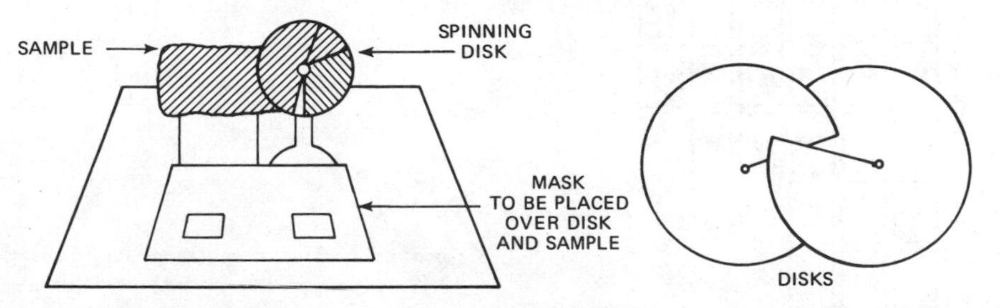


**Figure 11.11.** Diagram of a disk colorimeter, with detail of disk. The disks can be slipped together, showing portions of each. By spinning the disks the colors are mixed additively in the eye and can be matched to the color of the sample.

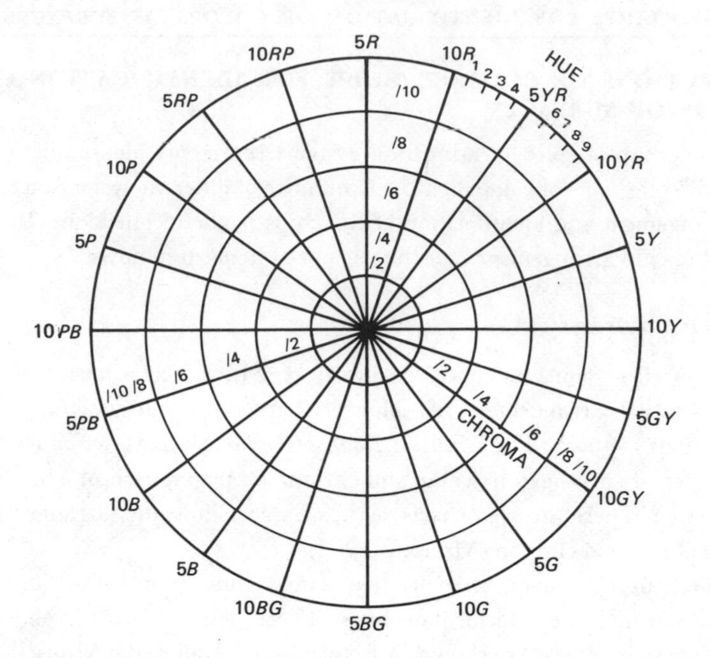


**Figure 11.12.** Designation system for Munsell hue. Five primary hues (red, yellow, green, blue and purple) and five intermediate hues (yellow-red, green-yellow, blue-green, purple-blue and red-purple) are each subdivided further and shown with steps of saturation (chroma) to the outside of the hue circle. From *Color in Business, Science, and Industry* by D. B. Judd and G. Wyszecki, John Wiley & Sons, 1963.

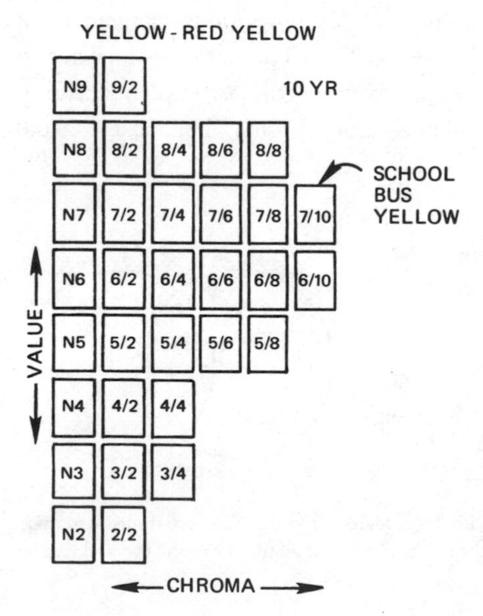


**Figure 11.13.** Arrangement for Munsell yellow-red yellow hue (10YR) by value and chroma. The chip closest to the color of the school bus is found on this page at value 7 and chroma 10. The Munsell designation for the school-bus color, therefore, is 10YR 7/10.

match to school-bus yellow is found on this page and can be designated as 10YR 7/10.

Munsell color designations are achieved by visual comparisons, either in the disk colorimeter previously described or by visual comparison with chips in the Munsell Book of Color. When trying to match a sample with a chip in the Munsell Book of Color, the page of chips with the closest visual hue is determined, and then the colors on that page of the book are examined visually in close juxtaposition to the object. One of the advantages of the Munsell system over some other systems is that the designations permit interpolation on all three scales so that any color can be specified in Munsell terms even if it falls in between the actual chips. Another advantage is that it is not restricted to available pigments. Should more saturated pigments become available, they can be assigned a higher chroma specification and take their place in the system.

In 1943 the chips in the *Munsell Book of Color* were measured on a spectrophotometer and designated with colorimetric specification by $X$, $Y$, $Z$ values (Kelly, et al., 1943). Recommendations to improve the visual uniformity of the spacing of the color chips were made at the same time by Newhall, Nickerson, and Judd (1943). The recommended system of color chips with improved spacing is called the "Munsell Renotation System" (see Chapter 8).

The history of the Munsell system has been carefully and interestingly documented by Dorothy Nickerson, who has been closely associated with it from its beginnings. Her articles are recommended to the reader who is interested in more information concerning the Munsell system and its unique position in the history of object-color identification (Nickerson, 1940, 1969).

## ISCC–NBS System of Color Names

The ISCC Color Names resulted from a comprehensive project designed to apply popular color names to the Munsell Colors and thus generate a standard verbal designation for color (Kelly and Judd, 1955). In all, 267 standard designations are possible in the system. Since the eye and brain can only distinguish about 300 colors by memory, this system is of about the right size for distinguishing basic colors. A series of "centroid" chips are available which represent each of the colors. The chips are identified by descriptive color names together with serial numbers. For example, school-bus yellow would carry the designation "Moderate Orange Yellow." This system is used for color identification in design, architecture, and art. (See Figure 11.14.)

When Kelly and Judd published the 1955 "ISCC–NBS Method of Designating Colors" in NBS Circular 553, they included in it a section coordinating

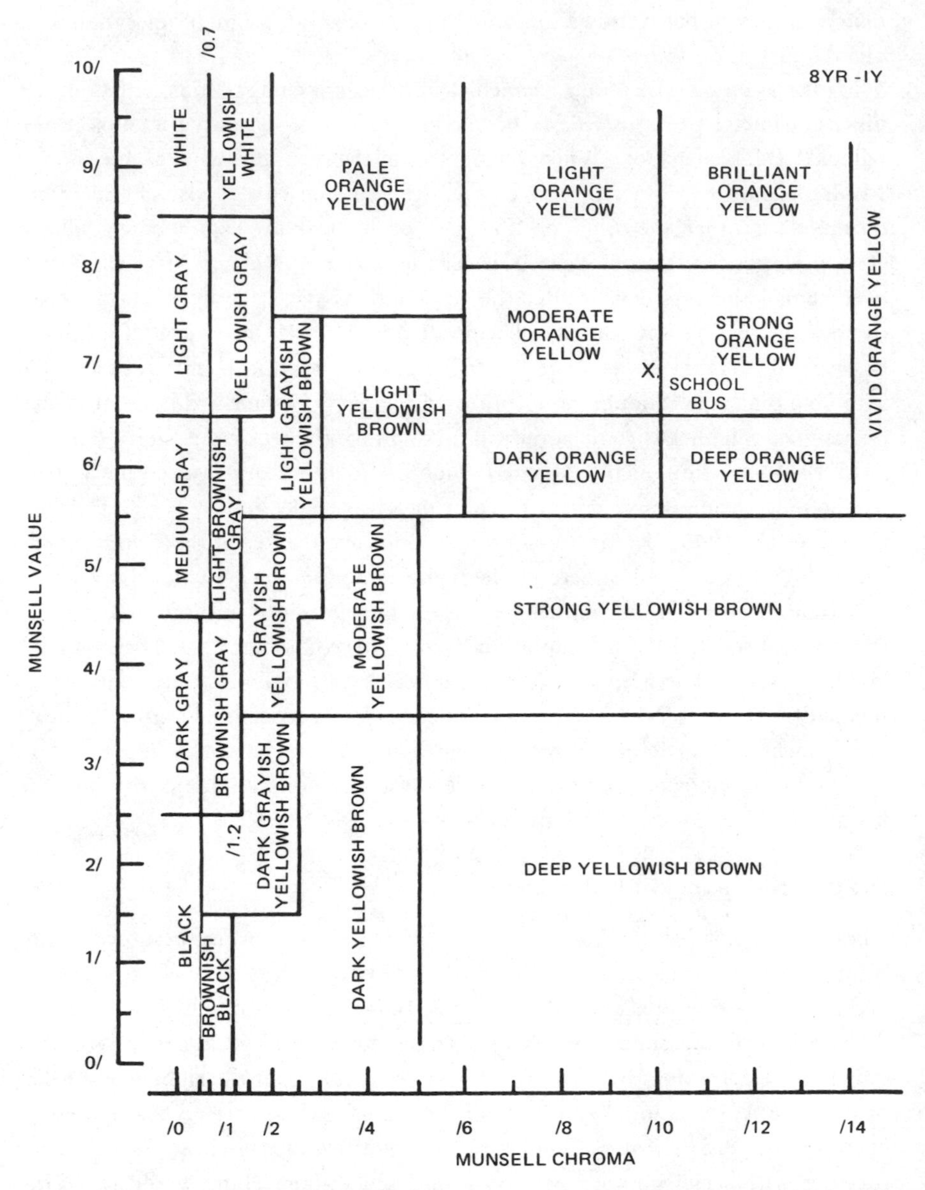


**Figure 11.14.** Page from ISCC-NBS Method of Designating Colors (NBS Circular 553) for 8YR-10YR (1Y) hues. The name for the yellow school-bus color is found by locating the plot of 10YR 7/10 on the page. It lies almost on the line between "Moderate Orange Yellow" and "Strong Orange Yellow."

color designations of many color-order systems then in use in various industries. Among those included were the Plochere Color System (Plochere, 1948), the Textile Color Card Association Standard Color Card of America, the U.S. Army Color Card, and the Federal Specification TT–C–595 Colors for Ready-Mixed Paint.

## Ostwald Color System

The Ostwald System (see Chapter 1) is based on a hue circle of most-saturated colors (full colors), spaced so that the differences between them represent approximately equal vision hue differences. (See Figure 11.15.) The rest of the color solid is built up by mixing each full color with white and with black. This is unlike the Munsell System where the colored chips were placed in all three dimensions so that differences between them matched equal visual differences. In the original Ostwald triangle, the amounts of black and white to be mixed with the full color were chosen by matching an additive color mixture arrived at by the spinning-disk method (Foss, Nickerson, and Granville, 1944).

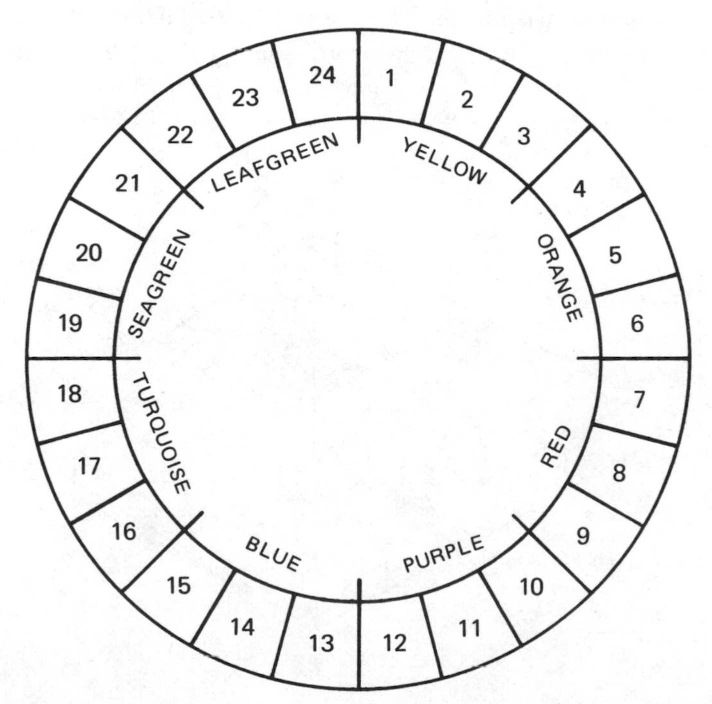


**Figure 11.15.** Designation system for Ostwald hue. Eight primary hues are subdivided into 24 intermediate hues.

Figure 1.5 (Chapter 1) shows a "leaf" from the Ostwald color tree. The upper edge of the triangle represents mixtures of the full color and increasing amounts of white, up to white at the top. The lower edge represents the full color mixed in the same manner with increasing amounts of black down to black at the bottom. Colors in between are mixtures of the full color with varying amounts of black and white. At the outer point of the triangle is the full color, with the greatest color content. Each hue has a separate leaf, and when arranged around the gray scale axis the leaves form the solid shown in Figure 11.16.

An Ostwald specification gives the number of full-color hue, and its white or black content on scales that run from the full color up to white, and from the full color down to black. Figure 11.17 shows the leaf for hue 3 (an orange yellow) and the numerical scales for white and black content. The Ostwald designation for school-bus yellow might be 3pc.

The Ostwald Color System is used by color technologists and color stylists because the dimensions of color change in the system are those met in producing colored materials. Technologists learn to visualize colors in terms of the pigments needed to make them. The Ostwald system is limited, however, in that there is no place in it for a color more saturated than the full color of that hue, which means that the basic colors must be changed as more saturated colors become available. The Container Corporation of America made chips

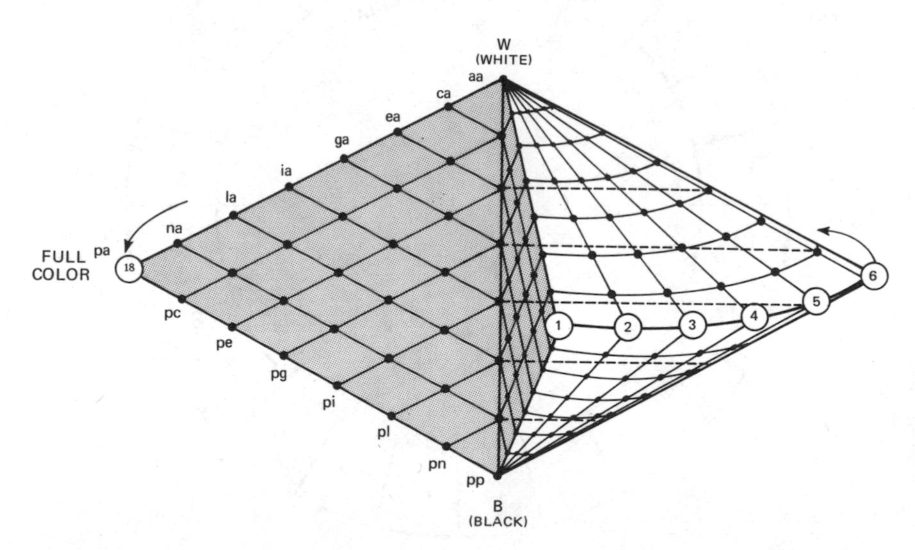


**Figure 11.16.** The Ostwald Color Tree. Hues are arranged around the axis as indicated by the arrows. The full-color hue is at the apex of the triangle with lightening of the hue occurring on the upward diagonal, and darkening on the lower diagonal (see also Figure 1.5).

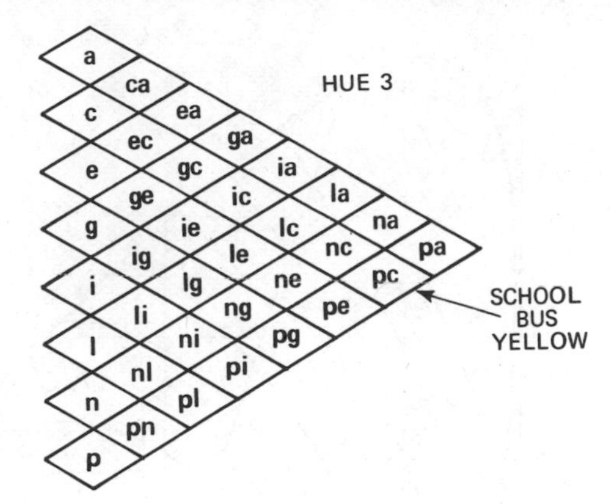


**Figure 11.17.** Leaf for hue 3 from the Ostwald Color Tree. The school-bus yellow designation is found from its CIE *x, y* specifications to be 3pc, shown on this page.

representative of this system available in its *Color Harmony Manual* (Jacobson, 1942), and colorimetric specification of the 680 color chips of the manual were supplied in 1944 (Granville and Jacobson, 1944).

## The DIN Color System

This system, developed in Germany by Dr. Manfred Richter, and used primarily in Europe, is specified in the German Standard DIN 6164. The spacing of colors in this system was done by making visual judgments of equal-saturation differences among colors of constant dominant wavelength. The results are shown in Figure 11.18. The DIN system is defined in terms of CIE units and is somewhat similar to dominant wavelength and purity with regard to chromaticity specification.

The coordinates of the DIN system are *Farbton* (hue), *Sättigungsstufe* (saturation), and *Dunkelstufe*, which relates to the darkness of a color. Farbton and Sättigungsstufe specifications do not change with lightness level but bear a fixed relation to the *x, y* specification of a color. Thus the system does not provide for the decreased visual chromaticity perception of differences in *x* and *y* as one goes to darker colors.

The darkness scale, which runs from 0 for white to 10 for black, was set up by making visual estimates of equal differences as was the saturation scale. These two dimensions, therefore, are visually uniform. However, the DIN hue lines radiate straight out from Illuminant C and do not curve as do lines of visually constant hue, as shown in Figure 11.19.

The DIN system suffers somewhat from being difficult to work with. There is no simple transformation from it to the CIE system. In order to obtain the DIN color coordinates from CIE measurements, one must plot the *x, y* values

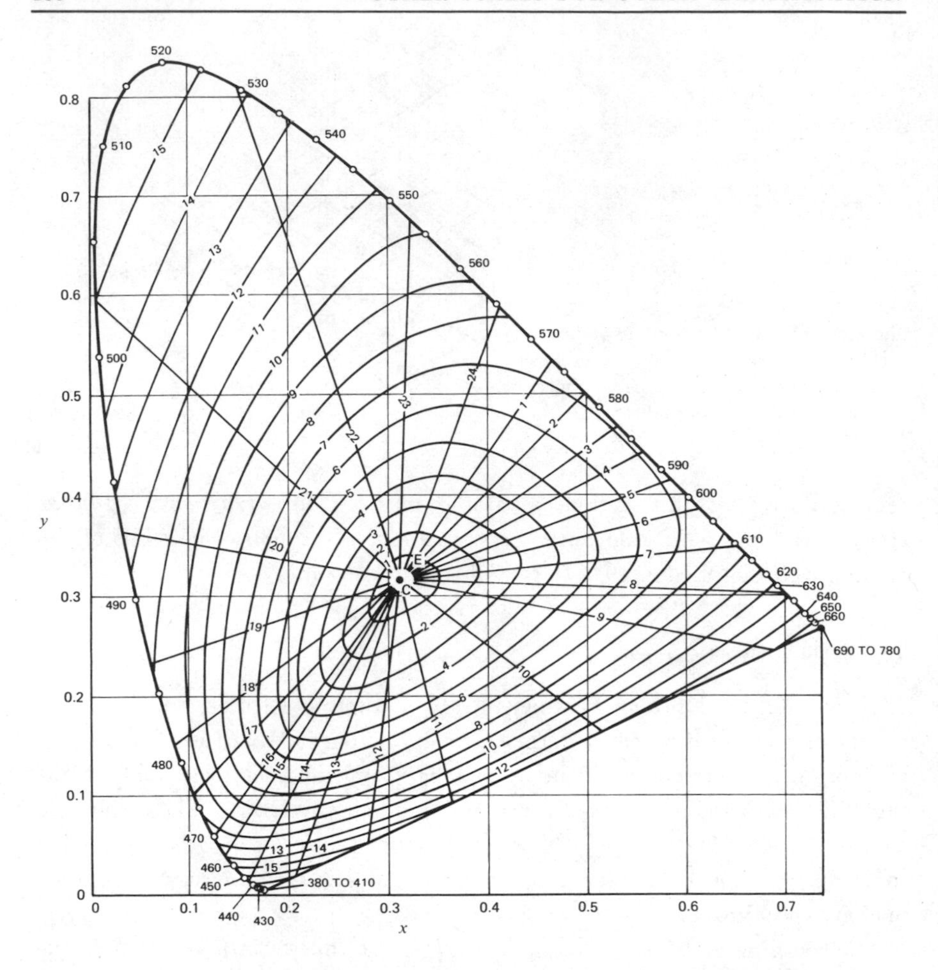


**Figure 11.18.** Chromaticity diagram with straight lines of constant DIN–farbton and ovals of constant DIN–Sättigungsstufe. The 24 straight lines radiating from the C point to the spectrum locus designate constant hues (farbton), and the ovals indicate the locations of object colors of the same saturation or chroma (sättigungsstufe) with saturation increasing with distance from the illuminant point. The farbton designation is similar to dominant wavelength, and sättigungsstufe resembles a purity specification. From *Color in Business, Science, and Industry* by D. B. Judd and G. Wyszecki, John Wiley & Sons, 1963. Used with permission of *J. Opt. Soc. Am.*

on the diagram shown in Figure 11.18 and then read the DIN hue and saturation from it. Color chips have been prepared for use in visual comparison.

## The Natural Colour System

The Natural Colour System (NCS), recently developed in Sweden as an outgrowth of the *Hesselgren Colour Atlas*, is interesting because it attempts to combine a number of features of other color systems. The NCS basic specification of color is in terms of color content of each of the six opponent colors of the Hering theory: red ($r$), yellow ($y$), green ($g$), blue ($b$), black ($s$), and white ($v$). The total "content" of any color is 100% of the various components so that a typical specification might be

$$y_{20}r_{20}s_{35}v_{25}$$

This color is equally red and yellow. It is 40% saturated (20 + 20) and has more black content than white. This color can also be specified in terms of hue, chroma, and blackness (rather than lightness) as follows: $y50r/40 - 35$, where 40 represents the total chromaticity component, 50 the percent that is red, and 35 the percent blackness. In terms of color content, black content, and white content, the system correlates with the Ostwald arrangement of colors.

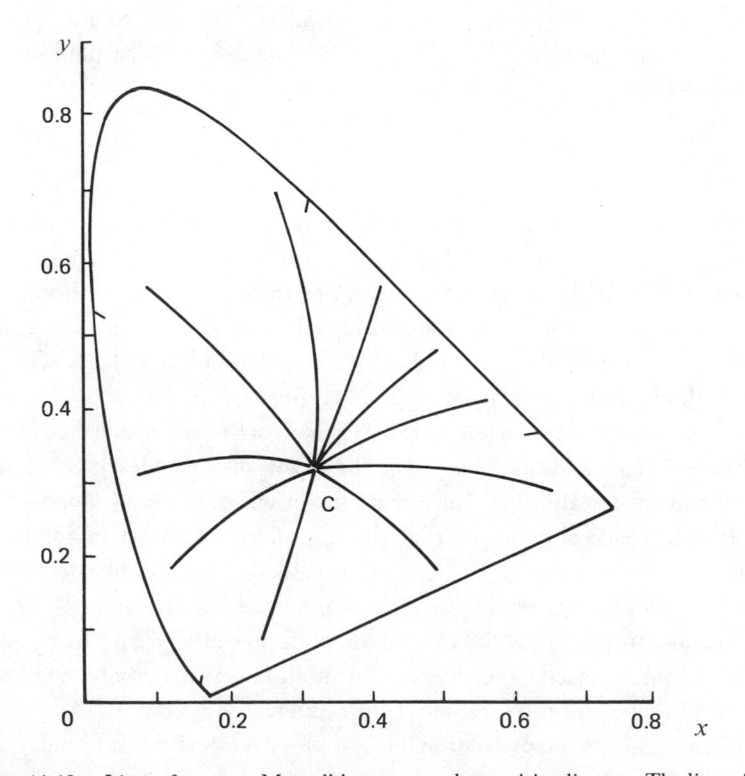


**Figure 11.19.** Lines of constant Munsell hue on $x$, $y$ chromaticity diagram. The lines of constant Munsell hue curve slightly on the $x$, $y$ diagram, in contrast to the straight lines of constant hue (farbton) shown in Figure 11.18. This indicates that the DIN hue specification is not visually uniform. Reprinted with permission from J. M. Adams, *Optical Measurements in the Printing Industry,* 1965, Pergamon Press Ltd.

### The Ridgway Color Standards and Color Nomenclature

The Ridgway charts were one of the first systems of ordered colored chips, having been developed in 1886 and subsequently modified (Ridgway, 1912). The system is based on mixtures of colorants with white, and it contains charts with names for 1115 colors, divided into 36 hues. A Ridgway designation for school-bus yellow might be Capucine Yellow III 15b. This system has enjoyed substantial use in agricultural and horticultural color identification.

### Maerz and Paul Dictionary

The *Maerz and Paul Dictionary of Color* (Maerz and Paul, 1930) is the most intensive dictionary of color names. It contains 7056 different colors arranged into seven hue groups and is used in art and industry for identification of colors by name. This is an example of a color chart produced by printing. The colors were created by variable-density overprints of inks of different colors. Munsell specifications of the red-to-yellow Maerz and Paul colors were published by Nickerson (1947).

## SUMMARY

Among the miscellaneous scales for color identification are single-number color scales used primarily for grading clear-to-amber oils, resins and chemicals, and certain foods. The methods for measurement of the clear-to-amber liquids are old. Many are still based on crude visual comparisions, but steps are being made to substitute instruments for improved accuracy, precision, and speed. The special color scales for grading food are recent in origin.

Another special area of color measurement involves the use of subtractive color mixture rather than addititve color mixture. One of the oldest instrumental methods involves the calibrated Lovibond glasses which are superimposed one over another to yield color stimuli that visually match unknown samples. Also related to subtractive color mixture are the methods of color densitometry so widely used in the assessment of color photography and color printing.

The disk mixture method of measurement is essentially an instrumental elaboration of the Munsell Color System. Munsell papers are combined in disks to generate additive color mixtures matching unknown samples.

Finally, there are the many color identification systems based on visual matching to color chips. The Munsell System is the best known and probably the most used of these. Some of the color chips are mounted in books, others are based on card files, and each system has its own particular arrangement.

# PART THREE

# INSTRUMENTS FOR THE MEASUREMENT OF APPEARANCE

# CHAPTER TWELVE

# INSTRUMENT CLASSIFICATION AND COMPONENTS

Our knowledge of the behavior of light and its interaction with objects tells us how to design instruments that can make the same kinds of judgments that human observers do. Our investigation of how the observer's eye-brain combination responds to the light it sees suggests the proper numerical scales for the psychophysical instruments.

Since the objective in the use of appearance-measuring instruments is a meaningful correlation with visual perception, the instrument design tends to follow the lines indicated in Chapter 4 concerning standardized conditions of viewing. However, an instrument has nowhere near the versatility of the eye. It recognizes only a specific attribute of an object, whereas the observer perceives a number of different attributes simultaneously. To simulate the complex human visual process, one would need several different instruments, each designed to sense a specific attribute of the object. These instruments then provide numbers for specific attributes of appearance and operate under standard conditions that can be reproduced at other times and places.

In comparing these instruments with the eye, the following can be said:

1. The eye is more versatile primarily because of its sensitivity to geometric factors of direction, pattern, shape, and so on. However, its evaluations are subjective, disputable, variable with changes in viewing conditions, and variable from observer to observer.
2. Instruments are less flexible than the eye, but they quickly reduce appearance to numbers that usually correlate well with visual evaluations. They provide evaluations that are more repeatable than those of the unaided eye.

## INSTRUMENT TYPES

Generally, there are four distinctions that must be borne in mind as one considers which of the various types of instruments he needs and what characteristics are essential for these instruments.

The first important distinction has its basis in how we see objects in the world around us. As was shown in Chapter 1, when we analyze product appearance we differentiate between

1. Chromatic, or color, attributes
2. Geometric attributes such as gloss, diffuse reflectance, haze, and texture

Instruments are grouped first into these two categories.

The second distinction is between physical and psychophysical instruments:

1. Physical analysis instruments (spectrophotometers and goniophotometers) measure the essentially physical properties of the light distributed by objects. Chapter 3 described spectrophotometric curves, which indicate amounts of light from objects as a function of wavelength. Also described were goniophotometric curves, which measure the geometric distributions of light by objects. The spectrophotometer and the goniophotometer provide physical analyses of the light reflected or transmitted by objects, but they give little information about how the observer sees this light.
2. Psychophysical analysis instruments (tristimulus colorimeters, reflectometers, glossmeters, hazemeters) are designed to give measurements that correlate with the human eye-brain impressions. It is necessary to build into these instruments characteristics that simulate the operation of the eye and brain in judgments of appearance.

The third distinction involves accomodation to the optical nature of the specimens to be measured. In Chapter 1 we identified three optical modes:

1. Illuminant mode
2. Object mode
3. Aperture mode

Specimens seen in the object mode are of greatest interest and are further divided into

1. Diffusing surfaces from which color is seen by diffuse reflection, and gloss, if any, is seen by specular reflection
2. Metallic surfaces from which color is seen by specular reflection, and reflection haze is seen by diffuse reflection
3. Translucent objects from which color is seen by diffuse transmission
4. Transparent objects from which color is seen by specular transmission, and transmission haze is seen by diffuse transmission

Geometrically different instrument arrangements are required for the measurement of each of the modes and each of the optical types of objects.

A fourth distinction of growing importance is that between

1. Laboratory instruments
2. On-line or continuous-process instruments

Laboratory instruments must be accurate and trustworthy when used by skilled operators under favorable operating conditions. On-line instruments, on the other hand, will be in operation in hostile environments and must thus be rugged and subject to effective maintenance by persons having production skills. In general, it is not necessary that the on-line instruments have high accuracy, but only that they have very good precision and stability so that they can readily detect departures from production standard conditions.

## SPECTRAL AND GEOMETRIC FACTORS OF INSTRUMENT DESIGN

As we have seen, instruments for the measurement of object appearance can be classified by whether they measure primarily chromatic or geometric attributes. Actually, both spectral factors (such as the color of the light source in the instrument) and geometric factors (such as directions of the incident and viewing beams of light) are involved in every instrument design. With color-measuring instruments the spectral properties of the instrument will be the most important. When a geometric attribute such as gloss is to be measured, the angular dimensions and directional aspects of the instrument are the most important. However, for measurements to be accurate and reproducible, both geometric and spectral aspects always must be controlled. All rigorous written methods of appearance measurement identify spectral and geometric conditions of measurement.

The three elements of the visual object-observing situation are duplicated in any appearance-measuring instrument. Each instrument has a source of light, a place for the location of an object to be observed, and a light receiver. Most instruments employ photoelectric light receivers that convert light to electric current, which is then measured. In an instrument for measuring the color of opaque nonmetallic objects (Figure 12.1*A*) these photodetectors are placed so that they receive the light that is reflected diffusely by the object. This is because (as was pointed out in Chapter 1) we see the color of such objects by diffuse reflection. On the other hand, if the gloss of the object is the attribute to be measured, the photodetector will be placed where the specular beam of light reflects from the object, as in Figure 12.1*B*.

Thus in an appearance-measuring instrument, there are controlled conditions for the observation of the appearance attributes of interest together with a

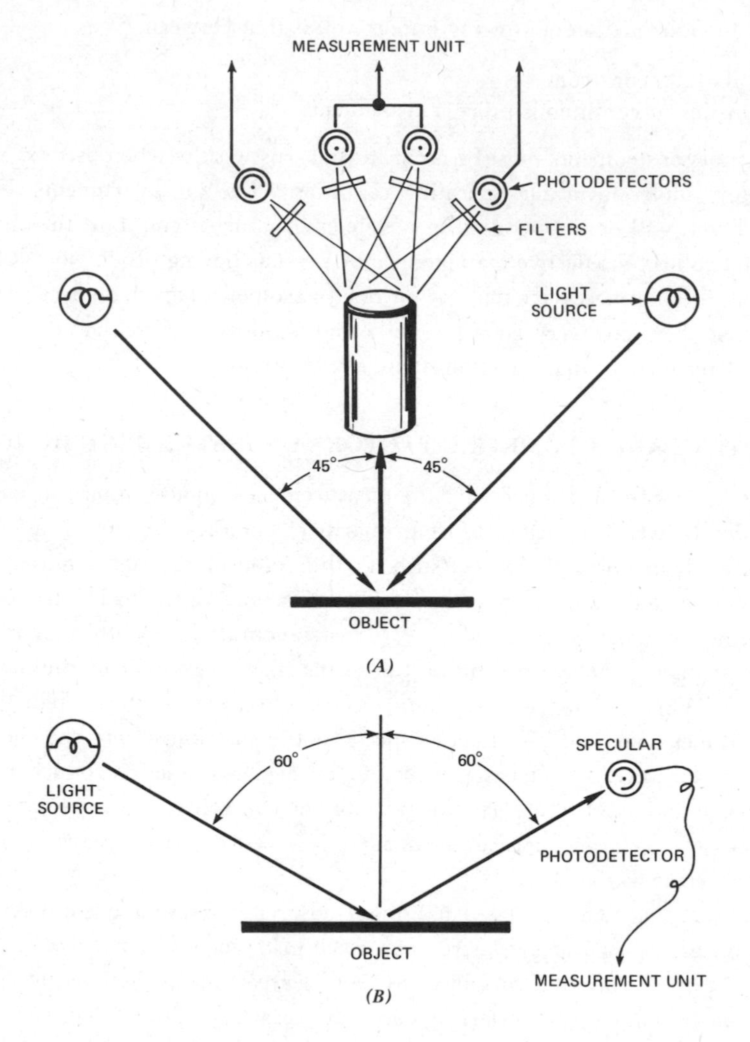


**Figure 12.1.** Duplication of the visual observing situation in the instrumental measurement of (*A*) color by diffuse reflection and of (*B*) gloss. (*A*) For the measurement of color the object is illuminated at 45° by two beams of light. Viewing is at 0°, to measure only the diffuse reflection. The light-diffusing element directly over the object uniformly mixes the light reflected by the object and carries it to the photodetectors. Filters modify the reception of the light so that the response of the detectors approximates that of the human observer viewing the object under a certain illuminant. The response signals are taken to the measurement unit. (*B*) For the measurement of gloss, if the light is incident on the object at 60°, the photodetector is placed at the same but opposite angle to receive the specular reflection, the light distribution that shows glossiness.

mechanism for measurement of the light signals received. Conditions of illumination, specimen placement, and light reception and measurement must all be controlled if measurements made by the instrument are to be standardized or reproduced from one instrument to another.

The components from which instruments are fashioned are

1. Light sources
2. Devices that control the directions, dimensions, and other characteristics of the illuminating and viewing beams of light, thus establishing the geometric designs of instruments
3. Photodetectors and wavelength-selection devices
4. Electrical-measurement devices and signal-analysis systems
5. Data-display devices

## LIGHT SOURCES

The primary objectives in the selection of the light source for a color-measuring instrument are stability, directability, life, and spectral usability.

### Incandescent Light Sources

The incandescent lamp is the light source most commonly used in appearance-measuring instruments. It contains a conducting metallic filament within a clear bulb. Tungsten is normally used as filament material because it has strength and the ability to withstand high temperature without melting. The spectral characteristics of a tungsten lamp are shown in Figure 12.2. In the visible portion of the spectrum its curve is essentially the same as that of a blackbody radiator. This means that its spectral emittance is the maximum possible at a specified temperature.

The features of incandescent lamps that make them desirable for use in appearance-measurement instruments are

1. Continuous spectral distributions of energy, as characteristic of sunlight
2. Steady output of light with time
3. Availability of small coiled filaments to give concentrations of light flux that can be formed into beams and directed
4. Control of light output by control of power
5. Low cost

The preferred incandescent lamps are low-voltage tungsten lamps with

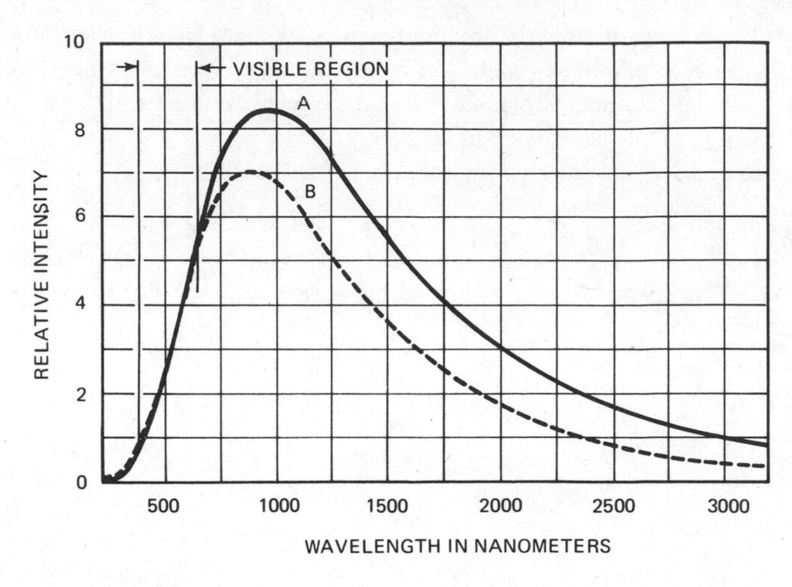


**Figure 12.2.** Radiating characteristics of tungsten. (*A*) radiant flux from 1 $cm^2$ blackbody at 3000 K. (*B*) radiant flux from 2.27 $cm^2$ tungsten at 3000 K. The curve for tungsten is equal to the blackbody curve in the visible region of the spectrum. From the *IES Lighting Handbook,* 5th ed. Courtesy of Illuminating Engineering Society, 1966.

small, heavy-duty filaments that provide bright, relatively small concentrations of light. Such lamps give a maximum of controllable light with a minimum of undesirable heat. Accurate positioning of the filament in relation to the geometric-control components of the instrument is essential if light beams are to be properly directed. Methods to achieve this are shown in Figure 12.3. Reproducible positioning of the filament is achieved in the first lamp shown by means of a focusing ring soldered to the base. The second lamp is a quartz-halogen light source mounted on a small printed circuit card so that it can be accurately located relative to the instrument. In the third, a small lamp used in a portable glossmeter, the lamp is immersed in potting plastic within a mounting ferrule so that the filament position can be duplicated with respect to the dimensions of the ferrule.

The recently developed tungsten halogen-cycle lamp has an important advantage over the older tungsten lamp. The filament in any tungsten lamp evaporates during burning. Normally this tungsten deposits on the glass bulb, darkening the lamp. In the halogen-cycle lamp, however, a halogen gas is present within a quartz envelope that is so close to the filament that the envelope reaches a temperature of about 250°C. At this temperature, the tungsten that evaporates from the filament combines with the halogen (normally iodine or bromine gas) to form tungsten halide. This gas continues to circulate within the quartz envelope until it strikes the filament. The filament causes the tungsten halide to separate, with the tungsten redepositing itself on the filament. The halogen again circulates as a gas. This counter-evaporation advantage of the halogen-cycle lamp gives it a much longer lamp life (by a factor of about 2½ or 3 times), and eliminates the darkening of the bulb from deposits of tungsten. It is well to remember, however, that a halogen-cycle lamp does not give stable light output if the bulb is not hot enough to prevent condensation of the tungsten halide gas.

The properties of incandescent lamps vary importantly with age and voltage (or current). Figure 12.4 shows that light output varies more rapidly than lamp voltage. For this reason, light-source power must be precisely regulated in any instrument reflecting source output in meter indication. It is only within recent years that power regulators have achieved the stability required for stable single-light-beam measurements of appearance attributes. Both stable regulated power and stable electrical contacts to the lamp are essential for stable single-light-beam instruments.

## Fluorescent Lamps

These are used where it is desired to simulate in the measuring instrument the fluorescent light sources regularly encountered in stores and homes. So far,

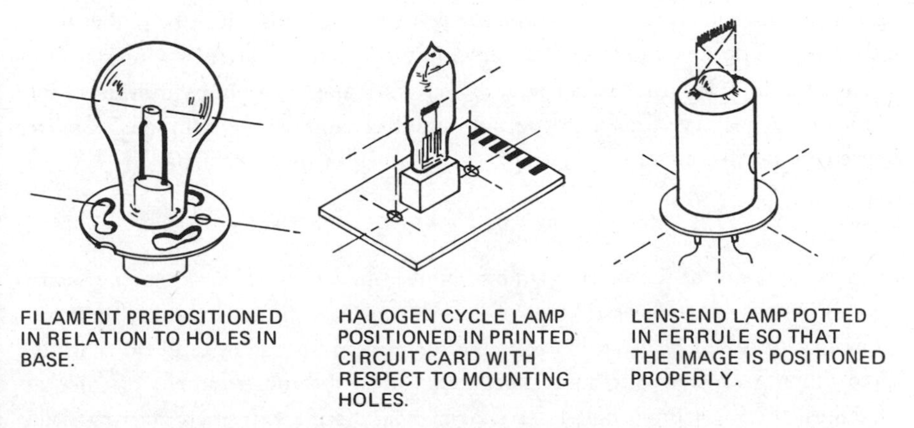


**Figure 12.3.** Mechanisms employed to locate filaments of light sources used in instruments.

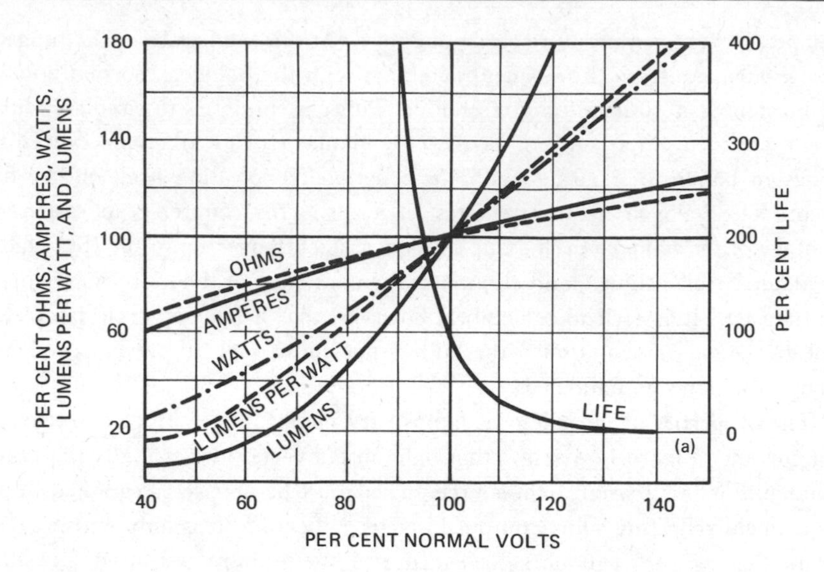


**Figure 12.4.** Effect of voltage variation on the operating characteristics of tungsten lamps. Both life and light output (lumens) vary more rapidly than lamp voltage. From the *IES Lighting Handbook*, Illuminating Engineering Society, 1966.

fluorescent lamps have been limited to use in experimental instruments, primarily because they are large-area sources (not concentrated) and are therefore difficult to incorporate into controlled optical systems.

## Xenon Arc Lamps

Another light source used in appearance-measuring instruments is the xenon arc lamp. The xenon arc has the advantage of being spectrally closer to daylight than is the incandescent source. As with most arc lamps, however, it is difficult to achieve the light-output stability and the continuous line-free spectrum needed for precision and accuracy in color measurement.

## Lasers

Lasers have not at the present time been used in appearance-measuring instruments because of their restricted ranges of spectral output. However, their narrow parallel light beams seem to make them ideal for use in uniformity testing and other geometric attribute instruments. For the measurement of color attributes, the most important laser characteristic is the extremely narrow-width monochromatic beam produced. Frequency-variable lasers are being tried in limited-range spectrophotometric instruments, replacing source-monochromator combinations.

## GEOMETRIC DESIGN—ILLUMINATING AND VIEWING BEAMS OF LIGHT

The primary objectives in selecting the elements of an instrument that establish its geometric design are accurate identification and realization of the required directions of sample illumination and view. For this one must know not only the axial directions of the illuminating and viewing beams, but also the beam width, primarily defined by the aperture stop, and the sizes of the field angles defined by the field stops. Also important are the auxiliary stops for eliminating stray light.

### Directions of Light Beams

Traditionally, light is considered to be traveling either in a single direction such as can be represented by a vector, or in all directions in a diffuse manner. Figure 12.5 identifies the directions of object illumination and view most frequently used for measurements of object-appearance attributes. Note that

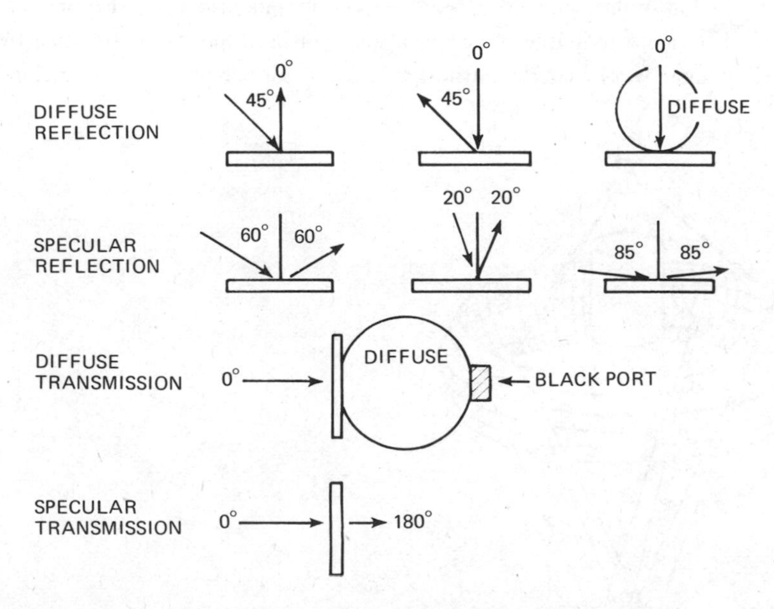


**Figure 12.5.** Axial directions of illumination and view commonly used.

white-lined diffuse-integrating spheres are used to gather or project light in all directions.

Over the years there has been considerable discussion concerning the relative advantages of the geometries of 45°, 0° and diffuse, 0° for the measurement of surface color. In general, it can be said that the 45°, 0° geometry is superior for correlation with visual estimates of color and color differences. However, for color-formulation computations, the diffuse, 0° geometry is preferable since it handles the specular (uncolored) reflection uniformly and without bias, whether the surface is glossy or matte.

In the measurement of diffuse reflection or color by 45°, 0° geometry, some specimens are directional; that is, their instrument readings change as they are rotated within the plane of the surface. This sensitivity of light reflected to surface orientation with respect to incident light is often found in textiles, textured plastics, paper tissues, and some foods. To reduce sensitivity to specimen orientation, 45° incidence and 0° viewing are used together with a circumferential arrangement of incident beams as shown in Figure 12.6.

### Spatial Dimensions of Light Beams

In measurements such as those for gloss, where geometry is critical, it is essential that more than the axial directions identified in Figure 12.5 be specified. It is not possible in any real beam of light to restrict the directions of travel to exactly one angle, although a laser source provides conditions that

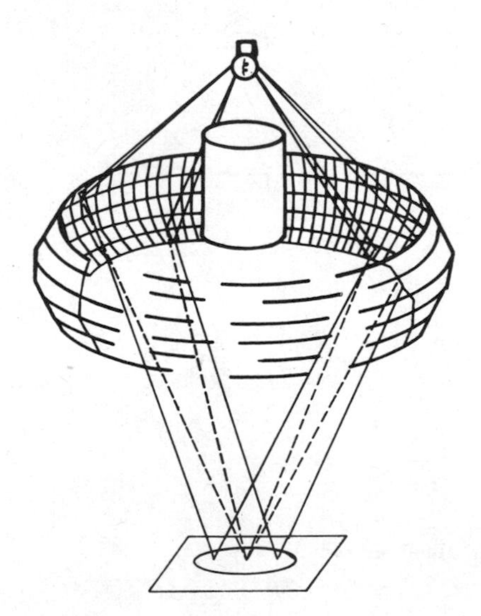

**Figure 12.6.** Circumferential illumination of a specimen.

come close to this. In real instruments, light sources of finite size are used. The light from these, in passing through an instrument, always contains rays with directions that diverge from each other. The angular extent of divergence of these rays in any illuminating or viewing beam is primarily identified by what is called the "field angle." These field angles, in addition to the axial directions of the beams, must be controlled and specified. The size of the field angle is controlled by a "field stop." As is shown in Figure 12.7, the field stop that sets the field angle should be centered on the axis of the light beam.

Identification of field angles is essential if the geometric conditions of measurement are to be completely identified. For high-gloss and transparency measurements, the fields are narrow and must be precisely controlled if the measurements are to be reproducible. In addition to the source and receiver field stops, there are other important stops designed for geometric control of the light passing through an instrument. The aperture stop essentially establishes the actual diameter of the beam permitted to pass through the instrument and determines the illuminance at the image. There should be only one aperture stop used in the entire instrument. (See Figures 12.8 and 12.9.)

The specimen areas exposed are established by the sizes of the components used in the instruments and the stops through which the light beams are allowed to pass. Other stops are used to eliminate stray, unwanted light and thus improve the geometric control of light flow through the instrument. The word "baffle" is frequently used as a synonym for stop, especially when the purpose of the baffle is merely to intercept stray light not associated with the desired light beams.

## Vignetting

Vignetting is a problem that arises through the improper use of stops. It occurs when two stops both serve to limit the size of the same beam of light. The second "stop" is often the edge of a lens or other optical element. Vignetting of rays produces nonuniform beams and makes the instrument sensitive to minute dimensional changes such as might be caused by temperature change or vibration. To avoid vignetting there must be only one aperture stop in each instrument and only one field stop in each illuminating or viewing beam.

## Parallel and Converging Beam Configurations

The geometric control of light as it proceeds from light source to test specimen to receiver is accomplished not only by stops or baffles, but also by the use of lenses, mirrors, optical fibers, and sometimes dispersers and light diffusers, all

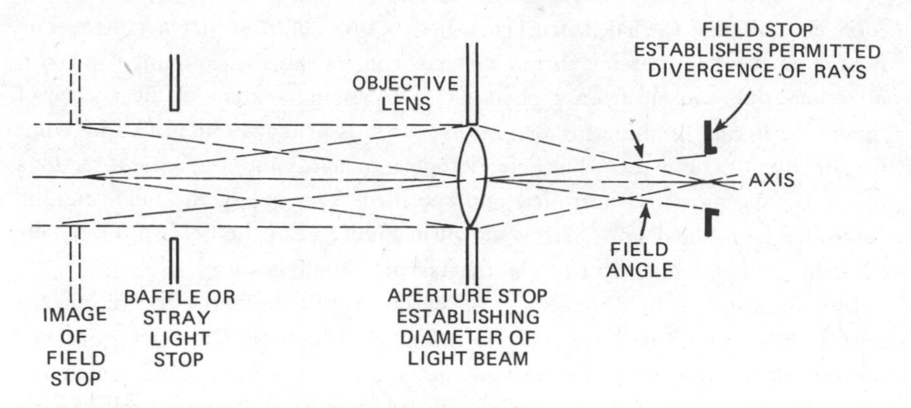


**Figure 12.7.** Spatial dimensions of a light beam. Field angles of illuminating and viewing light beams, as well as their axial directions, must be controlled.

within the lighttight instrument housings. Lenses are used in conjunction with stops to establish whether the principal rays are parallel, converging, or perhaps even diverging when they intercept the specimen. Figure 12.8 shows a parallel beam configuration of rays and Figure 12.9 shows a converging beam configuration. The two beam designs differ from each other in that with parallel beam design, the principal rays (light rays from the center of the source to the center of the receptor) are parallel to each other and to the corresponding axial ray when they impinge on and leave the specimen. In the converging

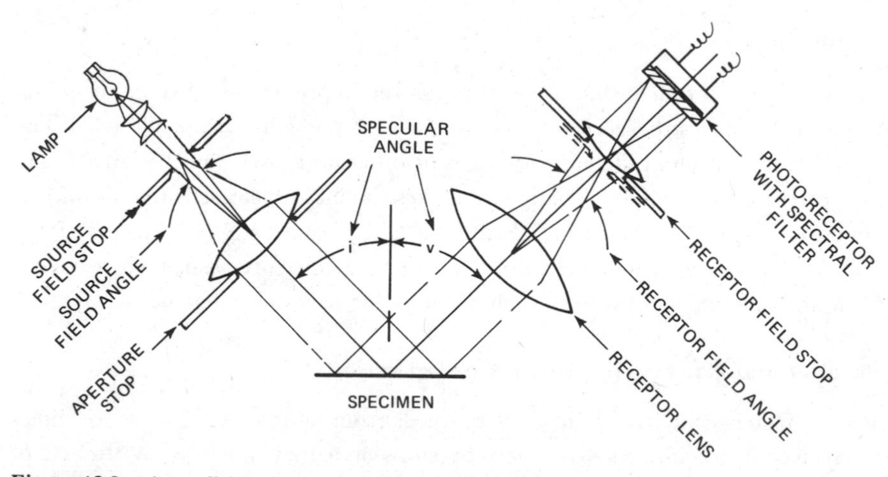


**Figure 12.8.** A parallel-beam specular reflection instrument.

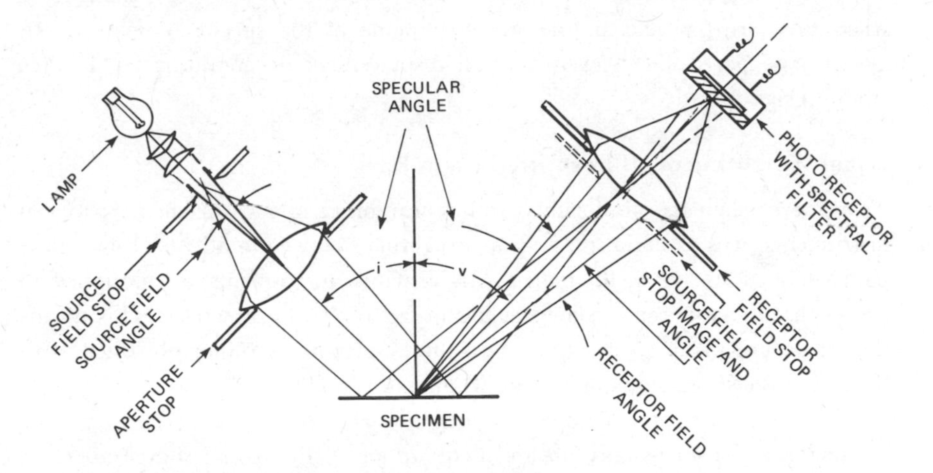


**Figure 12.9.** A converging-beam specular reflection instrument.

beam design, on the other hand, the principal rays are converging toward the center of the receptor field stop when they impinge on and leave the specimen. Both of these configurations are used in specular-reflectance instruments such as glossmeters, but the converging beam design of Figure 12.9 is more popular. It is less susceptible to vignetting since there is no receptor collimating lens. This eliminates the necessity of precisely aligning both a source and a receiver collimating lens.

## The Specimen and its Window

The specimen window itself is an often overlooked optical component. In fact, it represents an aperture stop if it is filled to its edges with light. If the specimen is translucent, or if light penetrates the specimen to any extent, vignetting of rays may occur and adversely affect instrument accuracy and reproducibility. This type of vignetting is a special problem in the color measurement of food. In a well-designed instrument for the measurement of food the specimen window is illuminated with a beam of smaller diameter than the diameter of the window itself. The difference between the diameters of the window and the beam should increase with depth of light penetration of the specimen.

The angles of illumination and view of the instrument are measured from perpendicular to the plane of the specimen. Specimens that are not flat, that bulge into the window or are held back from the specimen window by cover glass or some other apparatus, will not give the same measurements as they do when they are flat and in the specified plane of the specimen window. In glossmeters, particularly, even a small displacement or nonflatness will alter readings significantly.

### Equal Weighting of all Permitted Light Rays

To achieve geometric uniformity and repeatability of measurements between instruments, it is essential that equal weighting be given to all rays that follow paths defined by the field angles. High-quality instruments are designed to assure that the apparent geometries are in fact the real geometries. This means that steps are taken to see that the source aperture is uniformly filled with light, that the specimen is uniformly illuminated, and that all rays of light passing through the receptor window are equally evaluated.

Condenser lenses usually do an adequate job of filling the source aperture when properly used. However, in instruments with critical field angles, such as glossmeters, the lamp position must usually be carefully adjusted to make sure that the lamp filament image is centered in the source slit and fills the slit completely (See Figure 12.10.) For valid gloss measurements it is essential that the dimensions and positioning of the receptor and source field stops be accurate. In tristimulus colorimeters, which use signals from three or four photodetectors simultaneously, the light from the specimen must be distributed equally to each detector. In multiple-detector sphere instruments the sphere itself serves to distribute the light equally to each detector. In projected light beams, components such as diffusing sheets, prisms, and light fibers are used to transmit and mix the light. Figure 12.11 shows some methods for distributing light from a specimen to two or more photodetectors equally, or for distributing light to two or more specimens (such as sample and standard).

### Interchange of Directions of Illumination and View

According to the law of reciprocity first stated by Helmholtz, directions of illumination and view in a photometric instrument can be interchanged without

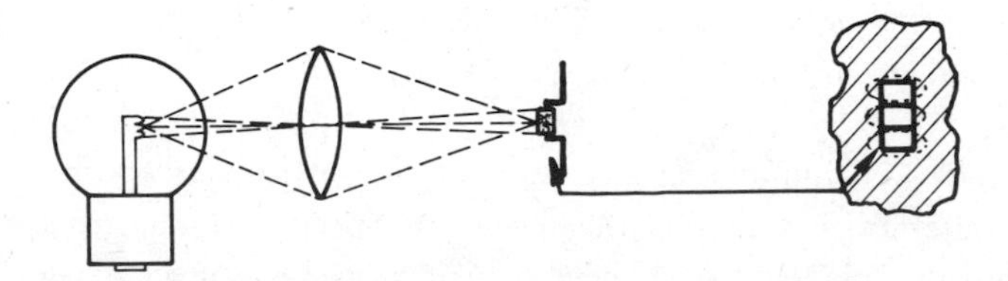

**Figure 12.10.** Source filament lamp filling the source field stop. If the source field stop is not uniformly filled with light, valid and repeatable measurements cannot be made.

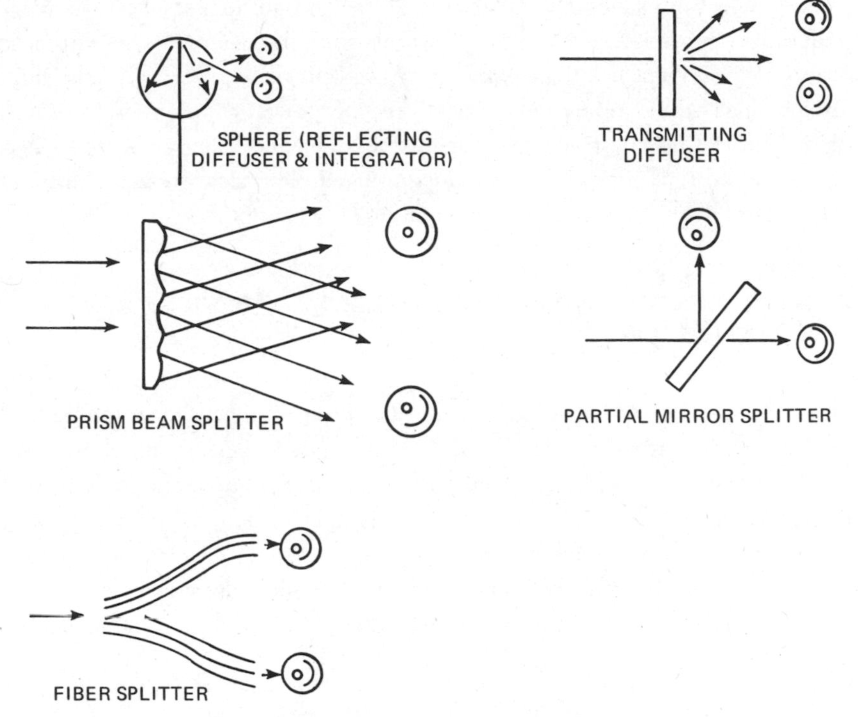


**Figure 12.11.** Distributions of light for equal reception at detectors. Light from the specimen must be distributed equally to each of the photodetectors.

changing the light-flow efficiency of the system (Helmholtz, 1909). According to the Helmholtz reciprocity relation, the 0°, 45° geometry is equivalent to 45°, 0°, and the 0°, diffuse geometry is equivalent to diffuse, 0° and so on. However, this is true only if care is taken to see that not only the axial angles of incidence and view are interchanged but also the field angles as well. Furthermore, care must be taken to avoid vignetting and to assure equal weighting of all permitted directions of travel of the light rays. Without this equal weighting, the reciprocal conditions cannot be said to be fulfilled. Several investigators have reported results challenging the validity of the reciprocal relation. It is believed that care was not taken in these experiments to assure equal weighting to all permitted directions of the light rays in both the original and reverse directions.

## Precision of One- and Two-Beam Configurations

In this chapter we have confined our discussion to one-beam instruments. There has been, however, considerable use made of a two-beam configuration in which light from a single light source is directed both to a specimen and to a standard. A comparison is made of the signals from the two readings. This type of ratio measurement has the advantage of high precision, even when the light source is not stable. However, the advent of power supplies of improved stability has made one-beam instruments feasible and often preferable. Not only is the cost of the second beam eliminated, but also greater flexibility results from having to provide for only one set of beam optics.

## SPECTRAL DESIGN—WAVELENGTH-SELECTION DEVICES AND PHOTODETECTORS

A tristimulus color-measuring instrument is designed to simulate spectrally the characteristics of both the illuminant and the observer involved in the normal color-observation situation. The measurements made by such an instrument are normally designed to duplicate visual observation under one of the standardized illuminants, usually Illuminant C or Illuminant D65, both of which represent standardized average daylight. Therefore the basic objective where spectral design is concerned is the selection of source-filter-photodetector combinations that have stable measurable responses to light relatable to the color-matching response functions of the human observer.

### Wavelength-Selection Devices

The elements of a color instrument are shown in Figure 12.12. The spectral energy of the light source ($E_i$) is adjusted by a filter ($T_i$) to duplicate the visual illumination to be simulated in the instrument. A second wavelength discriminating device ($T_v$) is placed in the beam viewed by the photodetector. Depending upon whether the instrument is a spectrophotometer or a tristimulus instrument, one of these devices isolates either a narrow band of wavelengths by prism, grating, or interference filter, or a broad band of wavelengths by means of a tristimulus absorption filter.

Where there is no fluorescence of specimens involved or no danger of damaging the specimen by heat, it makes no difference where in the optical system (before or after the specimen) the spectral modifying components $T_i$ and $T_v$ are placed. Thus in the Color Difference Meter developed by Hunter and produced from 1948 to the present, $T_i$ and $T_v$ are actually combined in single filters mounted at the receiving photocells (Hunter, 1958). Where fluorescence is present, however, it is important to make the wavelength selection (which will lead to appraisal of visual color) on the light which comes *from* the specimen. If

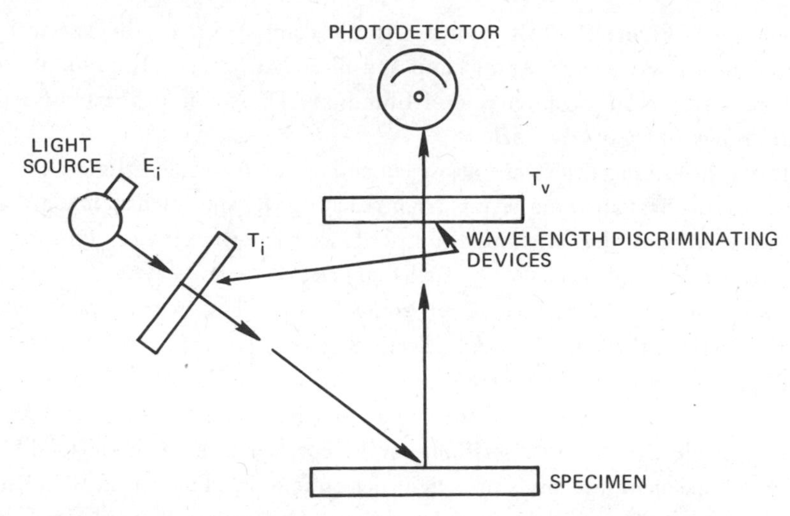


**Figure 12.12.** Wavelength-selection components of a color-measuring instrument. The spectral energy of the light source $E_i$ is adjusted by a filter $T_i$ to duplicate the visual illumination to be simulated. The filter $T_v$ selects the wavelength or band of wavelengths to be viewed by the photodetector.

wavelength selection is made on light proceeding *to* the specimen, the instrument will interpret as excessive reflectance the effect of specimen fluorescence at the wavelengths of light absorbed by the specimen, and color will be wrongly interpreted. Thus for color measurement of specimens that fluoresce, the light on the specimen should be spectrally like that in the observing situation being simulated, while the wavelength analysis to simulate the eye's response must be made on light leaving the specimen.

Incident-beam filters ($T_i$) have been developed to simulate daylight illuminants for instruments. The daylight-blue filters used in the past do not provide the fluorescence generating near-ultraviolet wavelengths required for most present-day evaluations. Therefore the CIE recommended in 1966 the addition of a new D-series of illuminants representing daylight over the spectral range from 300 nm to 830 nm. (See Chapter 4 and Figure 4.4.) Illuminant D65, with a correlated color temperature of 6500 K, is of special interest. Investigations have been carried out concerning artificial light sources and their approximation to Illuminant D65. Wyszecki (1970) reported for the CIE Colorimetry Committee E-1.3.1 on relative spectral irradiance distributions of filtered and unfiltered high-pressure xenon arcs, filtered tungsten-halogen lamps, and fluorescent lamps. Grum, Saunders, and Wightman (1970) also described work done with tungsten, tungsten-halogen, and xenon lamps and various filter combinations. Figure 12.13*A* shows Grum's comparison of the spectral irradiance of a low-wattage xenon lamp and filter with that of Illuminant D65. Wyszecki's results in a comparison of Illuminant D65 with a tungsten lamp T (3) are shown in Figure 12.13*B*.

Infrared-absorbing filters are also frequently used in incident light beams to prevent excess heat in these beams from reaching the specimen. They are also used to eliminate the unwanted infrared wavelengths that certain of the photoelectric cells (silicon cells, in particular) respond to.

*Narrow- and Broad-Band Wavelength Selection.* The two types of wavelength selection, the narrow-band selection of the spectrophotometer and the broad-band selection of the tristimulus colorimeter, are shown in Figure 12.14. In each case, both the ideal and the actual distributions are shown. In 12.14*A*, the ideal single wavelength discrimination is represented by a single solid black line in the spectrum for any single light determination. The actual wavelength response in any case will be broader than a single narrow band. In Figure 12.14*B*, the solid line represents the ideal curve of the $\bar{y}$ response function of the Standard Observer. The dotted line shows the actual approximation of the function as achieved in an actual tristimulus colorimeter.

Prisms, gratings, and interference filters are used to isolate narrow spectral bands in a spectrophotometer. Prisms and gratings both act to separate the wavelengths spatially. Interference filters are made either as individual wavelength-isolating devices or as continuous wedges that vary in the wavelength transmitted by location along the wedge. Thus a wedge interference filter is variable from point to point and can be used to separate wavelengths spatially, as a prism does. Uniform interference filters are used to isolate narrow spectral bands for abridged spectrophotometers or chemical colorimeters.

The spectral modifier ($T_v$) in a tristimulus instrument is normally an absorption filter that selects a broad line of wavelengths. Absorption filters may be fabricated of glass, plastic, dyed gelatin, or colored absorbing media in solution. In a tristimulus colorimeter, filters are used to duplicate directly the three spectral-response functions of the human eye as defined by the CIE $\bar{x}$, $\bar{y}$, $\bar{z}$ functions.

*Filter Combinations.* Developing filter combinations that achieve a desired spectral transmission function is both a science and an art. A major use of this science and art is in the design and construction of filters for tristimulus colorimetry. Successful tristimulus color measurements require that tristimulus signals from the photodetectors be accurately equivalent to those computed for the normal observer (represented by the CIE Standard Observer) when viewing the specimen under one of the standard CIE illuminants. To obtain accurate

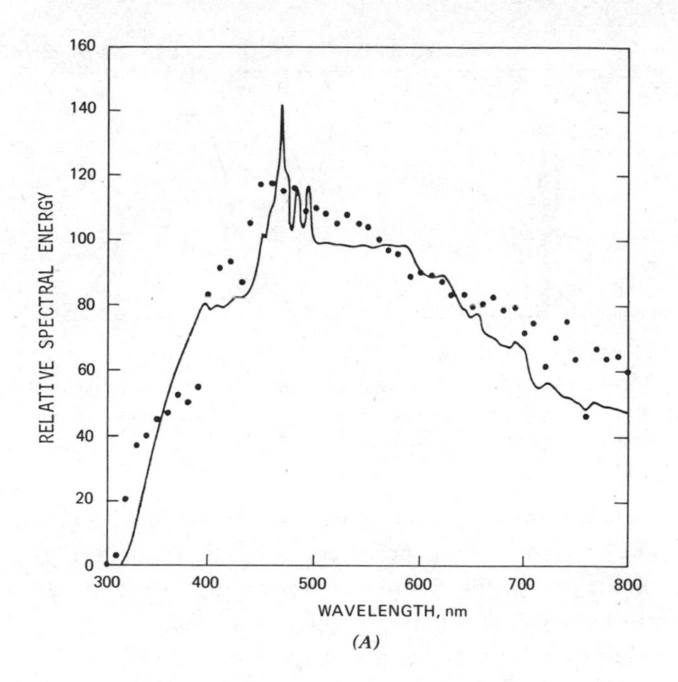


*(A)*

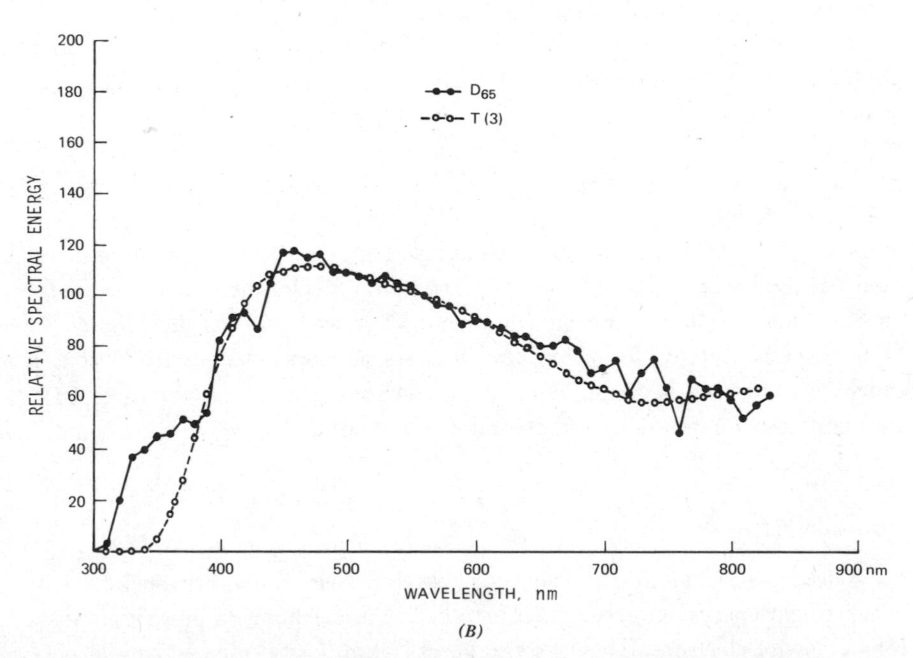


*(B)*

**Figure 12.13.** (*A*) Spectral energy distribution curve for filtered 75-watt xenon lamp (solid line) compared with Illuminant D65. (*B*) Spectral energy distribution curve for filtered tungsten source compared with Illuminant D65. From F. Grum, S. Saunders, and T. Wightman, *Tappi,* **53,** 1970; and G. Wyszecki, *Die Farbe,* **19,** 1970.

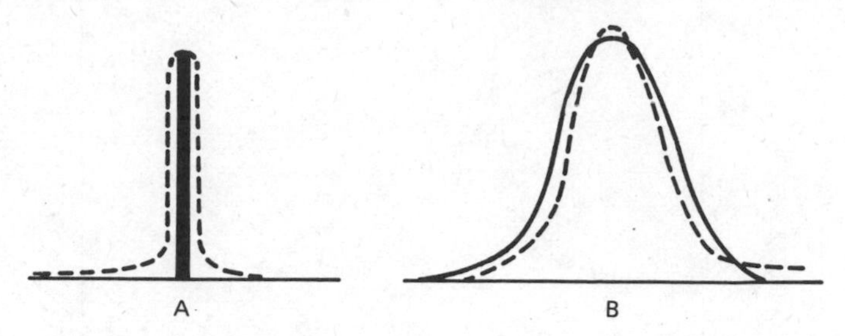


**Figure 12.14.** Narrow-band and broad-band wavelength selection. (*A*) Narrow-band selection characteristic of the wavelength selection device in a spectrophotometer. (*B*) Broad-band selection characteristic of filters in tristimulus colorimeters.

tristimulus signals the combinations of light source, tristimulus filters, and photoreceptor must be spectrally accurate. As explained in the section of Chapter 14 on tristimulus instruments, there have been a number of studies of the relations of spectral accuracy to accuracy of color measurements. The reader who is interested in further details of filter design is referred to articles by Gage (1937) and Davies and Wyszecki (1962).

The practice of correcting simultaneously for characteristics of sources and photodetectors is followed in the design of the majority of tristimulus colorimeters now in use. To date there have been few attempts to provide separately accurate spectral duplications of the CIE Illuminants C or D65 and accurate spectral duplications of the CIE tristimulus observer responses. However, efforts are now underway to achieve these spectral distributions separately because of the need for tristimulus measurements incorporating more than one light source. For example, in a tristimulus colorimeter equipped with an Illuminant A source and an Illuminant D source, such as the Hunterlab D38D Natick Labs Colorimeter, the light source adjustment is made at the source (Hunter and Christie, 1966). The adjustment filters for the three CIE standard observer functions are placed after the specimen, before the photodetector.

## Photodetectors

Photodetectors—phototubes, photocells, barrier layer cells—are devices that convert light energy into electrical current; i.e., they change a flow of photons into a flow of electrons. They are the "eyes" of appearance-measuring instruments. Photodetectors vary in their spectral sensitivity and in the methods of generating current flow. In terms of method of operation, there are four general types: photoemissive, photovoltaic, photoconductive (junction), and photoconductive (bulk).

*Photoemissive Detectors.* In the photoemissive process light strikes a specially surfaced metal called a cathode, knocking electrons free from the surface of the photocathode. If a voltage exists between the cathode and a small metal rod or plate called the anode, and if the electrons are free to move as in a vacuum, they will move to the anode, generating a current proportional to the light intensity. This process is illustrated in Figure 12.15.

Photoemissive devices include photodiodes and photomultipliers. Photomultipliers are photodiodes with built-in amplifiers (dynode stages) and are quite sensitive to very low light levels. They amplify the small photocurrents by factors as high as 10 million. They are used in situations where high sensitivity warrants the cost. Figure 12.16 shows the operation of a photomultiplier.

*Photovoltaic Cells.* Photovoltaic cells (barrier layer) are devices that actually convert light to electric power when illuminated. This is done in the absence of any applied voltage or current. In the photovoltaic process, as in the photoemissive process, incident light energy first knocks electrons free. In this case the electrons lie at a junction between two types of semiconductor materials. The important property of semiconductor junctions is that they carry a

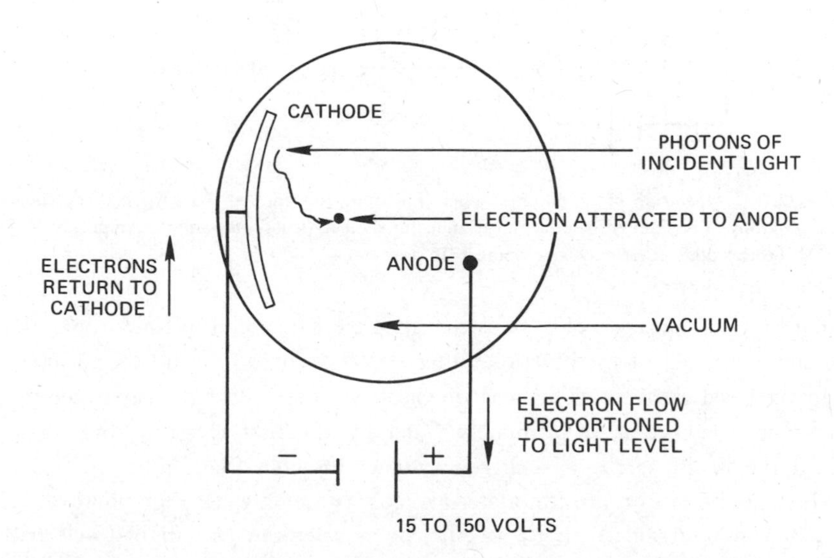


**Figure 12.15.** The photoemissive process. Light strikes the cathode, knocking free electrons which are then attracted to the anode, generating a current proportional to the light intensity.

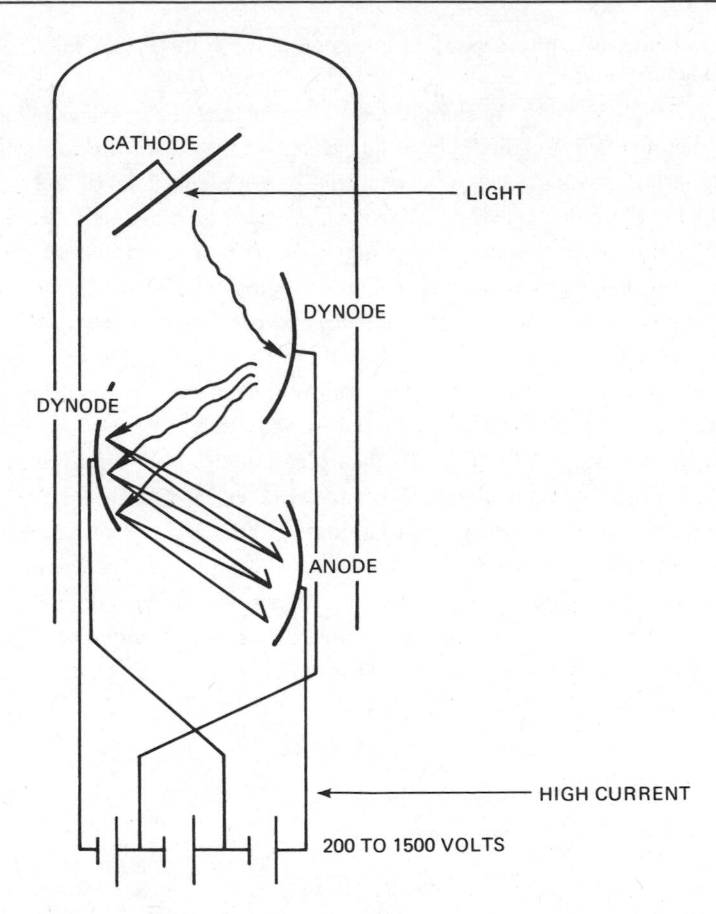


**Figure 12.16.** Operation of a photomultiplier. Each electron emitted from the cathode causes several electrons to be emitted from the next anode or dynode (small amplifiers). Amplification is repeated at each stage—there may be as many as 15 stages.

permanent local electric field. Thus the freed electrons tend to flow in a single direction. From across the junction, other electrons jump in to fill the vacancies left by the freed electrons. The electron vacancy (hole) works its way to the metallic contact where it is ultimately filled by the first electron that passes through the outside circuit as a current to reach the hole. (See Figure 12.17.)

Selenium and silicon are the materials most commonly used for photovoltaic cells. Silicon is about 20 times as efficient as selenium as a light-to-electric-power converter. Silicon cells were developed for and are used as solar power cells to power space satellites. Selenium cells, on the other hand, are spectrally close to the luminosity function of the eye. They are used, unfiltered, in light meters for camera work.

The high sensitivity of silicon cells to infrared radiation is a drawback in their use in instruments for appearance measurement. Infrared radiation is difficult to remove completely from instruments. When tungsten light sources are used there is a surplus of this radiation on the illuminant side, which compounds this difficulty. However, silicon cells are highly stable in their response to light, varying only with temperature. They are small, quite uniform, and require no power supply. There is a definite trend toward the use of silicon cells

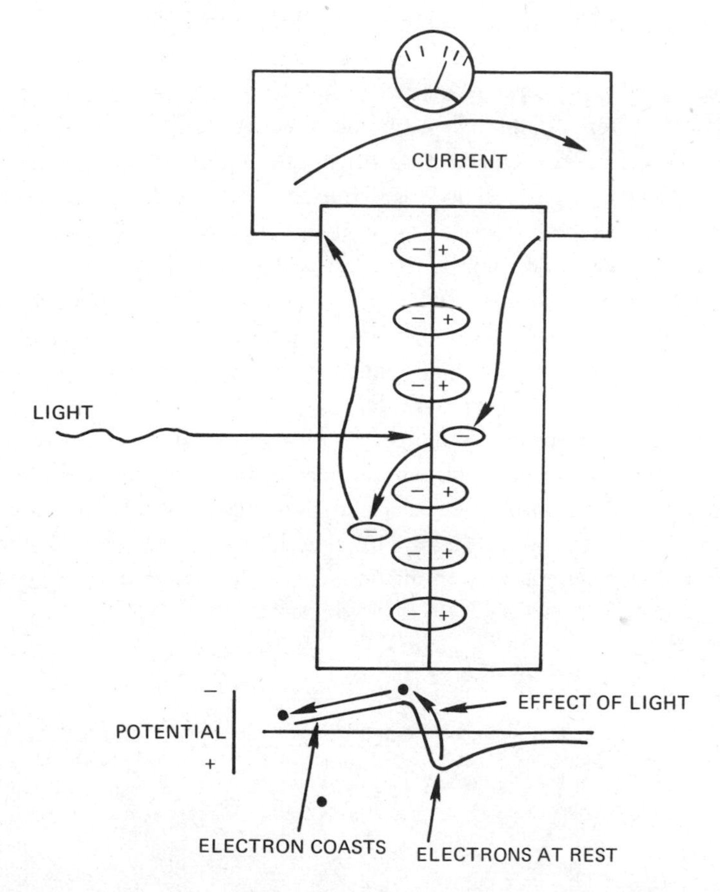


**Figure 12.17.** The photovoltaic process. Plus-minus pairs at boundary create a static field. Light energy pushes the electrons across the junction to a higher potential level. The exterior field causes the electrons to flow through the circuit producing a current.

for use in appearance-measurement instruments. The unwanted infrared energy is removed with special filters.

*Photoconductive Cells.* Some barrier (or junction) type cells can be operated in the photoconductive mode. Photoconductive cells (junction type) include both photodiodes and phototriodes (phototransistors). In operation a voltage is externally applied and resistance changes in proportion to the intensity of incident light. These devices have extremely fast responses but have high dark current and poor temperature stability compared to cells operated in the photovoltaic mode.

Photoconductive cells (bulk type) are composed of a homogeneous material that changes its electrical resistance with change in light level, the resistance dropping with increased light level. Although these devices are not linear in response to light, they are as sensitive to light as photomultipliers. Figure 12.18 shows the structure of a bulk type photoconductive cell.

There is variation in spectral sensitivity among all of these photodetectors. The sensitivity of a photocathode is defined by "S" number, which is based on the photoemissive characteristics of particular materials. Figure 12.19 shows the spectral response of some of the most widely used photocathodes and selenium and silicon photocells. It can be seen that by comparison with the sensitivity of the human eye (dotted line), photocathodes S4, S10, and S20, and selinium are all more blue or violet sensitive. Silicon and S1, on the other hand, have highest sensitivity on the red side.

The ideal photodetector for appearance instrumentation would respond only to light in the visible spectrum, would be constant with time in its response to light, and would be highly sensitive so that electrical detectors could easily measure small amounts of light. In actual practice all the photodetectors available depart significantly from these desirable conditions. All are sensitive

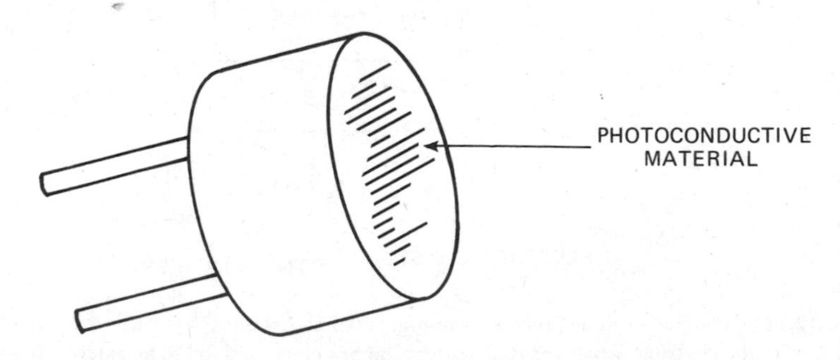


**Figure 12.18.** The photoconductive cell. A voltage is externally applied and resistance changes in proportion to the intensity of the incident light, dropping with increased light level.

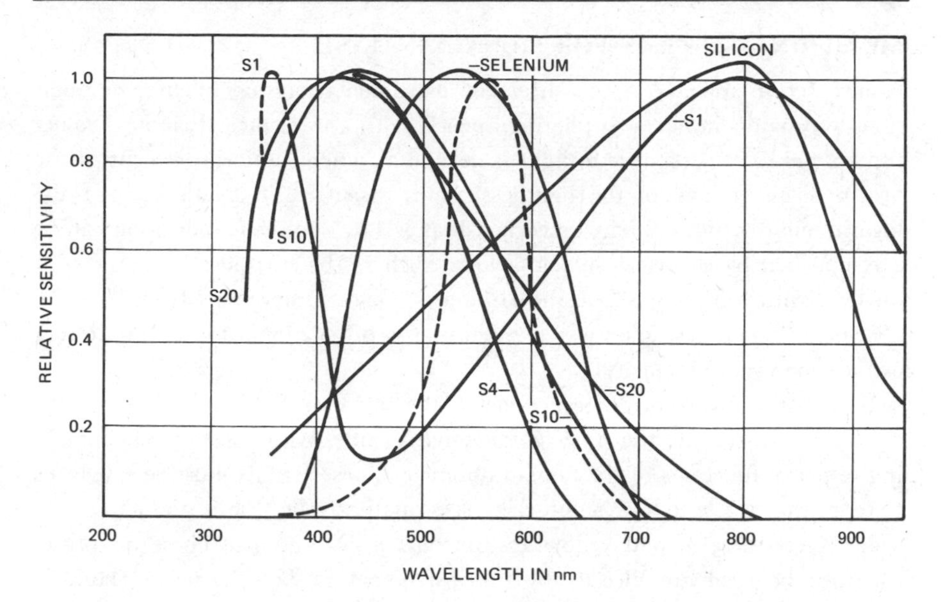


**Figure 12.19.** Relative spectral sensitivity of a silenium barrier-layer cell, a silicon barrier-layer cell and four photocathodes (S1, S4, S10, and S20). S1: Ag–O–Ce; S4: Ce–Sb; S10: (Bialkali) K–Ce–Sb; S20: (Multialkali) Na–K–Ce–Sb. The dotted-line curve is the CIE luminosity function of the human eye. Silicon and S1 have highest red sensitivity; all the others are more blue or violet sensitive than is the eye.

to changes in temperature as well as to changes in light. The majority of the photodetectors available are subject to some form of fatigue; that is, they are affected over a short or long term by light exposure. Silicon cells, however, have a major advantage in being relatively free from fatigue.

## ELECTRICAL MEASUREMENT DEVICES

After the light reflected or transmitted by the specimen in the instrument port is received by the photodetectors, signals proportional to the amount of this light are taken from the photodetector to an electrical measurement device. In the selection of the electrical measurement device, the objective is to achieve precise and accurate numerical measurement of the light reaching the receptor.

The electrical measurement device may be a galvonometer if photoelectric currents are being measured. If the initial photocurrents are converted by amplifiers into photovoltages, a voltage-measuring device such as an electrical bridge may be used. In this case, the current or voltage signal of interest is compared with a controlled signal adjusted, usually by a resistance bridge, until the two signals are equal, or "nulled."

## MEASURING SYSTEMS FOR SIGNAL ANALYSIS

A next step in appearance-measurement instrumentation may involve computations to provide numbers correlating linearly with appearance attributes. Sometimes a nonlinear response to light is desired as when the measurements made are in terms of any of the lightness scales discussed in Chapter 8, or the logarithmic density scale discussed in Chapter 11. Nonlinear indication can be accomplished by electrical analog devices, such as the current-adjusting circuit for the lightness function $L$ in the Hunterlab Color Difference Meter (Hunter, 1958). In this case a signal is generated that is proportional to the square root or the cube root of intensity.

In a tristimulus color difference meter, the signals delivered by the photodetectors are in terms of the $X$, $Y$, $Z$ tristimulus values based on the color-matching response functions of the human observer. These signals must be converted to terms that relate to how an observer sees qualitatively, that is, to lightness $L$, redness-greenness $a$, and yellowness-blueness $b$ evaluations. These quantities can then be read out directly by the instrument in $L$, $a$, $b$ terms (Hunter, 1958). With further modification, differences from standard, or from a stored memory standard, can be read out directly as $\Delta L$, $\Delta a$, and $\Delta b$.

The development of modern digital and analog computers and interfaces for computer and appearance-measurement instruments makes possible almost any kind of computation from the photosignals supplied by the instrument. The eye-brain combination is an exceedingly versatile detector-computer combination. There are obviously many opportunities for simulating this complex combination in modern instrument and computer technology.

## DATA DISPLAY DEVICES

Finally, the data that have been measured and converted electronically to numbers describing appearance attributes are displayed by one of a number of different devices:

1. Pointer position on a pivot meter
2. Position of dials connected to balancing potentiometers
3. Digital display or readout after analog-to-digital conversion
4. Position of a pen on a recorder
5. Position of an identifying spot on a cathode-ray tube

## SUMMARY

Psychophysical instruments are instruments designed to simulate the human observer in assessing the appearance attributes of objects. Because instruments are so much more limited than the eye in the number of attributes each can assess, many instruments must be built to equal the different evaluations made by one human observer.

An instrument is designed primarily to measure a specific chromatic attribute such as color, or a specific geometric attribute such as gloss. For this purpose each instrument duplicates the human observing situation for that attribute. Therefore each instrument contains a light source, a position for the placement of a specimen, and a receiver for the light from the specimen. There are a number of subsidiary components used in the instrument such as lenses, mirrors, stops, apertures, gratings, prisms, and filters, that contribute to the control of the geometric flow and spectral composition of light. Both spectral and geometric elements must be carefully selected to obtain valid measurements with either a color-measuring instrument or a geometric-attribute-measurement instrument.

# CHAPTER THIRTEEN

# INSTRUMENTS FOR THE GEOMETRIC ATTRIBUTES OF OBJECT APPEARANCE

This chapter describes the important characteristics of instruments used to measure those attributes associated with the capacities of surfaces to remit incident light in different directions. There are five types of instruments used for measuring these geometric attributes (Figure 13.1):

1. Goniophotometers, which measure the angular distributions of light reflected or transmitted by specimens
2. Diffuse-reflection meters, generally called reflectometers or opacimeters, which are used to assess diffuse-reflecting efficiency of surfaces and opacity by reflection
3. Specular-reflection meters, most of which are known as glossmeters
4. Direction-contrast meters for geometric attributes such as distinctness-of-image, luster, and reflection haze
5. Instruments for transmission measurements, both diffuse and specular, including instruments for opacity by diffuse transmission

## GONIOPHOTOMETERS

Goniophotometers assess the ways objects distribute light. They measure the amounts of light reflected or transmitted as the directions of incidence and/or view are changed. Goniophotometers provide physical analyses of the properties of objects responsible for geometric attributes. They are thus analogous, for geometric appearance, to spectrophotometers, which provide wavelength analyses of the properties of objects responsible for their color appearance.

### Historical Background

In 1922 L. A. Jones published a goniophotometric analysis of the reflecting properties of some photographic papers (Jones, 1922). Around 1930 H. J. Mc-

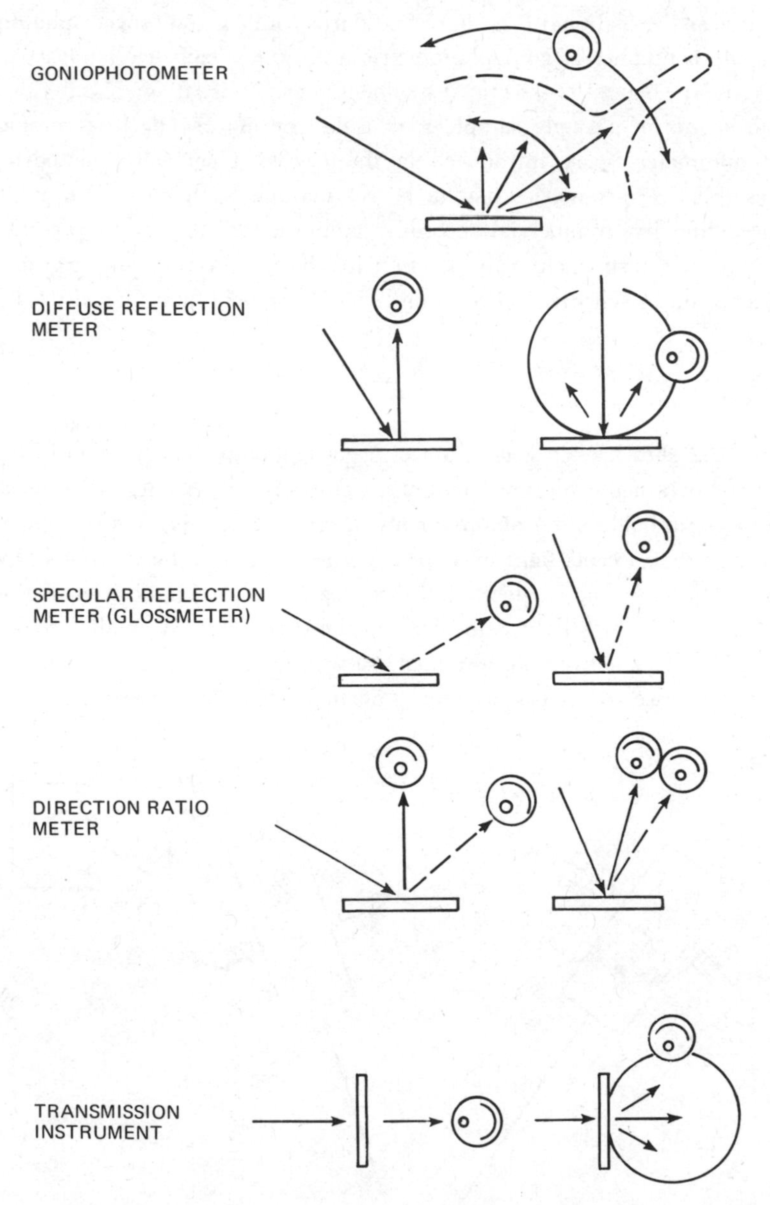


**Figure 13.1.** Geometries for five instruments for measuring geometric attributes.

Nicholas, at the National Bureau of Standards, built a goniospectrophotometer and made a number of goniophotometric and other geometric analyses of objects (McNicholas, 1934). His instrument was a visual one, and thorough measurements of a single sample took a day or more. The first commercial goniophotometers were introduced in the 1940s. They were photodetection devices, but the geometries had to be set manually. In 1953 Hunterlab introduced the first commercial recording goniophotometer. In its present configuration the instrument can be used for both reflecting and transmitting specimens and is sensitive to over a millionfold range of light intensity (Hunter, 1968).

## Goniophotometer Design

Figure 13.2 shows a block diagram of a goniophotometer. A beam of light is projected toward the specimen mounting position, where there is a device for mounting and holding test objects firmly in place. There is a light-receiving detector unit that accepts light in a specific direction from the specimen. Two of these components, and frequently all three, can be moved in position so that the directions from which light is incident on the specimen and the directions from which light is taken from the specimen are readily variable.

The light received by the detector is normally filtered to provide a $\bar{y}$ spectral

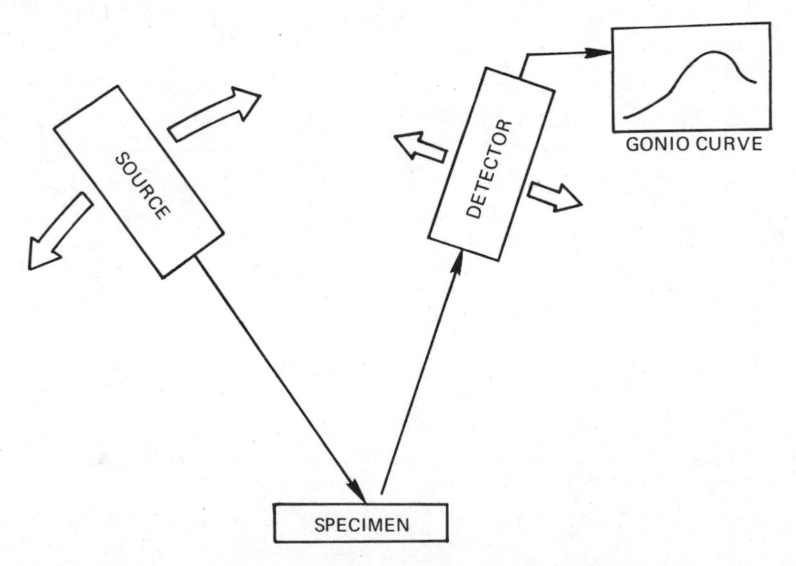


**Figure 13.2.** Block diagram of a goniophotometer. Reflectance is measured as a function of angle.

response. The detector response is converted to electrical signals that (if the instrument is equipped with a recorder) may be plotted as a continuing curve of the change in the amount of light reflected or transmitted by the specimen as either the angle of illumination or the angle of view is changed. Logarithmic rather than linear scales were used by Hunter (1968) for plotting reflectance and transmittance factors. Chapter 3 contains examples, descriptions, and discussions of such goniophotometric curves. In nonrecording goniophotometers light-intensity readouts are taken from meters or digital dials.

### Uses of Goniophotometers

Goniophotometers are used primarily to select the best fixed-angle conditions for subsequent measurements of gloss, textile luster, transparency, haze, and other geometric appearance attributes of specimens. Goniophotometers are restricted in application almost entirely to research work.

## DIFFUSE REFLECTION METERS

### Historical Background

One of the oldest reflectometers is the Macbeth-Taylor instrument originally developed to assess reflectance factors of surfaces involved in lighting (Taylor, 1920). The earliest appearance-measuring photoelectric reflectometer to enjoy significant use was the Hunter Multipurpose Reflectometer designed in 1937 and still in widespread use (Hunter, 1940).

The Hunter Multipurpose Reflectometer (see Figure 13.3) was used with amber, green, and blue filters to measure tristimulus values of color, and in addition it could be used to measure specular gloss. The photometer was an optical balancing device in which distance from the photocell to the sample was varied until the bridge circuit balanced. It was a sample-versus-reference comparison device in which the roles of the gloss and reflectance ports and photocells were interchanged when the operating mode was changed.

### Reflectometer Design

Diffuse reflection meters generally employ one of two geometries. They use either directed beams of illumination and view (usually 45° incidence and 0° viewing), or they use 0° illumination and hemispherical viewing. In the latter case a white-lined sphere is normally employed to collect the light from the specimen in all directions. As noted in Chapter 12, directions of illumination

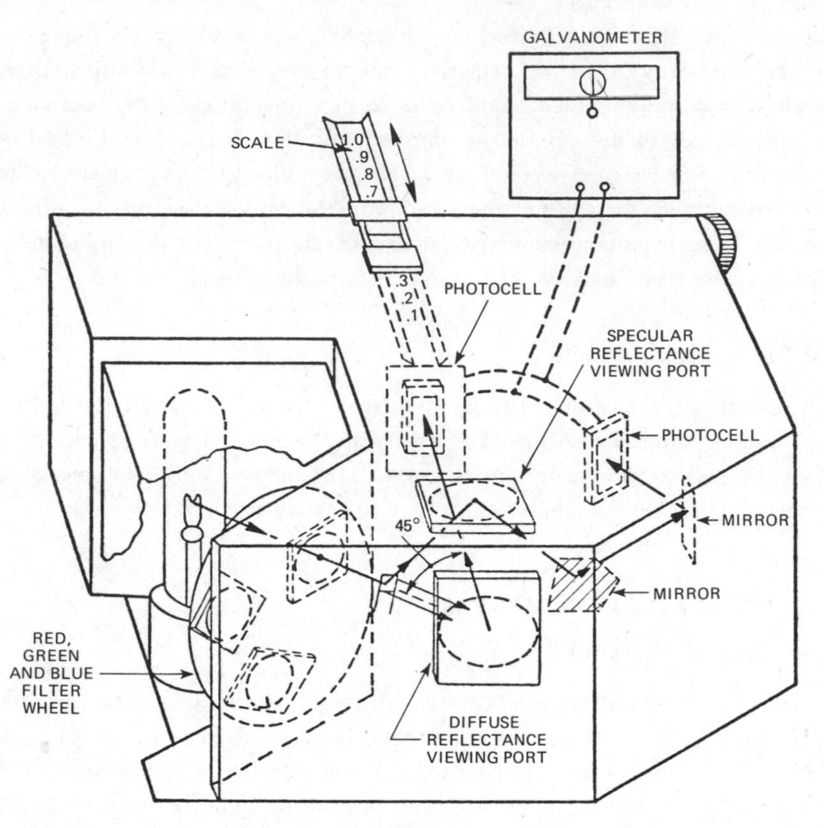


**Figure 13.3.** The Hunter Multipurpose Reflectometer. The 45°, 0° viewing port for diffuse reflection and the 45°, 45° viewing port for specular reflection serve as sample versus reference ports which are interchanged as the operating mode of the instrument is changed.

and view may be interchanged (Helmholtz reciprocity law) without change of measured quantities, provided care is taken to avoid vignetting, nonuniform weighting of light-beam fluxes, and interreflection errors.

Directed beam instruments measure diffuse reflectance factor only. Sphere reflectometers are used to measure total reflectance (diffuse plus specular), or diffuse reflectance factor only. The use of light traps to remove specularly reflected light from a sphere is described in Chapter 14 under spectrophotometers, with which instrument the technique is widely used.

A block diagram of the Hunterlab Reflectometer for Reflectance and Whiteness (Hunter, 1960) is shown in Figure 13.4 to illustrate a reflectometer with 0°, 45° geometry. It features a pair of phototubes receiving light through a single filter from sample and light source, respectively. Sample-light signal is measured against source-light signal with a voltage balancing bridge. A noteworthy optional feature shown here is an ultraviolet absorbing filter which may be located first in the incident light beam where it permits no ultraviolet radiation to reach the specimen, and then it can be positioned between the specimen and the photodetector where it permits ultraviolet light to reach the sample but not the photodetector. In this latter position any ultraviolet energy converted to visible by specimen fluorescence is added to that measured in the first position. Thus the contribution of fluorescent brighteners to blue reflectance can be measured with this instrument.

A reflectometer that employs a 30° illumination and diffuse viewing geometry by means of a sphere is shown in Figure 13.5. This is a diagram of

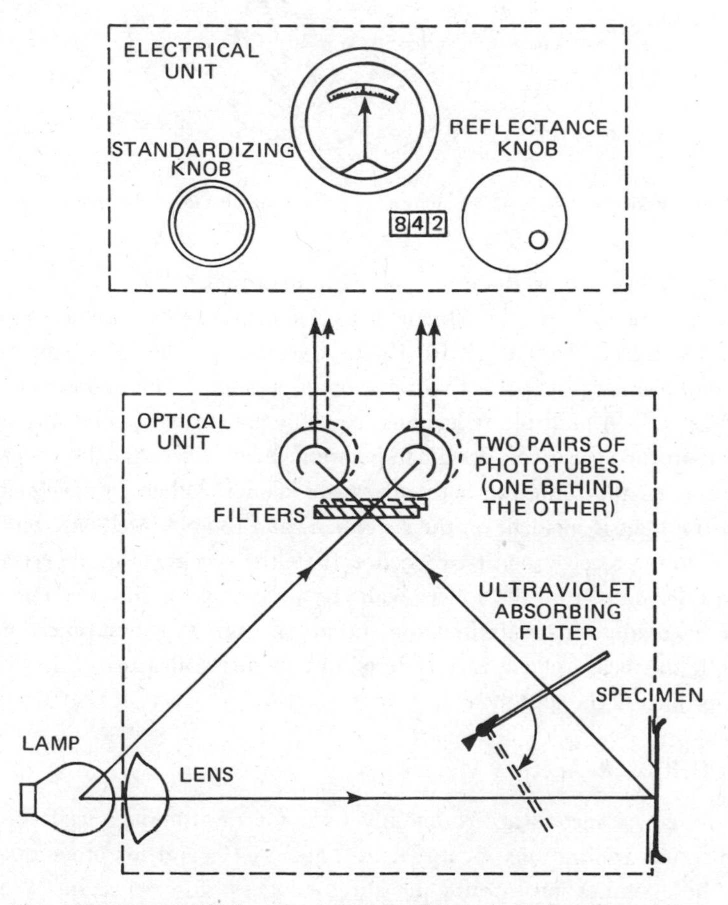


**Figure 13.4.** The Hunterlab Reflectometer for reflectance and whiteness. See text for explanation.

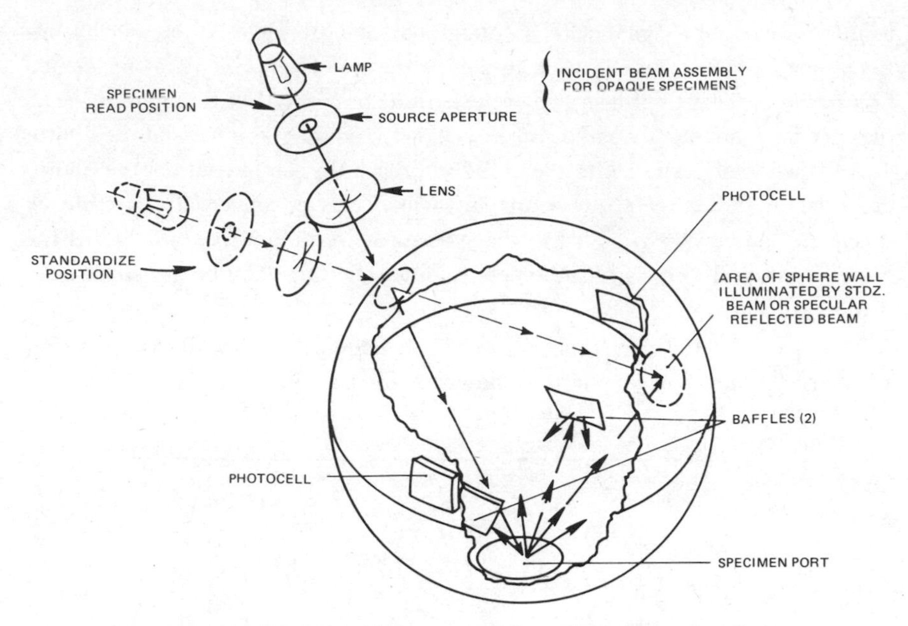


**Figure 13.5.** Principle of the Taylor-Baumgartner Sphere. See text for explanation.

the Taylor-Baumgartner Reflectometer (Baumgartner, 1937), which was the photoelectric successor to the Macbeth-Taylor visual reflectometer employing the same geometry (Taylor, 1920). In standardized position the light hits the sphere wall (which is as perfectly diffusing as possible). The photocells receive the first as well as multiple reflections from the sphere wall. The specimen to be read is in the specimen port during standardization so that the sphere efficiency remains the same for both operating modes. When in specimen-read position the light is incident on the specimen, and the photocells are baffled so that they do not receive the first reflection from the specimen but do receive the multiple reflections from the sphere wall. By accepting the first reflection from the sphere wall during standardization and not accepting the first reflection from the sample the measurement is considered to be a direct measure of the absolute total reflectance of the specimen.

## Uses of Diffuse Reflection Meters

Diffuse reflection meters are commonly used for the luminous reflection and blue reflection applications identified in Table A.10, and for other purposes. Among these applications are the measurements of diffuse reflectance factors of painted surfaces, laundered textiles, papers, porcelain enamels, and other materials. These measurements are used to evaluate the efficiencies of surfaces as reflectors of light either in lighting units, or of interior building walls. They are used to evaluate laundering and cleaning procedures by the measurement of black or soiling materials removed from textile swatches. In the ceramic, paint, and paper industries, reflectance factor ratios for black and white backgrounds are used to measure opacity. In the food industry and home economics laboratories the browning of food by roasting and baking is often assessed by a reflectometer.

In their uses as assessors of surface reflection efficiency, diffuse reflection meters employ a $\bar{y}$ (green) function response. For paper and textile blue reflectance factor measurements they are equipped with a $\bar{z}$ or other blue function. Diffuse reflection meters, when equipped with tristimulus filters, are used to measure color and as such are described in Chapter 14.

*Densitometers (Reflection and Transmission).* For printed or photographic reflection reproductions, diffuse reflection meters equipped with density scales and filters are used to assess ink or dye density (concentration) on reflecting substrates. For colored reflection prints, these reflection densitometers employ the subtractive primary red, green, and blue densitometric filters. Transmission densitometers are used for similar measurements of transmission prints. (See Chapter 11 for color identification from densitometer readings.)

## SPECULAR GLOSS AND REFLECTION METERS

Most of the instruments designed to measure specular reflection directly are called glossmeters. A specular glossmeter is an instrument that measures specular reflection under conditions designed to yield numbers that correlate with surface shininess or gloss. The specular angle of view is always equal to and opposite to the angle of incidence. A $\bar{y}$ luminous spectral response is normally used. Specular reflection is always involved in gloss measurement, but for some gloss measurements off-specular reflection is also involved, as in direction-contrast instruments for luster and distinctness-of-image gloss.

### Historical Background

The first commercial glossmeter was probably the Ingersoll Glarimeter developed in 1914 (Ingersoll 1914, 1921). It was actually a contrast-gloss meter and proved well suited for the measurement of gloss of white paper. Pfund (1930) developed the earliest specular glossmeter, which was a visual instrument as was the Ingersoll Glarimeter.

In the early 1930s Hunter designed a visual glossmeter (Hunter, 1934), and some years later he designed the first photoelectric glossmeter, now known as the Gardner Glossmeter. In 1937 Hunter pointed out that not merely two, as Pfund suggested, but five or six different types of optical measures are needed to analyze all types of glossy appearance (see Chapter 6).

The instrument associated with TAPPI Method T-480 for 75° gloss of paper in 1937 was initially called the "Oxford 75° Glossmeter" (Institute of Paper Chemistry, 1937). As a result of studies at the Institute of Paper Chemistry, several modifications to the instrument were suggested. Bausch and Lomb Optical Company then manufactured the Modified Oxford Glarimeter until World War II forced its discontinuance.

In 1939 ASTM Method D523 was adopted for the measurement of 60° gloss of paints and plastics, as the result of work done in ASTM Committee D-1 on Paint, Varnish, Lacquer, and Related Products (Hunter and Judd, 1939).

In 1945 Harrison gave a comprehensive review and evaluation of the state of knowledge concerning gloss (Harrison, 1945). He contributed an element of perspective to the subject of gloss analysis by suggesting that the difficulty lay not in getting goniophotometric curves of a material, but in knowing what to do with the curves.

## Glossmeter Design

In looking at a glossmeter it is helpful to consider two parts:

1. The optical unit, containing light source and incident light beam, specimen port, and viewing beam with receptor photodetector
2. The measurement or readout unit where the electrical signals from the optical head are converted into instrument readings

With some glossmeters these two components are housed in a single unit. In others they are housed separately and connected by cable. Separation permits flexibility in placing the optical unit on the specimen.

For a glossmeter to give measurements on one of established scales (Table A.1), accurate control of the effective directions of illumination and view is essential. All of the elements that make up the source lamp, condenser lens, source field stop, and aperture stop must be positioned accurately, as must the elements that constitute the receptor. In order to avoid vignetting there should be only one aperture stop limiting the diameter of the light beams. (See Chapter 12.) No incident rays should strike the edge of the specimen window. The illuminated area of the specimen must not be covered by film or glass and it must be perfectly flat, clean, and in the defined specimen plane. Gloss standards must be carefully cleaned and free of any film. Since gloss is the result of specular reflection at the object surface, phenomena of molecular magnitude within the skin of the specimen can significantly affect measured gloss.

The optical diagram of a Hunterlab Specular Glossmeter is shown in Figure 13.6. The receptor of the glossmeter consists primarily of the receptor field stop, which defines the viewing field angle, and a photodetector. The source image, as reflected by a flat, polished surface in the defined specimen plane, must be centered in the receptor window. (See Figures 12.8 and 12.9.) All the light passing the receptor field stop must be equally weighted by the photodetector. With high-gloss measurements, viewing field angles are narrow; when this is the case, any small displacement of the test surface from the required plane of measurement will introduce significant errors into measured specular gloss. In the high-gloss Hunterlab D47 Dorigon distinctness-of-reflected-image instrument, screw adjustment of the direction of incident light is used for every instrument setting in order to assure that the reflected beam of light is properly centered in the very narrow light-receiving pickup. Accuracy in

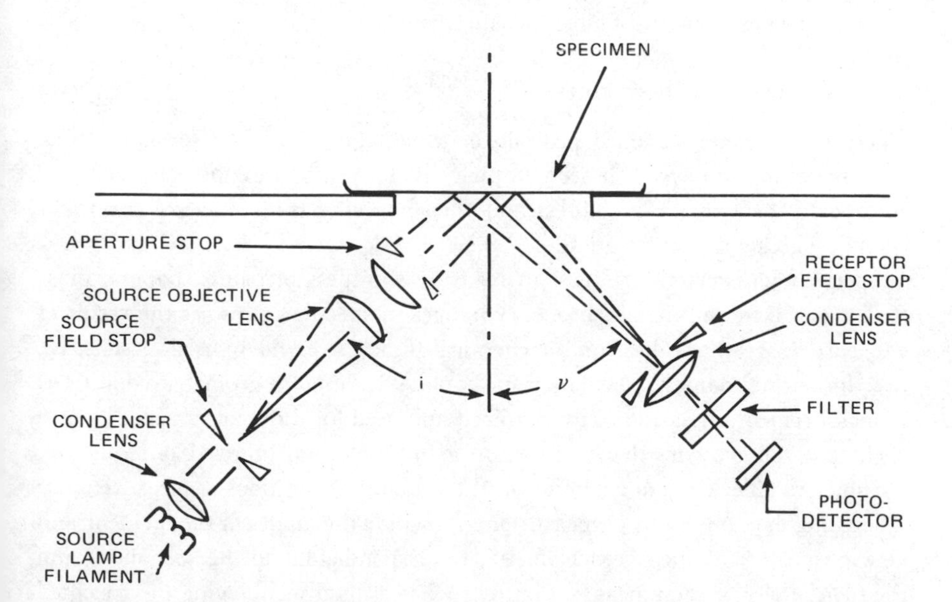


**Figure 13.6.** Optical diagram of Hunterlab Specular Glossmeter. The light from the source lamp, controlled by stops and lenses, is incident on the carefully positioned specimen. A photodetector placed at a viewing angle equal to the angle of light incidence receives light reflected from the specimen after the light has passed through a field stop, a lens, and a filter.

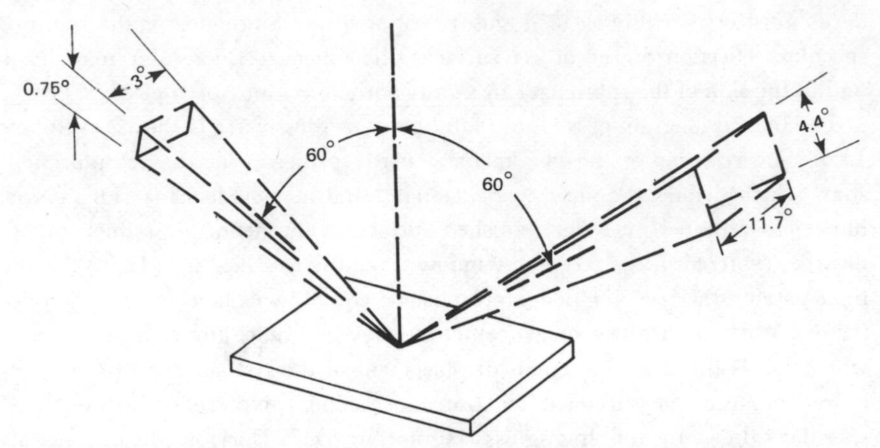


**Figure 13.7.** Specular and field angles called for by the 60° method of ASTM D523. The field angles are an important part of the specifications for any gloss measurement.

reproducing geometric conditions of measurement is normally the limiting factor in achieving accuracy of gloss measurements.

## Uses of Specular Glossmeters

There is no single standard procedure or Standard Observer for gloss. The gloss scales in use have each been empirically derived by ranking specimens according to their visual attributes, and then applying the numerical scale and geometry giving best correlation.

Specular glossmeters are used to measure shininess of paints, papers, plastics, and waxed and metal surfaces. Changes in specular gloss (as the result of exposure to weather, abrasion, or chemical fumes) are widely used as sensitive tests for surface damage. As shown in Table A.1, different geometric conditions of measurement are customarily required and used for different materials, since each material may involve a different manifestation of gloss. Consequently, glossmeters differ from one another in their angular conditions of measurement. These angular conditions of measurement include the angles of the incident and viewing beams (as measured from 0°, or perpendicular to the specimen) and the field angles of these beams. Figure 13.7 is a diagram showing the specified 60° specular angle and two field angles called for by the 60° specular gloss procedure of ASTM D523, which is commonly used for gloss measurements of paints and plastics.

## DIRECTION-CONTRAST METERS FOR CONTRAST GLOSS, DISTINCTNESS-OF-IMAGE GLOSS, AND REFLECTION HAZE

Within recent years specialized instruments to measure the contrast between reflections in two directions have been developed and used to measure geometric appearance attributes such as contrast gloss or luster, distinctness-of-image gloss, and reflection haze. The geometric conditions for the measurement of these different geometric properties are selected, in each case, to provide best numerical correlation with the specific appearance attribute of interest.

### Lustermeter for Measurement of Contrast Gloss

Figure 6.4 in Chapter 6 illustrates the particular kind of gloss called luster, or contrast gloss. This attribute can be most important in the appearance of textile fabrics, human hair, certain vinyl fabrics, and diffuse-finish bare metals. Luster can be defined as the contrast between the specularly reflected light and adjacent diffusely reflected light from the same surface. The formula developed to express this relationship is

$$\text{Luster} = 100\left(1 - \frac{\text{diffuse reflectance factor}}{\text{specular reflectance factor}}\right)$$

Actual numerical values of luster depend not only on the axial angles of the specular and diffuse light beams, but also on the receptor field angles.

The Hunterlab Lustermeter, a special glossmeter designed specifically to measure contrast gloss, employs both specular and diffuse viewing detectors and computes luster electronically using the foregoing formula. Angles of specular and diffuse reflection as well as field angles can be selected to give the best analysis of a specific product. Figure 13.8 shows the beam axes of a lustermeter.

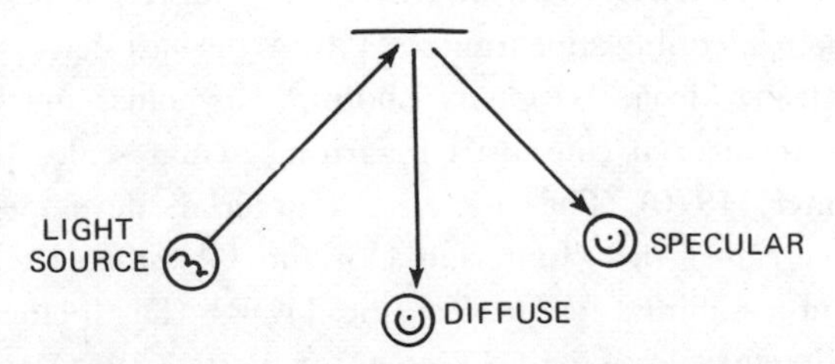


**Figure 13.8.** Diagram of Hunterlab Lustermeter. The angular conditions can be selected for the best analysis of a given product.

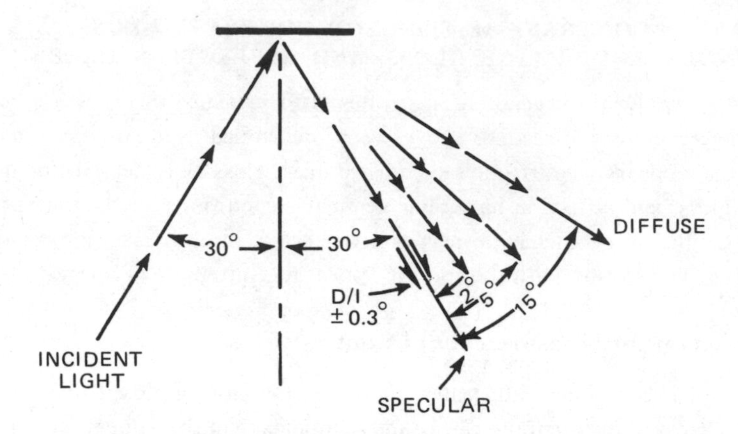


**Figure 13.9.** Light-beam angles of Hunterlab Dorigon Glossmeter.

## Distinctness-of-Reflected-Image and Reflection-Haze Meters

The image-reflecting capacity of a surface cannot be assessed by a measurement of specular gloss only. More sophisticated methods of measurement are required. Meaningful measurements are difficult for two reasons. First the angular resolution of the eye in image recognition is so high that it is difficult to duplicate in photometric instruments. Second, as in luster, it is the contrast with surroundings, not the image intensity itself that is responsible for distinctness.

Middleton achieved high angular resolution by projecting the image of a source slit onto the spoke-shaped narrow slits of a rotating disk with light receiver behind. Signal amplitudes were proportional to shininess. Rapidity of signal change was proportional to image sharpness (Middleton, 1953).

Recently the increased use of bare metal as an exterior architectural surface in buildings and as exterior trim in automobiles has created a need for gloss scales to evaluate metallic appearance. Within the last few years techniques developed by Alcan, Alcoa, Allegheny Ludlum, Reynolds, and Hunterlab have begun to move towards a consensus regarding actual scales for metallic appearance (Hunter, 1970). For this use, Hunterlab developed an abridged goniophotometer called the Hunterlab Dorigon Glossmeter (Christie, 1970). This instrument combines features of the DORI (Distinctness-of-Reflected-Image) meter designed by Tingle (Tingle and Potter, 1961) and the abridged goniophotometer designed by Tingle and George (Tingle and George, 1965). The Dorigon (Distinctness-of-Reflected-Image Goniophotometer) has the capability of measuring 30° specular reflectance factor, distinctness-of-image (at ± 0.3° from the specular angle), two kinds of haze (at 2° and 5° from the specular), and diffuseness (15° from the specular angle). Figure 13.9 shows the light-beam angles of the Dorigon Glossmeter.

## INSTRUMENTS FOR TRANSMISSION MEASUREMENTS

Instruments for the measurement of light transmitted through specimens fall into two categories:

1. Diffuse transmission meters for haze, turbidity and translucency, and opacity by transmission
2. Specular transmission and clarity meters for clarity and for chemical analyses by absorption

### Haze Meters

The sphere haze meter is used to determine the transmission haze of plastic films and sheets. It is designed to measure either essentially all of the light transmitted (specular plus diffuse) or the diffusely transmitted light only. Figure 13.10*A* shows the setting of the instrument for measuring specular plus diffuse light. The light source is pivoted 8° from its normal straight-through path, and the specular beam is collected by the white sphere wall. In Figure 13.10*B*, the specular beam of light goes straight through the sphere and into a light trap where it is absorbed. Only the diffusely transmitted light is measured. Haze is measured, according to ASTM Method D1003, as the ratio of the diffuse to total transmission according to the following equation:

$$\text{transmission haze} = \frac{\text{diffuse transmittance factor}}{\text{total transmittance factor}} \times 100$$

### Clarity Meters, Turbidimeters, Transmission Opacimeters

A clarity meter employs a 0° incident, 180° viewing geometry, and thus measures only the specular beam of light as it is transmitted by the specimen. A distinctness-of-transmitted-image instrument would be a more elaborate instrument of the same type, the transmission analog of the Hunterlab Distinctness-of-Reflected-Image instrument.

Instruments for the measurement of smoke as well as liquid cloudiness are important tools in the control of environmental pollution. These turbidimeters

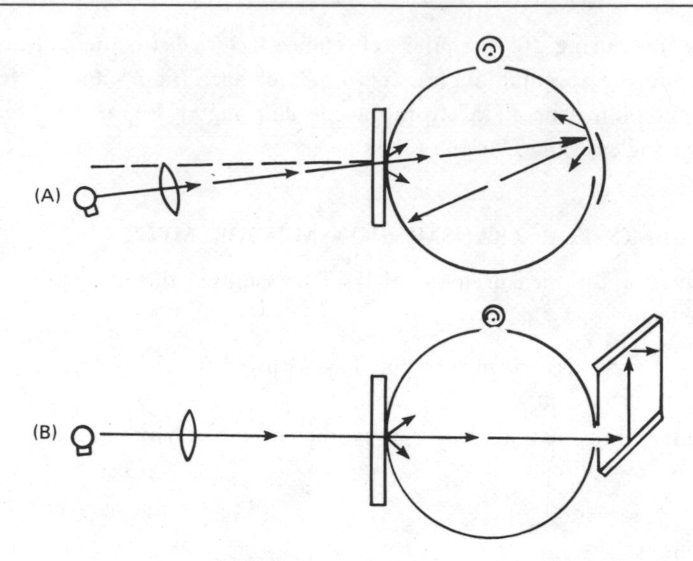


**Figure 13.10.** Optical diagram of the Hunterlab Sphere Transmission Hazemeter. (*A*) Position for measurement of specular plus diffuse; (*B*) position for measurement of diffuse only.

and other light-scattering photometers may employ the 0° incident, 180° viewing geometry of the clarity meter, or they may be more sophisticated devices that can view at 90°, 270°, or some other off-specular angle.

Transmission opacimeters are instruments that use diffuse transmittance to measure opacity of light-diffusing sheets, especially sheets of paper. As explained in Chapter 6 and illustrated in Figure 6.12, fractional increments of diffuse transmittance are roughly six times as sensitive to opacity as are fractional increments of contrast-ratio by reflection. Because of this significant advantage in precision, opacity is increasingly being measured by diffuse transmission, especially in on-machine applications.

## SUMMARY

The instruments in general use for the measurement of geometric attributes, such as gloss or haze, are goniophotometers, reflectometers, glossmeters, and transmission meters. Reflectometers measure diffuse reflectance factors used to evaluate reflecting efficiency, dirt removal, opacity, and darkening of surfaces. Specular glossmeters measure surface shininess and are used to assess appearance and surface change due to exposure. Other attributes of gloss can now be measured instrumentally but the use of these measurements is not well established.

Optical surface uniformity is a characteristic for which measurements are needed, but not available. Haze, turbidity, and opacity are regularly measured by transmission. Other transmission attributes analogous to the gloss attributes are the subject of little attention at the present time.

# CHAPTER FOURTEEN

# INSTRUMENTS FOR THE CHROMATIC ATTRIBUTES OF OBJECT APPEARANCE

Color measuring instruments differ from one another mainly in the method of wavelength selection and type of evaluation device. They can be classified as follows:

1. Spectrophotometers, which measure reflectance or transmittance factors one wavelength at a time. An abridged spectrophotometer is limited to a number of specific wavelength bands for its measurements.
2. Tristimulus instruments which can be filter photometers giving tristimulus reflection or transmission values (such as *X, Y, Z*), or tristimulus *L, a, b* colorimeters supplying direct readouts in opponent-colors dimensions.

## SPECTROPHOTOMETERS

Spectrophotometers measure the wavelength distributions of light responsible for color appearance. When used to measure spectral distribution of light by objects, spectrophotometers give percent reflectance or percent transmittance through the visible spectrum. The outputs of recording spectrophotometers are spectral curves such as those shown and described in Chapter 3. These curves are the physical analyses from which tristimulus color values can be computed. They also provide the data used in color formulations by computer. In addition to their use in color appearance analysis and color formulation, spectrophotometers are used for chemical analyses and ingredient identification. Accurate chemical analyses are made from spectral curves through knowledge of the specific spectral light-absorbing properties of chemical materials.

Although research spectrophotometers had been in existence for many years, the first spectrophotometer to have a major impact on color technology was the instrument developed by Professor A. C. Hardy and introduced by the General Electric Company in 1928. This instrument was designed specifically for color measurement (Hardy, 1935). Despite some component redesign, it has remained essentially the same basic instrument for the last 45 years, and for that period has enjoyed the position of chief reference instrument for color measurement.

Of the many spectrophotometers on the market, most are intended primarily for chemical analysis, usually by measurements of transmission of solutions in cells of standard light-path length. Some of these offer reflection attachments and other features to adapt them for use also in color appearance analysis. Few, if any, of these chemical analysis spectrophotometers adapted to color measurement have the photometric accuracy and precision of a Hardy spectrophotometer in good adjustment. Only the Hardy was designed specifically for color appearance measurement.

For some spectral analysis problems an abridged spectrophotometer is adequate. The abridged spectrophotometer is so called because its monochromator can select only certain wavelengths, usually by means of selected interference filters, each of which isolates a different band of wavelengths. A typical abridged spectrophotometer uses 16 filters at 20-nm intervals through the spectrum to provide a 16-point spectral curve. For many applications, specifically for computer-formulated color-match predictions, this has proven to be adequate, although the accuracy of abridged instruments decreases for specimens that have spectral curves with steep slopes and sharp peaks.

## Design Features of Spectrophotometers

The selectivity of wavelength isolation varies considerably among spectrophotometers. The most selective are the highly sensitive chemical analysis instruments. The least selective are the spectrophotometers used in the analysis of color appearance, where the narrow spectral-band isolation is not necessary. The Hardy spectrophotometer and the Kollmorgen KCS-18 abridged spectrophotometer both normally employ a wavelength-band pass of about 10 nm. The actual wavelength-band pass involved in any one value of reflectance or transmittance factor is quite a bit wider because the 10-nm width is measured at 50% of peak transmission. By contrast, the Bausch and Lomb 505, a spectrophotometer designed primarily for chemical analyses, has a minimum band pass of 0.5 nm. Several other chemical analysis spectrophotometers on the market have even narrower band passes.

*Wavelength Isolation and Identification.* The basic arrangement of components in a spectrophotometer is shown in the block diagram of Figure 14.1. Wavelength isolation, the most critical operation in spectrophotometry, is performed by means of gratings, prisms, or filters. In many instruments light is

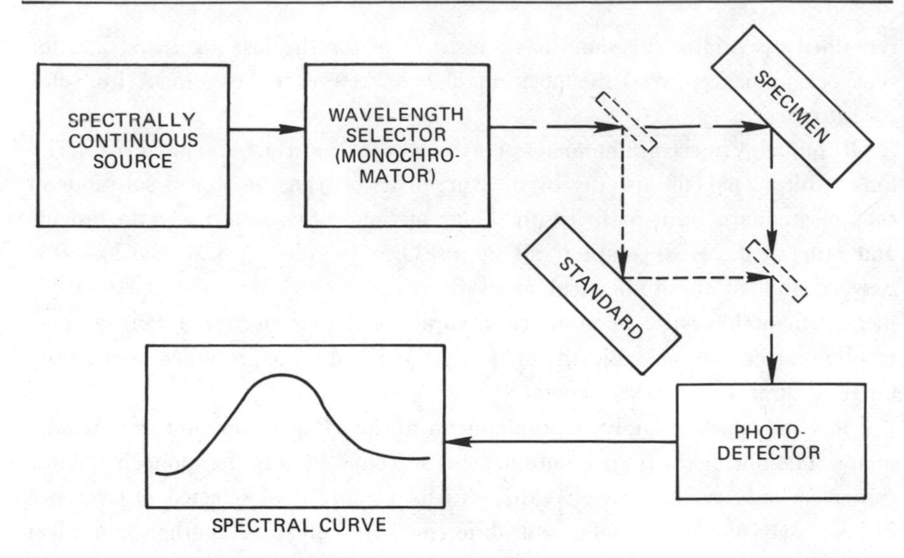


**Figure 14.1.** Block diagram of basic components of a spectrophotometer. It is necessary to place the monochromator between the specimen and the photodetector when measuring fluorescent specimens.

directed through duplicate wavelength dispersing elements in order to improve resolution and reduce stray unwanted wavelengths. The wavelength selectors and slits together define the monochromator of a spectrophotometer. Figure 14.2 shows the two-prism monochromator of the Hardy spectrophotometer. Note that the prisms remain fixed and the emerging beam is always in the same location. Wavelengths are selected by changing the position of the mirror-slit element that lies in the beam between the two prisms.

When fluorescent specimens are being measured with a spectrophotometer, the results differ depending upon whether the monochromator is ahead of or beyond the specimen. A fluorescent sample absorbs light at one wavelength and emits it at another, longer wavelength. When the monochromator is placed before the specimen (see Figure 14.2), only one wavelength band reaches the specimen at a time. Light emitted by specimen fluorescence is improperly attributed to the shorter wavelength exciting the fluorescence. When the monochromator is placed between the specimen and the photodetector, however, the fluorescence is correctly attributed to the longer wavelengths where it is emitted (and seen by an observer). In this instance the spectral character of the illumination on the fluorescent specimen is a matter of importance and will directly affect the shape of the spectral curve. Thus spectrophotometers with different illuminants will draw different curves for the

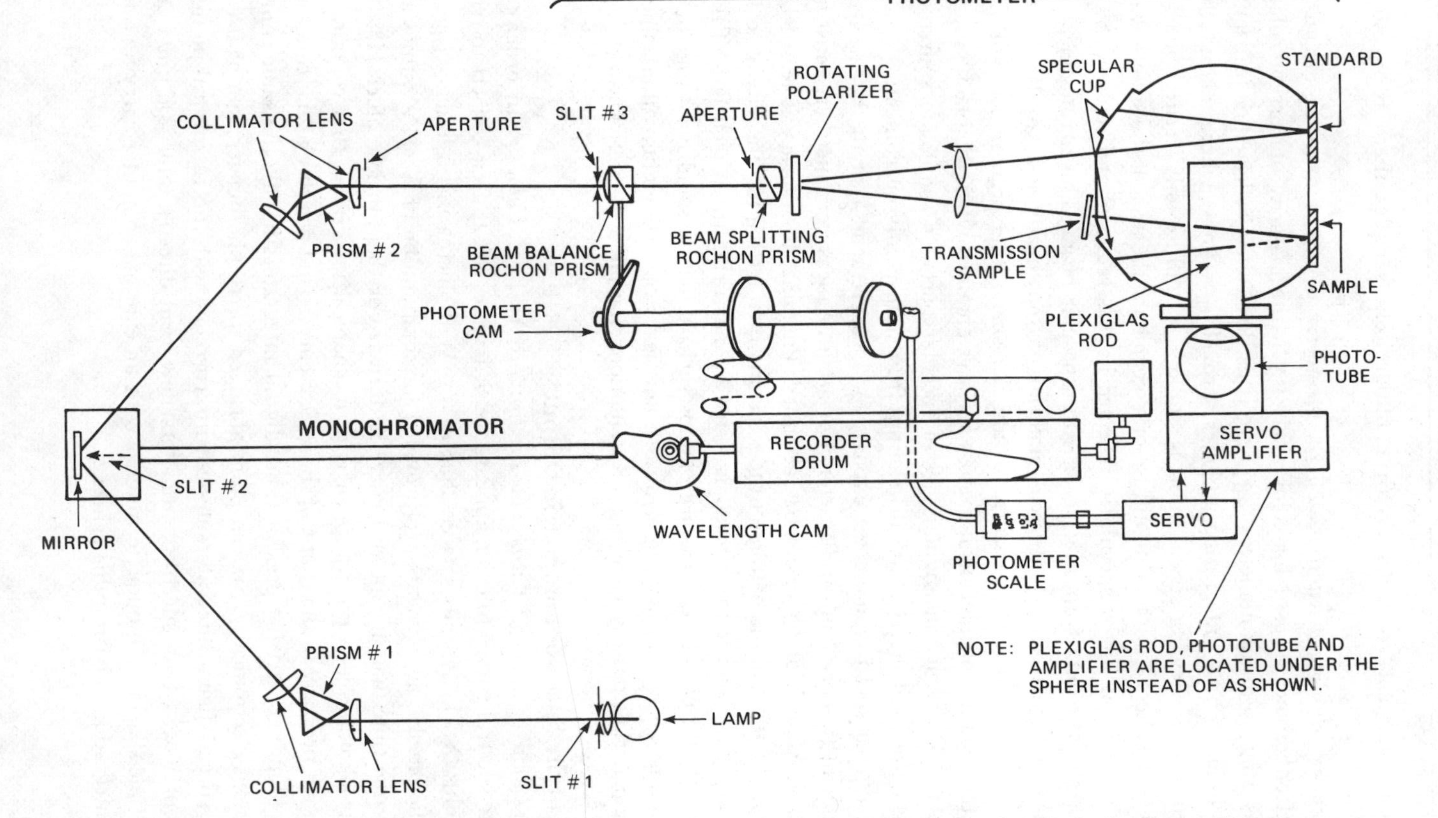


**Figure 14.2.** Design of the Hardy Spectrophotometer. This instrument employs two fixed prisms. The wavelength selection is accomplished by changing the position of the mirror-slit element.

same fluorescent specimen. With nonfluorescent samples there is no such problem.

*Specimen Illumination and Viewing.* Geometry is important in the design of spectrophotometers used for color analysis. In the majority of current color-measuring spectrophotometers, a sphere is used to collect reflected or transmitted light. These integrating spheres provide readings where either the illuminant or viewing specification is "total" or "diffuse only" reflection. To convert a sphere from total to diffuse-only geometry, the area of the sphere wall that is normally hit by specularly reflected light is removed and replaced by a black plug, or light trap, that absorbs the specular beam. This arrangement is shown in Figure 14.3. Since real materials do not neatly separate light into specular and diffuse components, the angular size of the specular port will affect the readings of intermediate-gloss surfaces obtained when the black plug is in place. Difference in angular size of specular port is a significant source of measurement "disagreement" in readings from diffuse-only instruments.

A typical spectrophotometer exposes for measurement a small flat area of specimen, roughly 1 in. in diameter. For accurate measurements there must be both wavelength and photometric accuracy. Photometric accuracy is tested by setting the top and the bottom of the scale and then testing for linearity of scale in between. Wavelength accuracy is tested by measuring the wavelength of the spectral lines of gaseous discharge lamps such as mercury and neon together, or by measuring the maximum absorption peaks of a didymium filter, since these materials possess wavelength distributions that are accurately known.

## Use of the Spectrophotometric Output

*Conversion of Spectral Curves to Tristimulus X, Y, Z Values.* In order to relate spectral curves to color appearance, it is necessary to first convert to tristimulus $X$, $Y$, $Z$ values (see Chapter 7). With increasing frequency spectrophotometric attachments are provided for computation of $X$, $Y$, $Z$ values at the same time that the spectral curve is drawn. For this purpose older Hardy spectrophotometers were equipped with mechanical integrators that contained cams cut precisely to the shape of the $\bar{x}$, $\bar{y}$, $\bar{z}$ curves, and possibly also a cam for each illuminant. At the present time, digital integrators handle the same computations electronically. Sixteen-point abridged spectrophotometers can be used with special digital integrators designed to correlate with the 16 spectral points they provide. Many spectrophotometers now read directly into tape units or into time-sharing computer terminals, with an external computer carrying out the integration to tristimulus values.

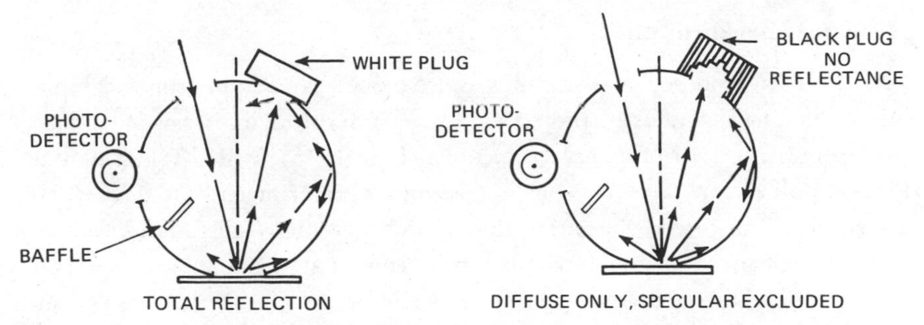


**Figure 14.3.** Converting a sphere from geometry for measurement of total reflection to that for diffuse only, with specular excluded. To exclude the specular reflection, the part of the sphere wall which is normally hit by the specularly reflected light is removed and replaced by a black plug, or light trap, which absorbs the specular beam.

Where the integration is not supplied by a spectrophotometer, the manual multiplication of spectral curve by the Standard Observer-Illuminant $\bar{x}E$, $\bar{y}E$, $\bar{z}E$ functions, once such a common color laboratory activity, is now largely done by advanced calculators or computers.

*Color Formulation by Spectrophotometry and Computer.* Color formulation of paints, plastics, and dyed textiles has developed rapidly during the last decade. A spectrophotometer and a mathematical model of the optical functions of the component ingredients are required for formulation of color. The Kubelka and Munk equations for scattering and absorption of light within light-scattering layers (Kubelka, 1948) and Beer's law for absorption of light in films are two of the simple equations used as mathematical models for computer formulation. Stearn's description of color matching by this method, with actual examples, is recommended to the reader (Stearns, 1969).

After the initial formulation of product to an approximation of the desired spectrophotometric curve, a tristimulus colorimeter often provides a less expensive, faster way than spectrophotometry to adjust to final desired product color.

## TRISTIMULUS INSTRUMENTS—REFLECTOMETERS AND COLORIMETERS

Tristimulus instruments are designed to respond to spectral distributions of light in the same manner as the human eye. Duplication of the human eye response to light is accomplished by adjusting with filters the spectral responses of photodetectors so that they are spectrally equivalent to the CIE Standard Observer functions.

## Historical Background

In 1915 Ives proposed a tristimulus reflectometer that approximated human color-matching functions by projecting a spectrum through a series of templates cut to duplicate a set of observer functions (Ives, 1915). In 1928 Twyman and Perry applied for a patent on a tristimulus colorimeter. In 1938 Perry described a four-filter colorimeter that would give direct tristimulus values of colors by means of source-filter-phototube combinations that were spectrally similar to the Standard Observer functions (Perry, 1938). The accuracy of such an instrument was shown to depend directly on the accuracy with which the Standard Observer functions were realized.

In 1937 Hunter developed the Multipurpose Reflectometer (see Figure 13.3), which incorporated amber ($A$), green ($G$), and blue ($B$) filters approximating the CIE functions (Hunter, 1940). The closeness of the spectral fit of these filters to the Standard Observer is shown in Figure 14.4. To obtain tristimulus $X$ values, the amber reading and a fraction of the blue reading were

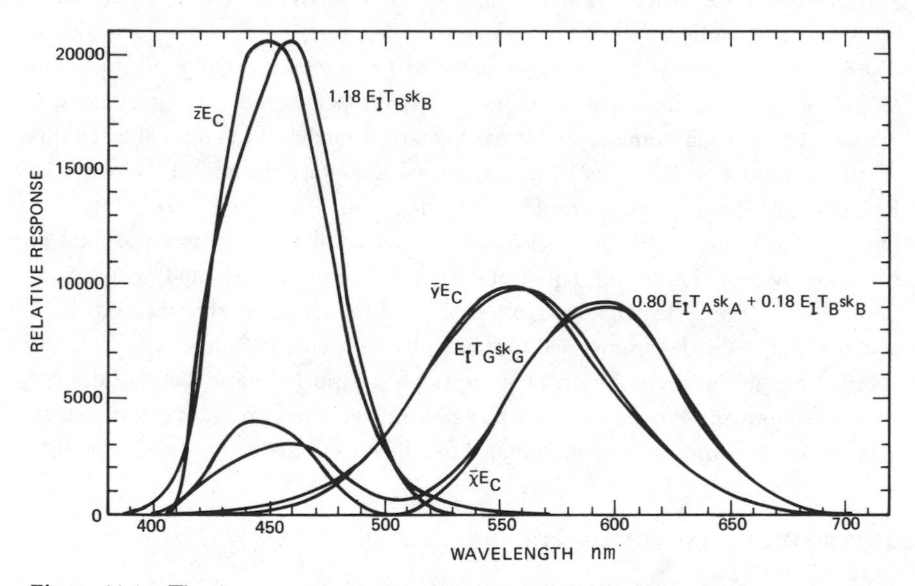


**Figure 14.4.** The closeness of duplication of the CIE Standard Observer functions by the source-filter-photocell combinations of the multipurpose reflectometer (Hunter, 1942). The Standard Observer functions combined with Illuminant C are represented by the curves $\bar{x}E_c$, $\bar{y}E_c$ and $\bar{z}E_c$. The responses of the instrument source-filter-photocell combinations are indicated by the other three curves, where $E_I$ is the spectral energy of the source; $T_A$, $T_B$, and $T_G$ are transmissions of amber, blue, and green filters; $sk_A$, $sk_B$, and $sk_G$ are spectral responses of amber, blue, and green photocells, with $k$ constants of proportionality.

added together by subsequent calculation. However, the recommended operations for color measurements was to convert the *A, B, G* reading to Hunter $\alpha$, $\beta$ values directly without first computing *X*.

By 1942 additional research and publications by Guild (1934), Winch and Palmer (1937), Dresler and Frühling (1938), Barnes (1939), Gibson (1936, 1939) and Van den Akker (1937) had laid a clear foundation for the future development of tristimulus colorimeters, with the major criteria for accurate colorimetry having been identified and reported. In 1942 Hunter described the variety of tristimulus measurements possible with the Multipurpose Reflectometer and similar instruments (Hunter, 1942). The errors inherent in such measurements were also studied in some detail. In the late 1940s Hunter developed a photoelectric instrument with circuits that gave Ohm's law analog values and readout in $R_d$, *a, b* scales (Hunter, 1948). A later development was the Hunter D25 Color and Color Difference Meter equipped with *L, a, b* scales (Hunter, 1958).

## Tristimulus Colorimeters

As defined above, tristimulus instruments are those that have source-filter-photodetector combinations that simulate the Standard Observer functions. In this book, we make a further distinction between a tristimulus reflectometer and a tristimulus colorimeter. A tristimulus reflectometer (such as the Hunter 1940 Multipurpose Reflectometer) gives reflectance factor readings only, with amber, green, and blue filters. These readings are called amber (*A*, or $X_{A\%}$, indicating it to be the same as the amber portion of the CIE $\bar{x}$ function), green (*G* or *Y*), and blue (*B* or $Z_{\%}$). A refined reflectometer will have a second blue filter, usually called $X_B$, to simulate the blue part of the CIE $\bar{x}$ function, making four filters in all. Van den Akker and others have shown that the four-filter tristimulus instrument is inherently more accurate than a three-filter instrument (Van den Akker, 1937).

A tristimulus colorimeter has, in addition to the three or four source-filter-photodetector combinations for Standard Observer simulation, a device that computes directly chromatic dimensions of color, such as *a* and *b*, or *x* and *y*. Such an instrument with direct-reading $R_d$, *a, b* or *L, a, b* scales is properly called a tristimulus colorimeter.

*Design of Colorimeters.* Figure 14.5 is a block diagram showing the light-viewing elements of a tristimulus colorimeter. There are several different geometries of light-beam incidence and view that may be employed, depending

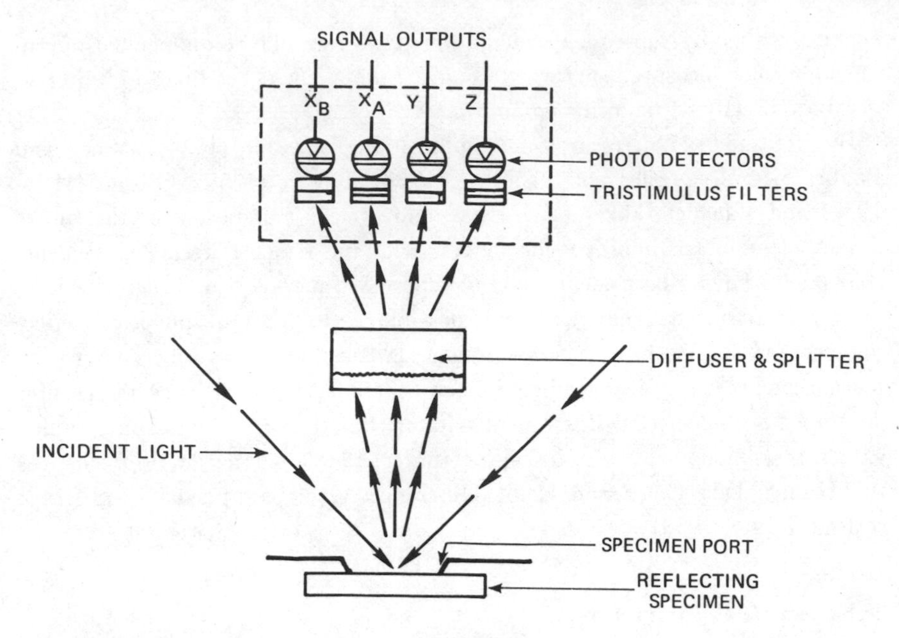


**Figure 14.5.** Block diagram of a tristimulus colorimeter. The most accurate tristimulus colorimeters employ two separate filters for the $X$ function, as shown here, or an $X$ filter that is a two part amber-blue filter to more adequately represent the double-humped $\bar{x}$ function of the Standard Observer.

upon the kind of sample being measured. Some of the geometries that can be used with the Hunterlab D25 instrument are shown in Figure 14.6, with a description of the type of sample for which each geometry was designed.

Tristimulus instruments have more flexibility than spectrophotometers in the specimen areas that can be viewed. The spectrophotometer is usually limited by dimension and cost considerations to a viewing area of 1 in. or smaller. Tristimulus instruments with unidirectional incidence and view can be readily equipped with larger specimen-viewing areas. In Hunterlab instruments, for example, 2-in. specimen ports are standard in D25 two-incident-beam (Model A) and annular ring (Model M) configurations. In the two-incident-beam instrument, small spot lenses can be used to reduce the specimen area to about 0.5 in. in diameter. In the annular ring configuration there is available a second optical unit (Model L) with a 4-in. diameter specimen area.

*Light Sources.* A normal tristimulus instrument employs a tungsten incandescent lamp with lenses and mirrors to project the light onto the specimen. A filter to absorb infrared heat is usually placed in the incident beams of light.

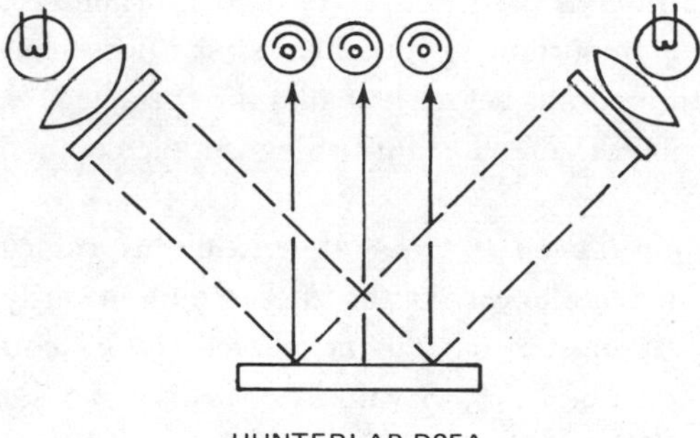

HUNTERLAB D25A

45° 2-INCIDENT-BEAM / 0° VIEWING, WITH 2" SPECIMEN VIEWING AREA (1/2" AVAILABLE WITH INSERTED LENS).

FOR FLAT SURFACES OF PAINT, PAPER, PLASTIC, CERAMICS, PRESSED POWDER, OPAQUE LIQUID IN CELLS.

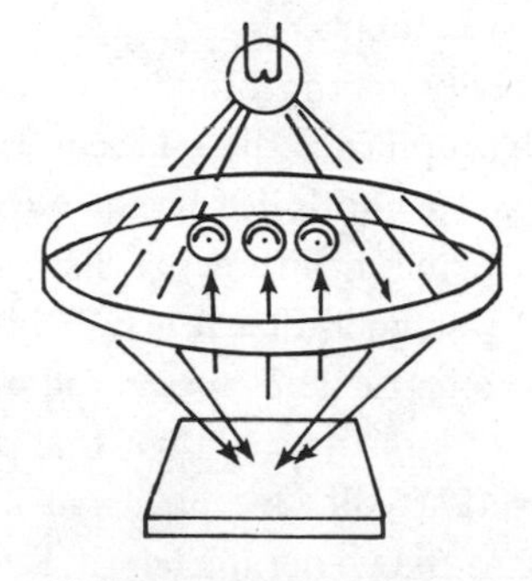

HUNTERLAB D25M, D25L

45° CIRCUMFERENTIAL INCIDENCE/0° VIEWING, WITH 2" (D25M) OR 4" (D25L) SPECIMEN VIEWING AREA.

FOR SWATCHES OF FABRICS, WADS OF FIBER, CARDS OF WOUND YARN. PEAS, BEANS, BERRIES AND KERNELS.

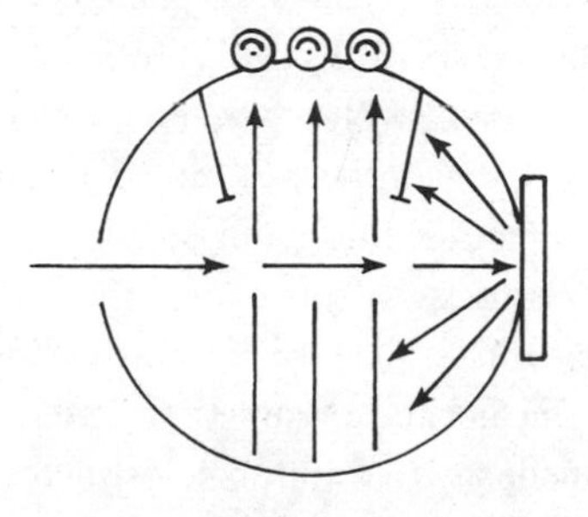

HUNTERLAB D25P SPHERE

0° INCIDENCE/DIFFUSE VIEWING, WITH 1 1/8" SPECIMEN VIEWING AREA.

FOR PLASTIC FILMS AND SHEETS, CLEAR GLASS, INSTANT TEA AND OTHER BEVERAGES BY TRANSMISSION, METALLIC SURFACES FOR COLOR BY SPECULAR REFLECTION.

**Figure 14.6.** Geometries of various optical sensors of the Hunterlab D25 Color Difference Meter.

Sometimes a bluish filter is used to convert yellow incandescent light to bluish daylight. Where it is important to correctly assess fluorescence, the ultraviolet light must not be filtered out before it strikes the specimen. Some colorimeters possess multiple sources, one for furnishing supplemental blue and/or ultraviolet light.

*Light Receivers—the Photodetectors.* In tristimulus colorimeters the light from the sample is directed to each of the filters simultaneously to provide three tristimulus signals. It must reach all the photodetectors equally to result in balanced measurements of $X$, $Y$, $Z$ values. For this purpose good tristimulus colorimeters employ light distributors as shown in Figure 12.11 to mix thoroughly the light directed to the detectors. In reflectometers, however, the general practice is to employ sequential measurement. This means that the filters are turned one after the other into the light beam, thus providing the reflectance readings in sequence rather than simultaneously. Simultaneous measurement has advantages in speed and precision over the sequential arrangement in the reflectometers. However, it is now feasible, with electronic digital storage methods, to make fast sequential direct readouts.

Colorimeters differ substantially in their conformity to the Standard Observer functions. No commercial colorimeter exactly duplicates the CIE curves (generally the curves are intended to simulate $\bar{x}$, $\bar{y}$, and $\bar{z}$ multiplied by the CIE Illuminant C values), but steady progress has been made in improving instrument accuracy. In some cases the filters are spectrally so good that it is difficult to measure accurately by conventional spectrophotometric and spectroradiometric means the errors that remain. Hunter and Gibson have shown that a good modern tristimulus colorimeter kept in calibration will give precision in color measurements of the same order of magnitude as the General Electric Recording Spectrophotometer (Hunter and Gibson, 1969).

For the best accuracy in colorimeters it is necessary that the temperature of the photodetectors be stabilized to prevent changes in spectral response characteristics with changes in ambient temperature.

*Instrument Read-Out.* Ten years ago instrument readings were made largely by the process of balancing calibrated bridges and reading position of the slide wire at which balance occurred. Now high-speed pulse counting digital readouts give the tristimulus color values directly.

*Uses of Tristimulus Instruments.* Although color measurements are frequently used for identification, sorting, and recording of color values, all the major uses of these instruments involve measurements of fairly small color differences. The most frequent uses of these small color-difference measurements are to establish closeness to standard and to give guidance for the adjustment of color mismatches. Color differences are also used in the study of product deterioration resulting from exposure and use.

Colorimeters should be used in temperature-stable clean surroundings. The instrument standards must be kept clean and should be used regularly to keep the instrument in calibration. Specimen preparation and presentation to the instrument is extremely important. To get repeatable valid measurements, test instructions and proper procedures should be followed. (See Chapters 15 and 16.)

## SUMMARY

Spectrophotometers and tristimulus instruments are used to measure color appearance. Spectrophotometers give wavelength-by-wavelength analyses of the light-reflecting or -transmitting properties of objects. Tristimulus instruments use filters that approximate spectrally the Standard Observer functions of the eye, and measure color in *X, Y, Z* terms, or in *L, a, b* values. Spectrophotometers are essential where color formulation is involved, and metamerism must be identified. However, tristimulus colorimeters and reflectometers provide generally more precise and less expensive means for the routine intercomparisons of similar colors and the adjustment of small color differences.

Color measuring instruments are less flexible than the eye, but, if conditions and scales of measurement are properly chosen, they quickly reduce color appearances to numbers that correlate well with visual evaluations. Instruments provide evaluations that are more repeatable than those of the unaided eye. Spectrophotometers provide ingredient and chemical information that is not available to (or from) the eye.

# CHAPTER FIFTEEN

# STANDARDS, STANDARDIZATION, AND MEASUREMENT TECHNIQUES

Standards provide the permanent bases against which appearance attributes are measured. Regardless of whether the instrument used is single beam (sample substituted for standard) or double beam (sample compared directly with standard), it is obvious that the accuracy and repeatability of measurement depend directly upon the accuracy and repeatability of the standards being used, and on the condition and cleanness of these standards.

## ACCURACY AND PRECISION

The term "accuracy" implies the existence of some standard values that are assigned the property of being "correct." The capability of an instrument to reproduce these standard values is a measure of its "accuracy."

When using appearance measurement instruments, one is more often concerned with precision than with absolute accuracy. This is because the instruments are used primarily as comparators rather than absolute devices. Normally one is interested in making measurements of small differences between similar specimens rather than in finding absolute values. Exact spectral and geometric conformity to specified conditions and photometric linearity are necessary for absolute accuracy. Because of the overall complexity of product appearance and the many aspects of its measurement, accuracy is generally not only difficult to achieve but also is frequently not as relevant to the problem at hand as is high precision.

"Precision" is the capability of an instrument to give the same results repeatedly. Precision is divided into repeatability and reproducibility.

1. Repeatability represents the degree to which a single instrument gives the same reading on the same specimen at different times. Even in highly accurate

instruments, low sensitivity can result in a wide spread of readings. Conversely, a highly repeatable instrument may be giving inaccurate data due to its failure to conform to the specified geometric, spectral, or photometric conditions of measurement. Repeatability is subdivided into short term and long term. Both short- and long-term repeatability are determinable only with stable specimens.

2. Reproducibility is an assessment of the ability of one laboratory to duplicate measurements made in another laboratory on the same type of instrument. It also is an assessment of the instrument manufacturer's ability to produce instruments that are alike. It represents the degree of agreement to be expected between two laboratories using the same type of instrument on the same samples.

## IDEAL OR PERFECT STANDARDS REFERENCES

There are two ideal standards referred to in reflection measurements: the perfect white for diffuse measurements, and the perfect mirror for specular measurements. There are corresponding standards of reference for transmission measurements:

1. *The perfect (ideal) white* standard, if it existed, would be a perfectly diffusing, totally reflecting surface. The term "reflectance factor" is used for the ratio of the light reflected by a sample being measured to that reflected by the perfect reflecting diffuser identically illuminated (CIE International Lighting Vocabulary, 1970). The scale associated with such a measurement is called a diffuse reflectance factor scale (formerly called "directional reflectance"). Comparison is with the ideal diffusing white, but only within the specific directions being measured. This is the reference normally used for color measuring instruments. Surfaces of pure barium sulphate powder approach this ideal.

2. *The perfect mirror*. The measurement scale having the imaginary perfect mirror as its reference is called a "specular" or "fractional" scale, since the concern is with the fraction of the incident light that is specularly reflected. This reference and scale are normally used when measuring specular reflectance factors and relative values of specular gloss. Surfaces of aluminum evaporated onto glass approach this ideal.

3. *Diffuse transmission white*. An imaginary, perfectly diffuse, totally transmitting white object is the reference for diffuse transmittance factor. There is no real-life ideal transmitting white, but it can be approached by goniophotometer comparisons with the ideal diffusely reflecting white surface as reference.

4. *Air* is the ideal reference for specular transmittance.

Table A.1 gives some specific values for certain of these primary reference standards.

## PRIMARY STANDARDS

A primary standard is one whose calibration is determined by measurement of parameters usually different from the parameter for which it will be used as a standard (ASTM Method E-284). For example, a primary standard for gloss measurement of nonmetals is a surface of polished black glass, calibrated by measuring its refractive index and then determining its specular gloss by use of the Fresnel equation (ASTM Method D2457). The black glass, with its specular gloss value, is then used as a primary standard for specular gloss measurements.

Similarly, a primary standard for measurement of color by diffuse reflection can be calibrated by measuring a pressed powder tablet of MgO or $BaSO_4$ and arriving at its absolute spectral reflectance values by the use of an auxiliary sphere (ASTM Method E-306). The white tablet, with the values assigned from this measurement, then becomes a primary standard for measurements of diffuse color by tristimulus instruments. The Taylor (1920) and subsequent Baumgartner (1937) Reflectometers (Figure 13.5) are also designed to measure absolute reflectance.

These primary calibrations are generally the concern of technical leaders in the field and of governmental agencies set up for this purpose. Some of these agencies are the National Bureau of Standards in Washington, D.C., for the United States; the National Physical Laboratory at Teddington, Middlesex, for England; the National Research Council in Ottawa for Canada; and the Bundesanstalt für Materialprüfung in Berlin, Germany.

Because primary standards are calibrated by elaborate techniques, they are used with great care and only for the calibration of secondary standards.

## SECONDARY STANDARDS

All standards other than primary standards may be classed as secondary standards and can be grouped into the following categories:

1. Master standards, used mainly as master references for calibration of other secondary standards. These other secondary standards are used in the daily operations of measurement laboratories. Master standards are calibrated by

reference to primary standards. When not being used to calibrate secondary standards, master standards are stored with care.

2. Working standards, calibrated against master standards and used as references for the calibration of laboratory instruments.

3. Instrument standards, used with a particular instrument only, for use in maintaining the calibration of the instrument.

4. Hitching-post standards, which, since they are similar in values to the specimens to be measured, can be used to minimize errors caused by scale-interval defects in the instrument.

5. Specific calibration standards, which are permanent specimens or standards used as calibrating standards for measurements of an impermanent product. A specimen of the impermanent product is first measured on a reference instrument. It then becomes a transfer standard and is used to calibrate a second instrument. On this second instrument, a permanent specimen is then read and assigned values from its comparison with the impermanent "transfer standard." This permanent specimen becomes the specific calibration standard and is used thereafter to calibrate the second instrument for measurements of the impermanent product. With this technique, numbers obtained on the second instrument will be comparable to those that the reference instrument would give.

6. Diagnostic standards.

A special category of standards, called diagnostic standards, is designed for analyzing and identifying the nature of instrument inaccuracies. Diagnostic standards are normally used in groups of two or more together. In each group there are differences in one optical aspect, but similarities otherwise. These differences can be put to diagnostic use in two ways. First, they can identify correct instrumental response to that optical aspect. For example, photometric linearity of an instrument may be checked by a set of tiles that are graduated in reflectance factor. Second, insensitivity to an optical difference may identify a source of some instrument inaccuracies. For example, one examines the readings of a tristimulus instrument from a pair of colorimetrically similar but spectrally different colored tiles, called color metamers. Any lack of correct tristimulus responses to different wavelengths is likely to result in erroneous readings. Similarly, a pair of tiles chosen so that they distribute light in two different ways can help determine whether the instrument is properly sensitive to differences in geometric distribution of light. A pair composed of an opaque and a translucent standard will indicate how the instrument responds to specimens that are not opaque. Table A.12 gives a listing and description of some of these diagnostic standards.

## STANDARDS MATERIALS

Materials for working standards for regular use are chosen for their permanence and resistance to change with use and cleaning. The majority of working and instrument standards in everyday use are of ceramic materials.

The most difficult standards to realize in practice are those for intermediate-scale gloss. Depolished black glass, partially delustered porcelain enamel, etched glass, ceramic tile, and oxidation-resistant metal have all been used. None completely meet the requirements of permanence and cleanability and at the same time have the required medium uniform gloss. Many gloss standards in use today vary in gloss from area to area on their surfaces. For this reason, gloss standards should always be centered in the same way for measurement. In addition, it is difficult in many laboratories to clean these gloss standards properly because of residues from the ordinary cleaning materials used. Gloss standards must be cleaned with special nonresidue materials and with special care.

Permanent standards for reflection color are made of porcelain enamel, opaque glass, and glazed wall tile. Glass filters are used for transmission color standards. While some of the ceramic colorants change with age and exposure, and some of the ceramic surfaces are attacked by chemical pollutants in the air, there are few alternative materials of comparable uniformity, durability, and color.

Many times specimens of the materials themselves are used as color standards. Also the painted, printed, or plastic color chips that are used for visual evaluation are often used as standards with instruments. With these standards it is difficult to prevent degradation in storage and almost impossible to prevent damage in use. Freezer storage of such paint chips and paper standards is quite widely used in order to preserve such product standards over periods of time (Huey, 1965).

It is desirable with color standards to use materials that clearly separate reflected light into diffuse and specular components. One is apt to forget, when looking at the numbers on the back of a color standard, that those values may only be accurate for the geometry under which the standard was calibrated. Thus a standard calibrated on a sphere spectrophotometer may not carry the right numbers for use with a 45°/0° colorimeter.

## CARE OF STANDARDS

All standards must be handled carefully, and in a manner that will cause no change in their optical properties. The most common problems arise from the presence of dirt, moisture, grease, and oil affecting both gloss and color. Damage is also caused by minute scratches that can also affect both gloss and color. Other common causes of damage to standards are placing them on dirty, damp, or oily surfaces, and using them as back-up plates when measuring specimens. Dyes and chemicals in papers and textiles can materially affect the measurement values of a tile used as a back-up plate for specimens.

Standards should be stored in closed containers so that they remain clean and in the dark, since light may fade some colored pigments. As Francis Scofield cautioned some 30 years ago, "Standards should be kept carefully in velvet-lined boxes, like jewels." The recommended procedure for cleaning ceramic standards is to wash them with a mild anionic detergent solution, using a nylon brush. Follow this with a running hot-water rinse and blot dry with a clean, lint-free paper towel to remove water marks. With gloss standards the surface should be cleaned with a mild anionic detergent and a nylon brush, followed by a forced rinse in hot water (temperature near 150°). Immediately *blot* the tile dry with a clean paper towel. The tile must never be rubbed with the hand or a paper towel, either during washing or drying. Paint, plastic, and color chip standards are never cleaned and are discarded after short periods of use.

## MEASUREMENT TECHNIQUES

Techniques of measurement importantly affect the accuracy and precision of instrumental measurements. Best results are obtained from instruments that are thoroughly warmed up and stabilized. Rapid work on instruments generally results in higher precision than does slow work. It is always advisable to take more than one reading of a sample because of the possibility of the operator making an error in his tabulations.

With many of the products regularly measured for color and other appearance properties, the variations encountered within one sample are the major cause for loss of precision. To avoid this, duplicate specimens from the same sample should be measured. Normal statistical procedure in the analysis of all data should be used.

## SUMMARY

Tables A.13 and A.14 give in tabular form the important facts about standards.

# CHAPTER SIXTEEN

# SPECIMEN SELECTION, PREPARATION, AND PRESENTATION

The specimen to be measured, as it fills the instrument viewing aperture, is equally as critical as any of the other components of the appearance-measurement operation. For the specimen does indeed become a component, completing the chain from light source to object viewed to detector to measurement unit. As noted in the previous chapter, the variability of specimens is very often the limiting factor in the precision of measurement. Meaningful and repeatable appearance measurements are not possible if there is faulty specimen selection, preparation, or presentation.

Ideally all of the instruments in general use (excepting special product devices such as the Hunterlab D45 Citrus Colorimeter) are designed to accomodate specimens that are flat and remit light for measurement from only a single plane of measurement. Many of the actual specimens tested are not flat. Others possess optical properties that cause them to remit light toward instrument receivers from other than a single plane.

In practice, those who use measurements of reflection color frequently intercompare results from instruments with different geometries. In addition, they intercompare results from single instruments where care is not taken to orient specimen surface directionalities in any standard manner. Such intercomparisons of results are reliable only when the specimens are opaque, uniformly diffusing, and uniform in color from area to area. All but a few of the specimens measured in everyday practice depart in some degree from the foregoing ideal conditions of flatness, uniformity, opaqueness, diffusion, and specularity in ways that can significantly affect instrument measurements of them. Therefore the ways in which specimens are obtained, prepared, and presented for measurement importantly affect the numbers obtained.

## OBJECTIVES IN SPECIMEN PREPARATION

The objective of instrument measurement is to obtain repeatable numbers that correspond to visual estimates of the appearance attributes of the materials involved. To achieve this objective, one must obtain specimens that show, for the conditions of measurement, the attributes of concern. Whenever possible these differences should be enhanced, as is the case in ASTM Method E450 for the measurement of the yellowness of nearly colorless, clear industrial-chemical liquids, where sample thicknesses of 10 cm (rather than the more customary 1 or 5 cm) are recommended. Specimens must also be selected and prepared in a standard way so that they are repeatable. They must be presented to the instrument in a consistent and repeatable manner.

## EFFECTS OF PREPARATION ON SPECIMEN PROPERTIES

The appearance-measurement technologist must be aware not only of how his specimens depart from the ideal form for measurement, but also what effect preparation procedures may have on the optical properties of his specimen. A change in an optical property can bring about a change in measured appearance attributes. For instance, when air-dry semigloss paint panels are prepared for testing, differences in the humidity and temperature of the air in which the panels are dried may cause differences in gloss by as much as 20 units. The highest gloss will result from drying in warm, dry air; the lowest, from cool, humid air.

Consider how a change in the specimen surface can affect appearance attributes. Gloss is a direct result of surface smoothness and the amount of Fresnel reflection from the surface. A change in surface smoothness will not only affect gloss but will indirectly affect diffuse reflectance and diffuse color also, since the amount of light entering the light-scattering structure beneath the surface will be changed. As we saw in Chapter 3, when the surface is rough the Fresnel reflection is scattered in all directions and contributes colorless reflected light to both the observed diffuse reflection and the measured diffuse reflectance factor. Thus a matte surface contributes to an increase in lightness and a decrease in saturation when measured instrumentally.

Some of the preparation factors that can affect surface reflection are:

1. Temperature, humidity, and chemical conditions at the time of surface formation
2. Pressure used, tension applied, knife-blade technique employed, surface polishing, and cleaning techniques used, or other mechanical factors in specimen preparation
3. Presence of liquid, vapor, dust, wax, or other film on the surface
4. Interaction of surface with atmosphere and other elements of the environment after formation, but before measurement

The light-scattering properties of the diffusing structure beneath the surface of the object determine diffuse reflectance and color and are significantly affected by a number of factors in preparation. For example, improved dispersion of light-scattering components within a diffusing structure increases the effectiveness of those components. That is, improved dispersion of white powders increases diffuse reflectance, but improved dispersion of colored pigments tends to decrease diffuse reflectance and increase saturation. Grinding, mixing, milling, and blending are some of the techniques used to achieve dispersion within light-scattering mixtures.

Compaction within a specimen of granules or kernels of food, fibers in a pile fabric, fibers in a wad, yarn in a skein, or loose powder will usually increase light scattering from the specimen. This is true so long as the compaction does not cause the individual granules, kernels, fibers, or powder particles to effect optical contact with each other. Optical contacts between components of the structure will eliminate reflecting interfaces and thus decrease light scattering. The presence of water, vapor, or other liquids will also increase the tendency for optical contact between light-scattering elements in a structure and thus decrease light scattering. Consider the difference between the appearance of wet and dry sugar as an example. Any increase of depth of penetration of light entering the body of the specimen tends to increase absorption and therefore darken it and usually increase its chromaticity.

Enlargement of granules, kernels, or clumps of material within a specimen is the opposite of compaction. Increased cavities between lumps will cause entering light to be trapped and thus decrease light scattering from the diffusing body.

## REQUIREMENTS FOR SPECIMEN PREPARATION DEPENDING UPON TYPES OF MEASUREMENT TO BE MADE

Before discussing specific specimens and some of the techniques for handling them, let us consider the conditions for specimen preparation that are dependent upon the type of measurement to be made. For example, when measuring the color of a specimen that might pillow into the viewing aperture, it is permissible to hold the surface flat by using a cover glass taped over the aperture window. Other materials being measured for color may be chopped up and placed in a glass specimen cell, or made into a paste and applied to a glass plate. None of these techniques can be used when gloss, rather than color, is the attribute being measured. Here it is the first surface of the material that is important, and nothing should be done to alter this surface such as polishing, chopping, or coating. Sheets and films may be flattened by tension or by a vacuum, if necessary, for gloss measurements, but not by the use of a cover glass.

When measuring attributes that depend upon the transmission of light, such as the color or the cloudiness of iced tea, the liquid may be placed in special optical cells that have flat, parallel surfaces. The thickness of the specimen presented should be chosen to maximize the haze or color differences.

## TECHNIQUES FOR HANDLING SPECIMENS OF DIFFERENT FORMS*

There are a variety of techniques that can be used in handling the various forms of objects and materials so that the most valid and repeatable measurements of their appearance will result. The remainder of the chapter is devoted to instructions for handling materials of seven different forms.

1. *Solid opaque objects such as ceramics, plastics, metals, and some foods.*

   a. Select areas suitably flat and uniform, or form flat area by machining or scraping. Because surface characteristics affect color, machining must be carried out in a repeatable manner.

   b. When making measurement of color by reflection, use small light spot if:

   (1) Specimen is too small to cover viewing window.

   (2) Specimen is nonflat but has a substantially flat, small area that can be measured.

   (3) Specimen is highly translucent and uniform. (Use small spot but large viewing window.)

   c. Orient in same manner for each measurement. For measurement of color by reflection where orientation is undefined, use circumferential lighting or measure twice, turning specimen 90° between readings, then average results.

* Drawings in this section by Joe Plott.

2. *Film forming materials such as paints, lacquers, waxes, printing inks.*

   a. Apply films of desired thickness to flat panels in a standard manner, as with a knife blade of prescribed opening pulled at a prescribed rate of speed.

   b. Measure in one direction, turn 90° and measure again, average results.

3. *Films, sheets, fabrics such as paper, plastic, cardboard, painted metal, and textile fabrics.*

   a. Expose flat and carefully oriented.

   b. Where there is insufficient specimen rigidity, tape cover glass over viewing window to prevent pillowing of paper tissue or fabric into the instrument.

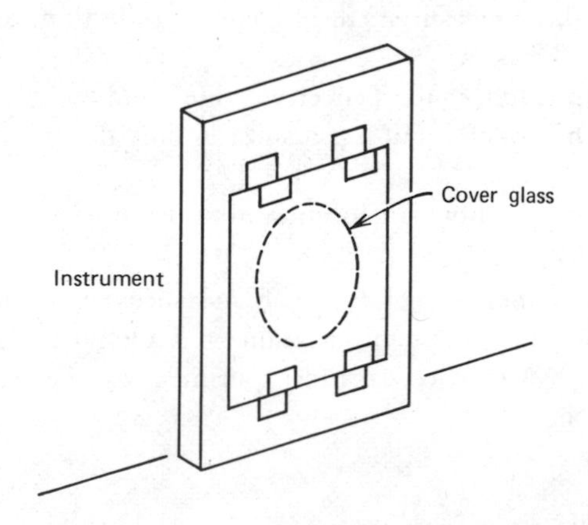

c. Where there is incomplete opacity, use several thicknesses of specimen or back specimen with an opaque object of nearly the same color. (For color measurements only.)

d. Where rotational or directional selectivity is suspected, measure in one direction, turn specimens 90° and average results, or for color, use 360° illumination.

e. Types of specimen holders:

(1) For color measurement of textile fabrics and plastic films, a telescoping ring or embroidery hoop.

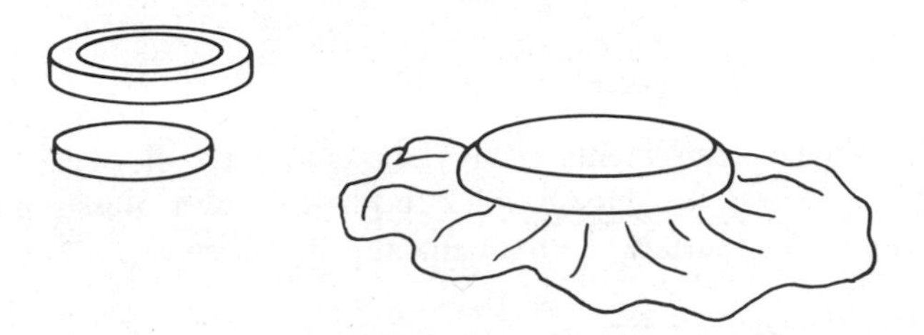

(2) For measurements of opaque plastic films and papers, a black metal plate with strip of two-sided pressure sensitive tape on underside.

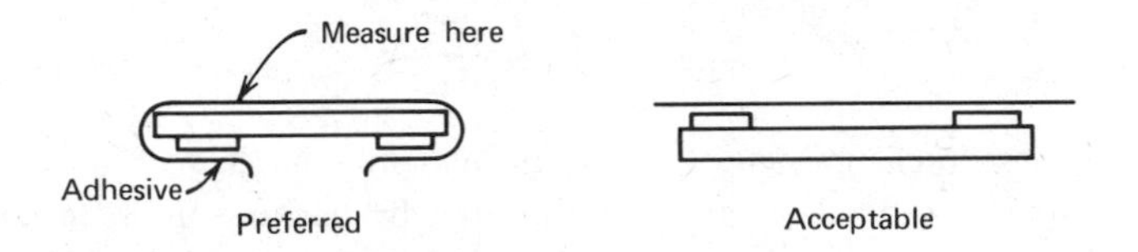


or a metal plate with vacuum ring groove in center.

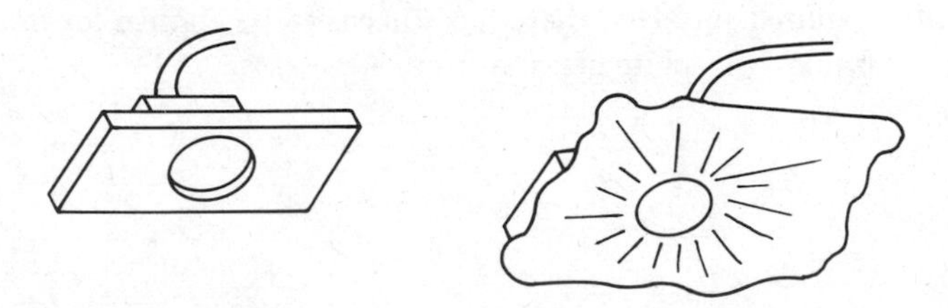

(3) For measurements of clear plastic films, where the specimen must be held flat while transmitted light is measured for haze, yellowness, or transmission factor, telescoping or supporting rings, depending upon whether the film is stiff or flexible.

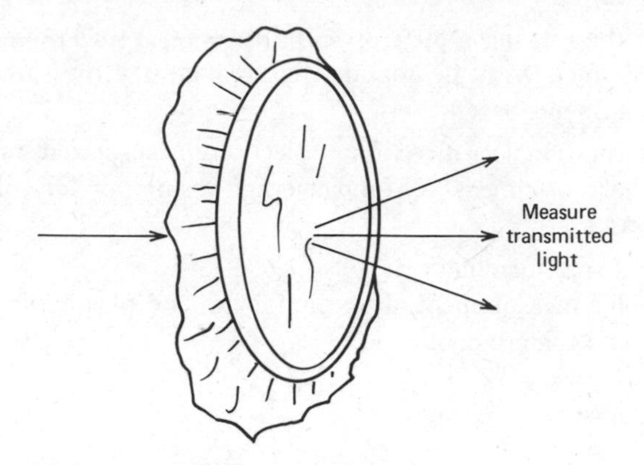


(4) For gloss measurements of vinyl films that are distorted by vacuum clamps, but have a tendency to stick to glass, a cover glass with the film applied to its *front* surface, with all entrapped air removed.

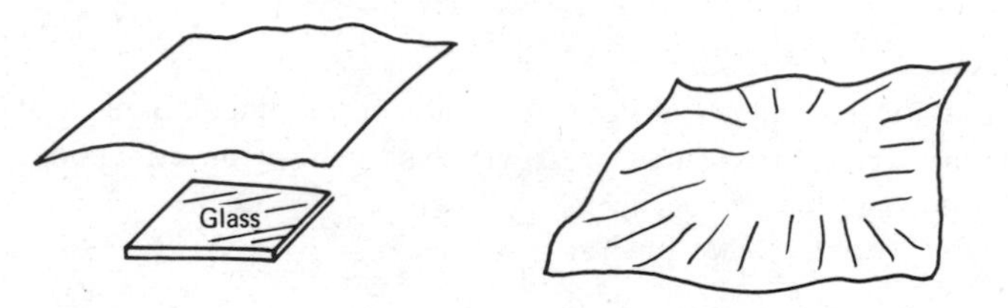


Where gloss measurement of top surface only of clear films is desired, second-surface reflection can be eliminated by backing the film with black glass that has been brought into optical contact with the second surface by means of wetting.

(5) For color measurements of sheet metal specimens, rigid plastic sheets, or nonflat painted surfaces where force needs to be applied for flattening, a clamping frame, plus plate glass in front, if necessary.

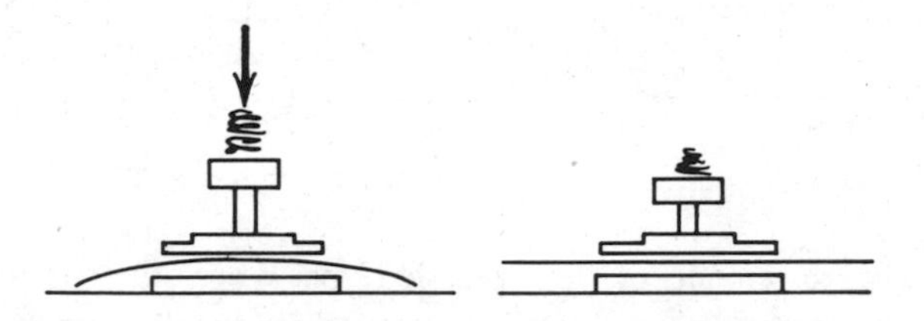

4. *Powders, granules, flakes, pigments, pellets.*

a. Sieve granules or pellets so that size will not vary from one measurement to the next. Shred flakes if too large to handle, (i.e., potato chips).

b. Pour into container with transparent window in bottom. Compaction and depth of material must be uniform. Measure through bottom window.

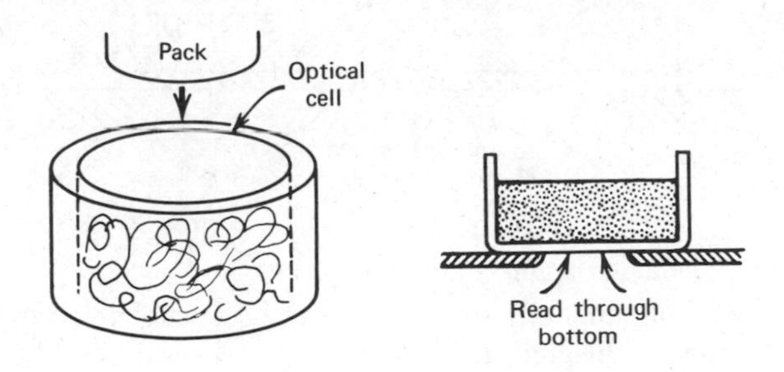


c. Measure from top using one of the following methods:

(1) Pour powder into container and scrape top flat with a straight edge.

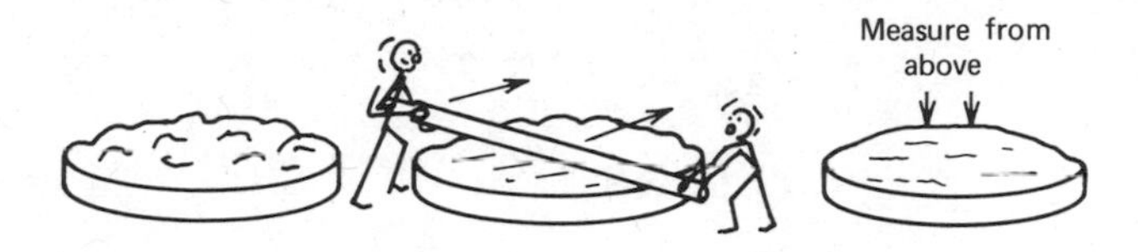


(2) Press powder into a cake.

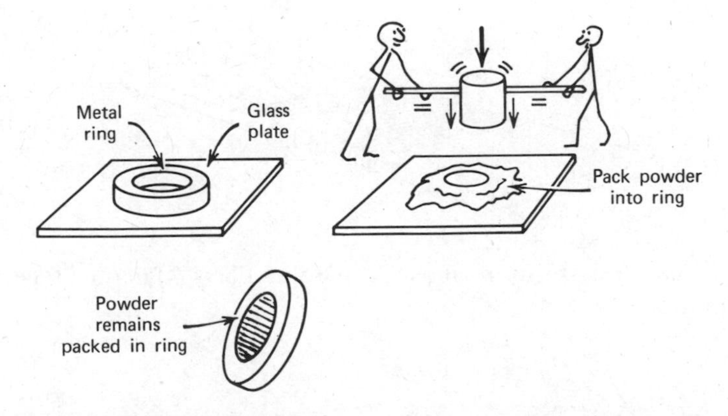


d. With bulk or lump products in which individual pieces are roughly ½ to 1 in. in diameter, pour into large container with transparent window in bottom. Arrange so that packing is nearly solid. Measure through bottom using a large-area viewing instrument.

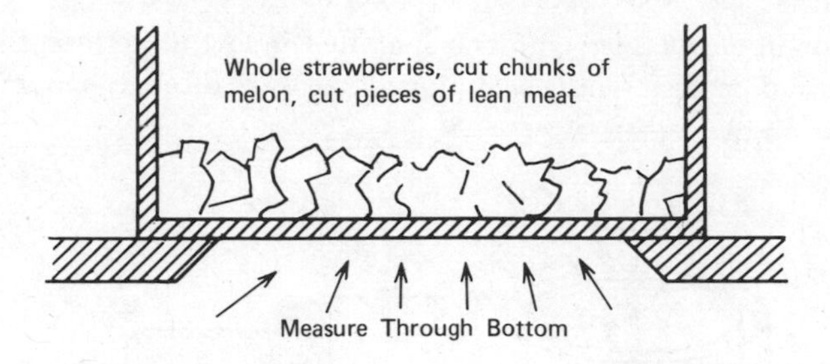


5. *Bulk cotton, wool, yarn, and loose fibers.*

a. Make wad of randomly oriented fibers large enough so that almost no light can penetrate through it. Place it against cover glass. Place a specified weight on top and measure through the glass.

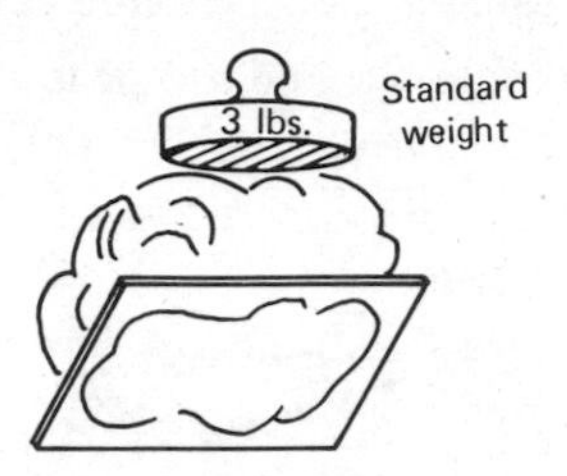


b. Wind yarn closely and parallel on cards to sufficient thickness to prevent show-through.

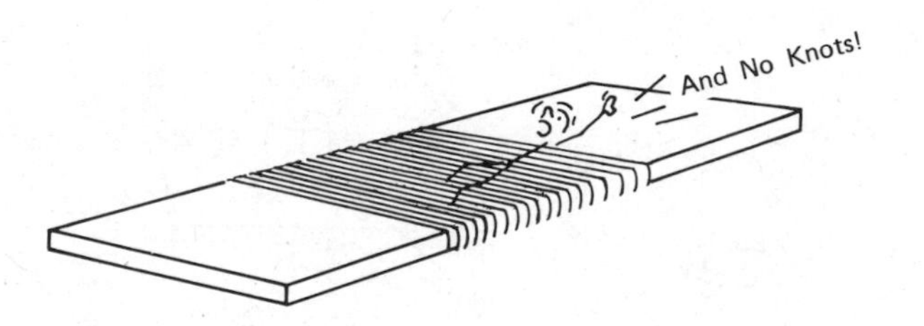


c. Comb fibers to form ribbon of parallel fibers, clamp or tape on flat support.

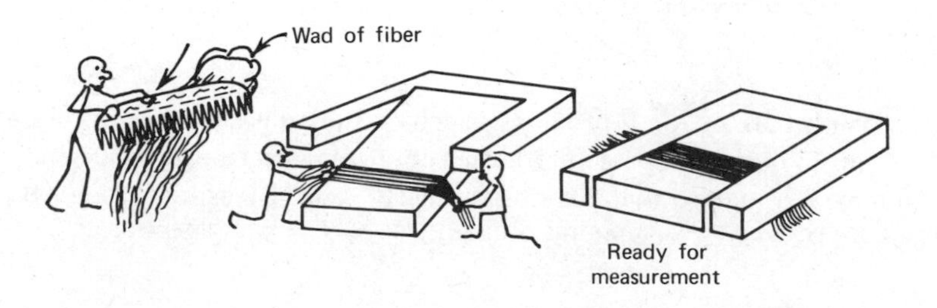

d. Use cover glass over viewing window if necessary to prevent pillowing into viewing area.

e. Use standard pressure to hold specimen against viewing window.

f. Measure in one direction, turn specimens 90° and measure again, average results, or use 360° illumination with beam splitter.

For contrast luster measurements (ratio of specular to diffuse reflection), yarns and fibers must be prepared as in b or c above.

6. *Translucent, turbid, and opaque liquids and pastes such as liquid paints, slurries, and many foods.*

a. Measure through bottom.

(1) Pour into plastic or glass container with clear bottom. Fill container to same depth for each measurement.

(2) Mount optical head of the instrument with viewing window facing upward.

(3) Place filled container on viewing window and measure through the bottom. If the specimen is translucent, the contribution of translucency to measured color will be minimal if an opaque backing of nearly the same color as the specimen is placed immediately behind it.

(4) Avoid air bubbles or "settling" by careful stirring.

(5) If lateral loss of light is a problem when measuring translucent specimens, the relative sizes of specimen area exposed and illuminated should be changed so that the area exposed is larger than that illuminated.

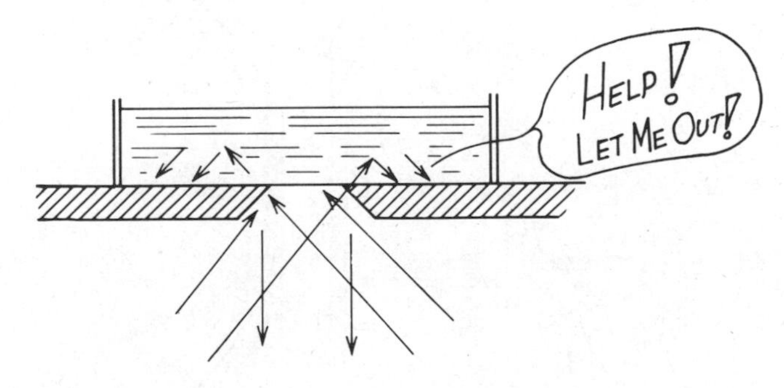


Loss of light when measuring translucent specimen due to small size of window.

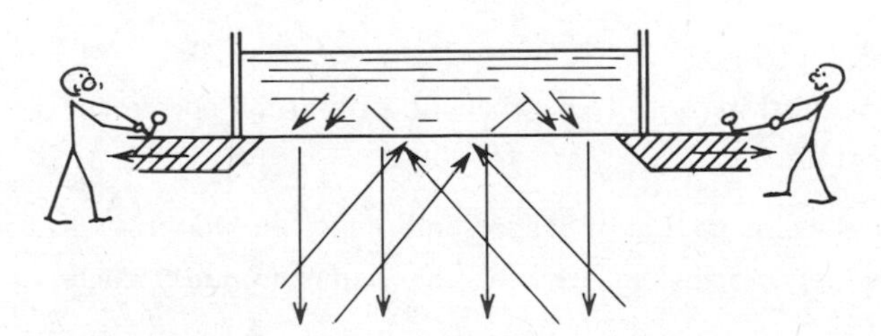

Window size larger than the size of the illuminated area prevents loss of light.

b. Measure top surface.

(1) Pour into shallow white container of standard depth and level the surface. Be sure to remove entrapped air.

(2) Measure top surface.

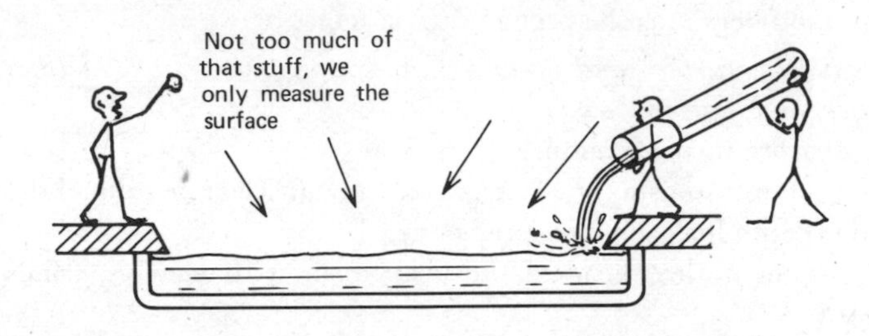


7. *Transparent liquids, many beverages, oils, varnishes, chemical and dye solutions.*

a. Pour liquids into container with flat and parallel optically clear sides of known spacing.

b. Make transmission measurements through cell.

## SUMMARY

The important things to remember in preparing specimens for instrument measurement are:

1. Choose specimens that are representative of the materials as used.
2. Prepare specimens in exactly the same manner each time they are

measured. When standard methods already exist (such as the ASTM and TAPPI methods), these should be followed.

3. Present the specimen to the instrument in a standard, repeatable manner.

4. Results measured will depend on the form of the specimen and its presentation.

If a method is established so that the same procedure is used each time specific specimens or types of specimens are measured, this will form a valid basis for comparison of measured results. It will also insure repeatability of results when measuring the same specimen.

# CHAPTER SEVENTEEN

# APPLICATIONS OF APPEARANCE MEASUREMENTS

The objective of making appearance measurements is to obtain numbers that are representative of the way objects and materials look. Although making total appearance analyses is generally not possible or practical, measurements of specific attributes of appearance can be exceedingly useful and economically important. Concerns for product appearance cross industry lines and touch almost every manufacturing industry where the consumer will ultimately judge the way the product looks.

## IMPORTANCE OF PRODUCT APPEARANCE

That the appearance of consumer products is becoming more critical every day cannot be overemphasized. The consumer wants and demands whiter linens. In the absence of other clues he may well select from the supermarket shelf the strawberry jam that appears freshest from visual examination. The quality of a product is certainly a factor in consumer selection of merchandise. Quality is associated with the appearance of products. The whitest objects will, in most cases, be preferred to near-whites that are yellowish or grayish. Bright colors will be preferred to colors that are muddy or dirty in appearance.

There seems to be another important factor in consumer judgments which Ralph Evans has termed "apparent achievement of designer's intent." This involves an assumption by the would-be purchaser that he understands the intent of the designer of the article under consideration. For instance, if all parts of an article are very nearly the same color, it is assumed that the designer intended them to be identical in color. Any failure to achieve uniform color is cause for rejection or criticism of the article. If the two sides of a seam in a dress do not appear to match completely, this is cause for rejection. It is to prevent such criticisms that color-difference measurements are so widely made.

Not only the appearance of products, but also that of their containers affects a person's selection. Market research people have found that brighter packages are considered to contain fresher products, and that color uniformity on the outside of a package conveys the idea of uniform quality of product inside the package. As competitition increases, the importance of quantitative measurement of the appearance of products in a precise and objective way becomes a matter of paramount importance to producers and entrepreneurs.

Appearance evaluation at the present writing is more of an art than a science. On the one hand the product technologist has tended to rely on empirical or anecdotal methods to explain the way things look and to adjust the appearance of a product by visual means. For many years it has been common in the textile, paper, and paint industries to rely solely on the visual judgment of the chief dyer, the machine foreman, or the production supervisor in grading the appearances of their products. There is no doubt that a skilled colorist over a period of years gains a tremendous amount of knowledge by observing the appearance of materials and making judgments to adjust or control a process to produce the desired product appearance. It should be recognized, however, that these visual observers are subject to human frailties. Some of the variables are the ability of the observer to remember colors over a space of time, the observer's age and change in visual accuracy over the years, precision among a group of evaluations, and variations in color vision among those making evaluations. Also, what happens when the colorist with 40 years of experience is no longer available? For example, in the case of one textile processor with a plant in a new location and with no experienced colorist available, it was necessary for the process line to stop until the plant general manager made a visual judgment on color match.

Most important of all, visual perception is a qualitative judgment and does not yield results that can be expressed numerically. Quantitative measurements of appearance differences are required if appearance technology is to be pursued as a science rather than an art. Only the use of instruments in measuring appearance can partially remove the many uncontrolled variables found when people are used exclusively to evaluate object appearance.

## PRODUCTS AND INDUSTRIES WHERE APPEARANCE ANALYSES ARE USED

Appearance has an optical basis; it is the result of the light-object interactions that were described in Chapter 3. The optical behavior of light is much the same in many paints, plastics, papers, textiles, ceramic products, foods, and other materials. Therefore the appearance attributes of these different products are generally similar. Where these similarities cross industry lines, the same approaches to appearance identification, the same terminologies, and similar techniques of appearance measurement can be used.

The products most frequently analyzed by instruments for color or other appearance attributes have diffusely reflecting surfaces and are either white or colored. Paints, papers, textiles, and many ceramic and plastic products contain as their major optical ingredients light-scattering (white) pigments or fibers and, if colored, dyes or colored pigments. Instrumental evaluations are made of these products in the process of attempting to achieve the desired final color appearance. Also, gloss, clarity, turbidity, and other appearance factors are adjusted by the selection of pigments, vehicles, and surface-forming ingredients.

In the food industry appearance measurements are used largely to assess quality rather than to guide formulation. Pure food laws tightly restrict the adjustment of color and appearance by formulation in food products. Measurements of food color are used to assess color quality of food ingredients and to assure maintenance of this quality through subsequent processing and storage.

The appearance of its product is becoming increasingly important for the metals industry, as exposed metal surfaces are used more and more to decorate as well as to protect. The metals industry, however, is behind the paint, plastic, and paper industries in controlling appearance. One reason for this is that metals are optically different from nonmetals and present unique problems in the measurement of their appearance attributes.

Signaling materials, designed to project light of specified colors and intensities in certain directions, must necessarily be evaluated for appearance. Also, the cleaning and polishing industry is concerned with the effectiveness of soaps, polishes, and cleaners, as well as the equipment used with them.

In addition to measurement of appearance properties of the products themselves in these various industries, evaluation of the processes and equipment used to produce or modify them is often performed. Thus not only are soaps and detergents evaluated for efficiency of dirt removal from fabrics, but also the washing machines used to clean them and the process cycles employed with these machines are studied to determine the most effective and efficient combinations for cleaning.

Table 17.1 shows how measurements of appearance attributes are used in the life of an industrial product. As can be seen, these measurements are utilized at different stages of product existence from initial conception to end use. Appearance analyses are used in production to assess the quality of incoming raw materials and the performance of the production process. Most important, they are used to analyze the appearance quality of the finished product before it is released to the market.

**TABLE 17.1**

SIX STAGES IN THE LIFE OF A PRODUCT WHERE MEASUREMENTS OF APPEARANCE ATTRIBUTES ARE USED

| Stage | Examples |
|---|---|
| A. Development | |
| 1. Before product is available<br>Product development<br>Research in methods and materials | Carotene in sweet potato seedling stems measured for color to assess quality of new genetic stock |
| 2. Designing product to have proper color<br>Color formulation (select pigments & dyes by computer) | Textile swatch from customer measured by dye manufacturer with abridged spectrophotometer to obtain values for input to colorant mixture computer |
| B. Production | |
| 3. Quality control (raw materials)<br>Tests of ingredients to be used in production | Before use in paper manufacture, pulp and clay pigment tested for blue reflectance to assure required freedom from blue-light absorption |
| 4. Quality control (in process)<br>Control of product during manufacture<br>a. Laboratory<br>b. Continuous (on line) | Porcelain enamel panels representing current production measured by frit manufacturer for difference from standard in color to assure conformity to tolerances<br>Color of painted continuous steel strip measured after emerging from baking oven to assure proper paint film thicknesses and oven temperature |
| 5. Quality control (finished product)<br>Inspection before release to customer | Cellophane sheet measured for ASTM D1003 haze and 20° gloss in quality control lab before shipment to customers |
| C. In Use | |
| 6. Performance evaluation<br>Tests after product is manufactured (durability, resistance to various service hazards) | Paint panel under exposure on test fence is regularly measured for color and 60° gloss in order to evaluate changes associated with fading |

## USES OF MEASUREMENTS OF APPEARANCE ATTRIBUTES

There are specific needs for appearance measurements common to many different industries since the economic reasons for their use are shared by these industries. Some of these uses are:

1. To identify a material for research, communication with others, or for the record
2. To test a product for conformity to specifications or tolerances for permissible differences from standard
3. To provide a score for samples graded for quality, maturity, or progress in treatment
4. To provide a shade-sorting classification so that all samples of the same shade number can be used together in an assembled garment, automobile or other product
5. To formulate a product color so as to achieve desired final appearance
6. To assess a product performance in service as measured by degree of change of appearance due to actual or simulated service exposure
7. To measure nonappearance variables that bear a known relationship to a measurable appearance attribute

### Identification

The first function of appearance measurements is to identify. Identification is involved in almost every use of measurements. In communication concerning a product, or in records for research or development, it is only by measurements that it is possible to identify a product adequately.

A food technologist working with beans and needing a colored standard for his work would use a numerical identification such as

$$L = 57.9 \qquad a = -15.7 \qquad b = +14.0$$

in order to insure getting a standard close in color to the beans.

### Tests for Conformity to Specifications

In terms of volume the major usage of appearance measurements is in testing products to see if they conform to specification. This testing may be necessary at several points in the production process. As indicated in Table 17.1, the products tested may be

1. Incoming raw materials
2. Materials in process of being manufactured
3. Finished materials ready for delivery

*Incoming Raw Material Testing.* A magazine printing house purchases paper for the magazine from several paper mills and issues specifications to which the suppliers must conform. These specifications provide tolerances for basis weight, thickness, burst and tear strength, and ink absorbency, as well as tolerances for appearance characteristics. For the appearance-measurement technologist, however, the tolerances of interest for a certain weight paper could be the following:

| | | |
|---|---|---|
| Color | *L* | 90.0 minimum |
| (Hunterlab) | *a* | $0.0 \pm 0.2$ |
| | *b* | $3.3 \pm 0.2$ |
| Gloss | Wire | $65.0 \pm 6.0$ |
| 75° | Felt | $65.0 \pm 6.0$ |
| Opacity | | 96.0 – 1.0 |
| Brightness | | 77.0 – 1.0 |

*Materials in Process of Being Manufactured.* The introduction of the computer to the production process has expanded the opportunity for automatic in-process continuous control. As the speed and volume of production increases, so does the need for this control of in-process material. Without almost immediate correction for defects in appearance, large volumes of off-specification product can be produced, resulting in correspondingly large losses. As an example of successful control of in-process material, a system of closed loop process control of the color of paper, utilizing a Hunterlab D44 On-Line Color Monitor in conjunction with an IBM 1800 computer has been in successful operation at Consolidated Papers, Inc., Wisconsin Rapids, Wisconsin, since 1968. In this system the Hunterlab D44 continuously measures *a* and *b* differences from standard of the paper run. The measured differences are converted by the computer into changes in flow of blue and pink dyes, thus substantially reducing grade change time and loss of off-color paper (Wickstrom and Horner, 1970). A diagram of this system is shown in Figure 17.1.

*Finished Material Ready for Delivery.* As an example of a specification developed for outgoing shipments, consider the following development involving a green paint supplied by a paint manufacturer to a coil coater for application to aluminum siding. This actual experience of the Sherwin-Williams Paint Company also reveals some of the hazards involved in attempting to meet a customer's tolerances. The coil coater initially restricted the paint manufacturer to tolerances of plus or minus 0.5 unit from standard in both the *a* (red-green) and *b* (yellow-blue) dimensions, as depicted by the solid circle in Figure 17.2.

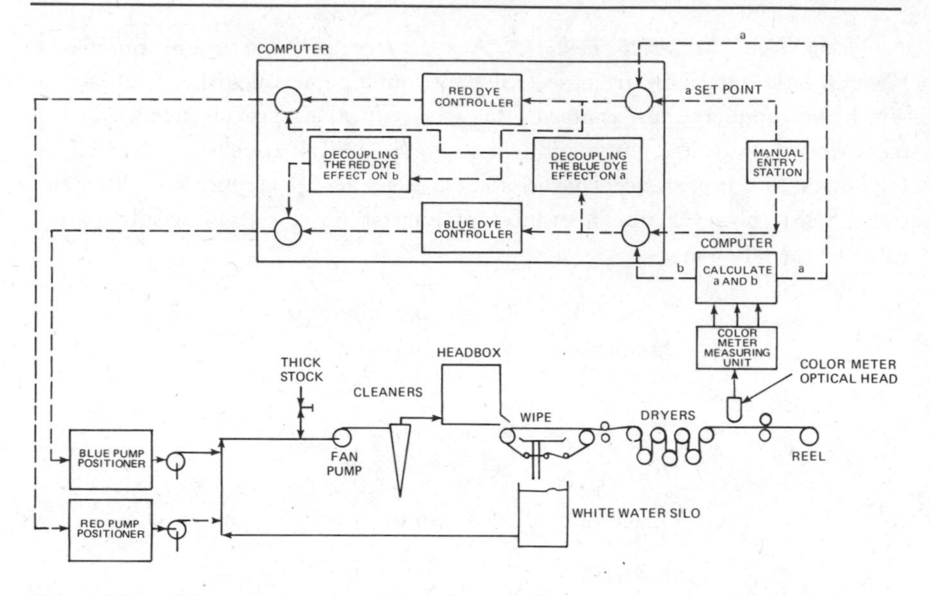


**Figure 17.1.** Schematic diagram of color control loop for printing papers. Color difference measurements made by the color meter optical head (Hunterlab D44 On-Line Monitor) are used to control the flow from the blue and red dye pumps to maintain correct color of paper. From Wickstrom and M. Horner, *Tappi J.*, **53,** 1970.

However, the first sample supplied to the customer with the full +0.5 range for the *a* value was rejected as being "too red." The customer therefore reduced the tolerance to +0.25 units on both the *a* and *b* dimensions. However, the sample supplied with this +*a* tolerance was also rejected as being too red, and the tolerance was cut to +0.15 units. The paint company's technologist described it thus, "This was the wrong approach because we found out much later that he would tolerate a minus reading of 0.5 *a* and a plus *b* reading of 0.5, but would not accept any variance from the standard in terms of +*a* and −*b*." (See dotted line in Figure 17.2.) "In other words, he wanted the green to be on the yellow side of the standard and more saturated. As a result of this, he kept tightening the entire tolerance down beyond the ability of the instrument to reproduce itself. We continually state that when plus-minus tolerances are used, the color involved must be considered so as to have numbers that agree with a visual observation [Sam Huey, Sherwin Williams Co., private correspondence, 1969]." This practical example illustrates vividly the wisdom of the rule that should be the foundation of any appearance-measurement application: "Never forget to use the eye so as to include visual observation of each specific example." The instrument is the tool to analyze a specific attribute; the eye is the final judge of overall appearance.

## Grading for Quality

In this area of application, measurements of product appearance are used to assign color grades, scores, or index numbers for purposes of product-quality identification. The food industry is the major area of use of color measurements in quality grading. Color is a primary indicator of quality of many food products. In the food industry, the processor does not have flexibility to adjust color by the modifications of ingredients. Rather he must achieve and preserve

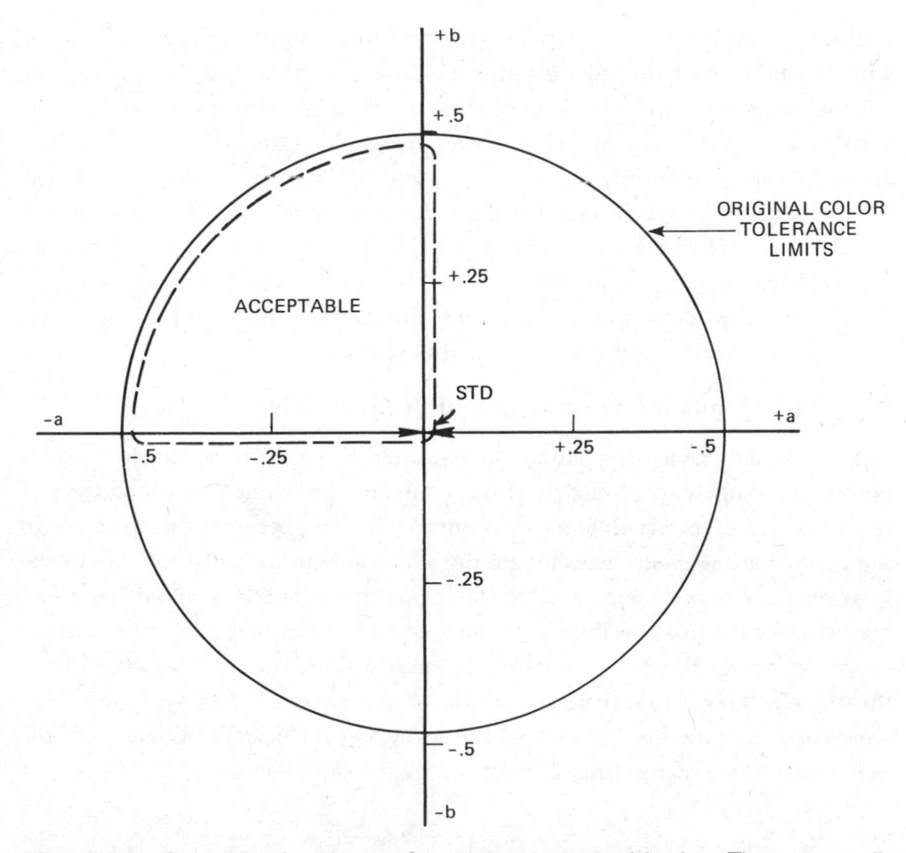


**Figure 17.2.** One user's color tolerances for a green paint in *a* and *b* values. The user originally gave the color tolerance limits described by the solid circle, 0.5 unit in any direction from standard. It eventually became clear, however, that only the area within the dotted line contained acceptable paint color. Any deviation from standard color in the red or blue direction was rejected by the user.

the color imparted by his raw materials throughout the processing and handling of the product. For example, the U.S. Department of Agriculture procedures for grading orange juice require that for the juice to be classified as USDA Grade A, it must be OJ4, or better, in color. Figure 17.3 shows how these United States government grades scores relate to equivalent color scores derived from Hunterlab Citrus Colorimeter measurements. (See also Figure 11.3.)

## Shade Sorting

Sorting for color-shade classification is commonly used in the textile and ceramic fields. Here one is dealing with products for which it is impossible or uneconomical to limit the color variability of the bolts of cloth, or the ceramic wall tile, to such tolerances that any part of the production run can be used with any other part to make a finished product. In this case, the production outputs are sorted into lots of items that are close enough in color to be used together in final use. Figure 17.4 shows an *a, b* diagram used in sorting blue denim fabrics. It identifies by a "shade number" bolts of denim that trained observers have selected as suitable for use together. Note that some errors in the visual estimate of shade seem to be revealed by the instrumental measurements made. In the majority of color-shading applications, most of the sorting is still done visually. Instrumental measurements are used to referee visual judgments, decide doubtful cases, and maintain stability of shade ranges.

## Guiding or Adjusting Formulation of Product Color

Color formulation of compounded light-scattering products by means of computers and mathematical models of the compound has come into use in the last ten years. Here, spectrophotometry is universally used because it is necessary in the mathematical model to combine the physical constants of the ingredients, wavelength by wavelength, and thereby generate the spectrophotometric curve predicted for the final product. Computer programs that aid in formulating the proper color to match a standard are mostly based on the Kubelka-Munk theory, which holds that there is a relationship between reflectance at any given wavelength and the absorption and the scattering coefficients ($K$ and $S$) of the ingredients. This relationship is described by the equation

$$\frac{K}{S} = \frac{(1 - R)^2}{2R}$$

where K is the light absorbed, S the light scattered, and R the reflectance.

## SCORING GUIDE FOR COLOR

## IN VARIOUS PROCESSED ORANGE JUICES

| ORANGE JUICE | USDA OJ 1 Equal to or better than | Not as good as OJ 1; better than OJ 2 | USDA OJ 2 Equal To | Not as good as OJ 2; better than OJ 3 | USDA OJ 3 Equal To | Not as good as OJ 3; much better than OJ 4 | USDA OJ 4 Equal to or slightly better | Not as good as OJ 4; better than OJ 5 | USDA OJ 5 Equal To | Not as good as OJ 5; better than OJ 6 | USDA OJ 6 Equal To | Not as good as OJ 6 |
|---|---|---|---|---|---|---|---|---|---|---|---|---|
| Frozen Concentrated<br>Pasteurized<br>From Concentrate<br>Canned Concentrate | 40 | 40 | 40 | 39 | 39 | 38 | 37 | 36 | 36** | 35 | 34 | 33 or less |
| Canned<br>Concentrated for Manufacturing | 40 | 40 | 40 | 39 | 39 | 38 | 38 | 37 | 37 | 36** | 36** | 35 or less |
| Dehydrated | 40 | 40 | 40 | 39 | 39 | 38 | 37 | 36 | 36 | 35 | 34** | 33 or less |

** *Limits for U. S. Grade A*

**CAUTION - Concentrate must be reconstituted correctly, before color evaluation.**

**Figure 17.3.** The relationship of United States government grade score for orange juice to equivalent color scores derived from citrus colorimeter measurements. Citrus Colorimeter CR (citrus red) and CY (citrus yellow) values are used to derive equivalent color scores (see Figure 11.3). These ECS values are shown with the U.S. Department of Agriculture grade scores from OJ1 to OJ6 related to them.

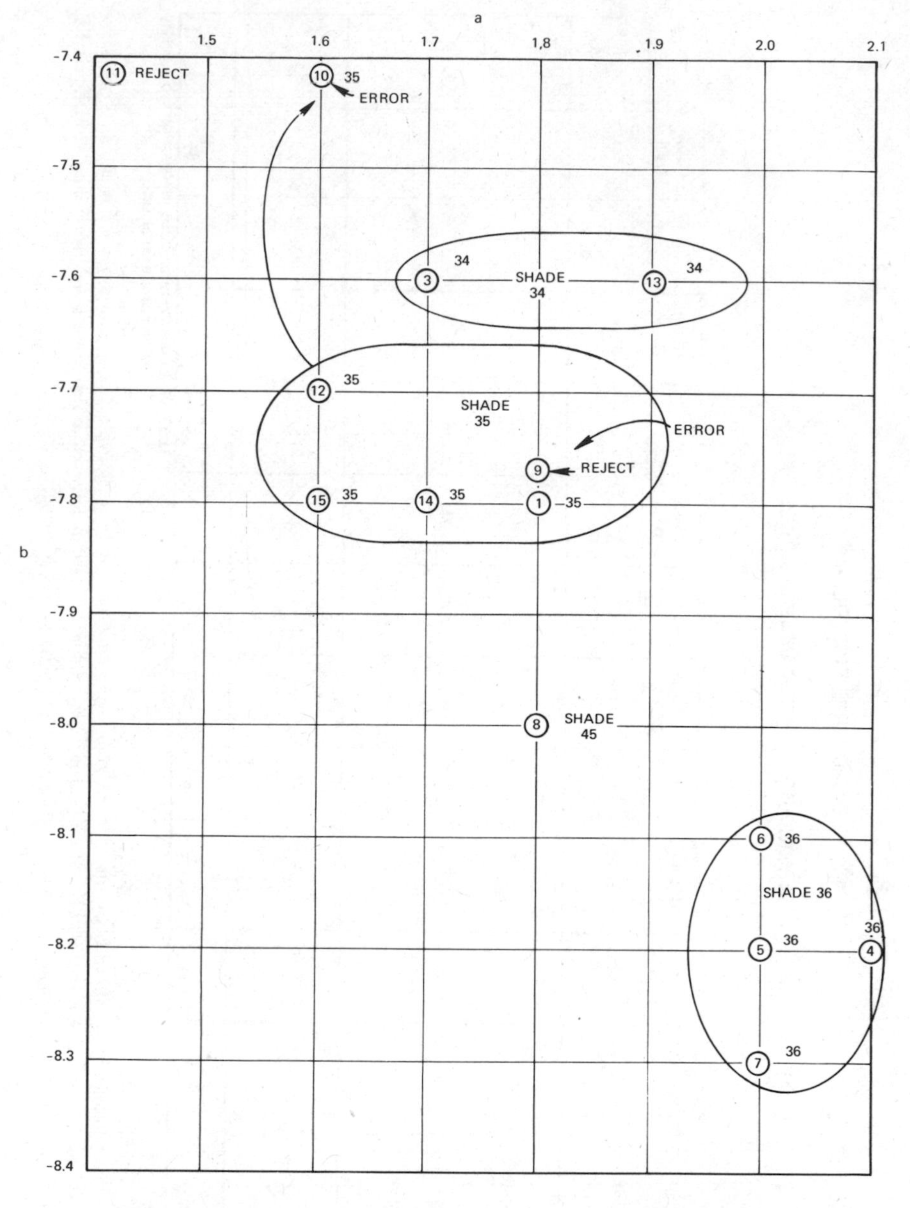


**Figure 17.4.** Measurements of shade-sorted blue denims plotted on an *a, b* diagram. The denims falling within the same group, as visually sorted, are circled after being plotted by their *a* and *b* measured values. The use of the *a, b* diagram shows clearly two errors made in visually sorting the denims (10 and 9).

The theory further holds that the $K/S$ value for a mixture of colorants is the sum of the $K/S$ values of the individual colorants. When several colorants are mixed together, the $K/S$ value of the mixture, therefore, is described by the equation

$$\left(\frac{K}{S}\right)_{\text{mixture}} = a\left(\frac{K}{S}\right)_{C_1} + b\left(\frac{K}{S}\right)_{C_2} + c\left(\frac{K}{S}\right)_{C_3} \cdots + \left(\frac{K}{S}\right)_{\text{Base}}$$

where $C$ is the colorant, $a$ the concentration of colorant 1, $b$ the concentration of colorant 2, and $c$ the concentration of colorant 3.

The use of the Kubelka-Munk theory for formulation of color assumes a mathematical model for the materials used, and the behavior of light within and without these materials. The method is not valid for some materials in which the mathematical model does not apply. It has nevertheless proved to be a very useful tool for developing first-approximation formulas for the colors of many compounded products. The method has been described and documented with examples (Park and Stearns, 1944; Allen, 1965, 1966; Davidson and Hemmendinger, 1966).

After the initial color formulation of a compounded product by spectrophotometer and computer, a precise colorimeter can be used to make small adjustments necessary to correct any shade that is off from the desired standard color. This can be done using color difference values alone. Figure 17.5 shows graphically the approach to adjustment by formula for small errors in color obtained by use of a specific formula. The three vectors pointing away from the standard represent the changes in color that occur with a 1% change in the concentration of each of the three colorants. This must be determined by experimentation.

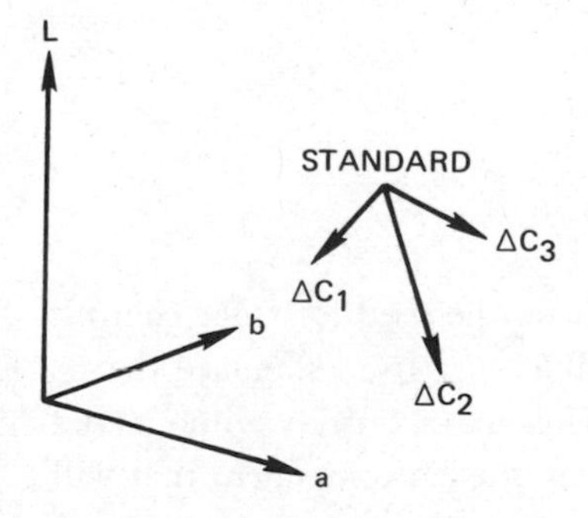


**Figure 17.5.** Vectors representing changes in color of a product with 1% increase in concentration of each of three colorants.

*Experimentally Determined Constants.* The experimentally determined relation between color change and 1% concentration change shown in Figure 17.5 is actually recorded as nine numbers. The values are represented with partial differential notation since the actual change in $L$, $a$, and $b$ will be the result of the combination of the changes induced by all three colorants.

$$
\begin{array}{llll}
\text{For } C_1 & \dfrac{\partial L}{\partial C_1} & \dfrac{\partial a}{\partial C_1} & \dfrac{\partial b}{\partial C_1} \\[2ex]
\text{For } C_2 & \dfrac{\partial L}{\partial C_2} & \dfrac{\partial a}{\partial C_2} & \dfrac{\partial b}{\partial C_2} \\[2ex]
\text{For } C_3 & \dfrac{\partial L}{\partial C_3} & \dfrac{\partial a}{\partial C_3} & \dfrac{\partial b}{\partial C_3}
\end{array}
$$

These constants are now combined into complete statements that summarize the color change resulting from *any* colorant change within the region near the standand.

$$\Delta \mathrm{L} = \frac{\partial L}{\partial C_1} \Delta C_1 + \frac{\partial L}{\partial C_2} \Delta C_2 + \frac{\partial L}{\partial C_3} \Delta C_3$$

$$\Delta a = \frac{\partial a}{\partial C_1} \Delta C_1 + \frac{\partial a}{\partial C_2} \Delta C_2 + \frac{\partial a}{\partial C_3} \Delta C_3$$

$$\Delta b = \frac{\partial b}{\partial C_1} \Delta C_1 + \frac{\partial b}{\partial C_2} \Delta C_2 + \frac{\partial b}{\partial C_3} \Delta C_3$$

Using standard matrix algebra, these formulas can be turned around to generate new formulas in which $\Delta L$, $\Delta a$, $\Delta b$, become the independent variables rather than the dependent variables. These formulas are specific for the standard color selected, and the three colorants selected.

$$\Delta C_1 = f_1(\Delta L, \Delta a, \Delta b)$$

$$\Delta C_2 = f_2(\Delta L, \Delta a, \Delta b)$$

$$\Delta C_3 = f_3(\Delta L, \Delta a, \Delta b)$$

With the formulas in this form the equations can be used for color control. The variables $\Delta L$, $\Delta a$, $\Delta b$ now represent the difference from standard that is desired to be corrected; $\Delta C_1$, $\Delta C_2$, $\Delta C_3$ are all determined quickly without trial and error and predict the change in concentration of each component that will bring about a match. If the color change to be made is large, that is if the present color is far from standard, successive application of the formulas may be required to achieve a match since the formulas assume a linear relation between the percent change in concentration and color change.

These experimentally determined relationships between color change and concentration of dye change can be shown in graph form. Figure 17.6 shows data for a red spray enamel, for which it was determined by experiment that addition of one gallon of red pigment or one gallon of yellow oxide resulted in the following average changes in *a* and *b*:

| | *a* | *b* |
|---|---|---|
| Red | −0.08 | −0.95 |
| Yellow oxide | −0.40 | +0.20 |

Plotting the $\Delta a$ and $\Delta b$ point for the red change and then drawing a line through that point and through the standard point gives us a grid on which can be marked off distances equal to the change in *a* caused by the addition of one gallon of red pigment. Doing the same for the $\Delta a$, $\Delta b$ point of the yellow oxide change gives us the comparable grid for the *b* dimension. By plotting on this graph the point representing an initial-match color, using its *a* and *b* differences from standard, the amounts of yellow and red corrections needed to bring it to the standard color can be read. In terms of *a* and *b* differences from standard color, $\Delta a = -0.80$, $\Delta b = +1.09$. From its plotted position on the graph, it can be seen that subtraction of 2.2 gallons of yellow oxide pigment and addition of 0.7 gallons of red pigment are needed to adjust it to the standard color.

One needs to know what to do when a negative amount of one of the components is called for, as with the yellow oxide ingredient in the example just cited. If, as frequently happens, the ingredients of the formula are already mixed so that removal of some of this one component is no longer feasible, then it is necessary to consider additional ingredients. For instance, one might find that a negative amount of white pigment is required to achieve the desired match. A negative white pigment is, of course, a black pigment, and so one would then have to go to a second calibration of the formulation involving the results of adding black rather than white. Similarly, a yellow pigment would be the negative of blue-pigment contribution to color. There are variations of the foregoing method of color adjustment (Dana, Bayer, and McElrath, 1965). Not only color, but also opacity of paper and gloss of paint, to name two specific geometric attributes, are susceptible to adjustment in this manner.

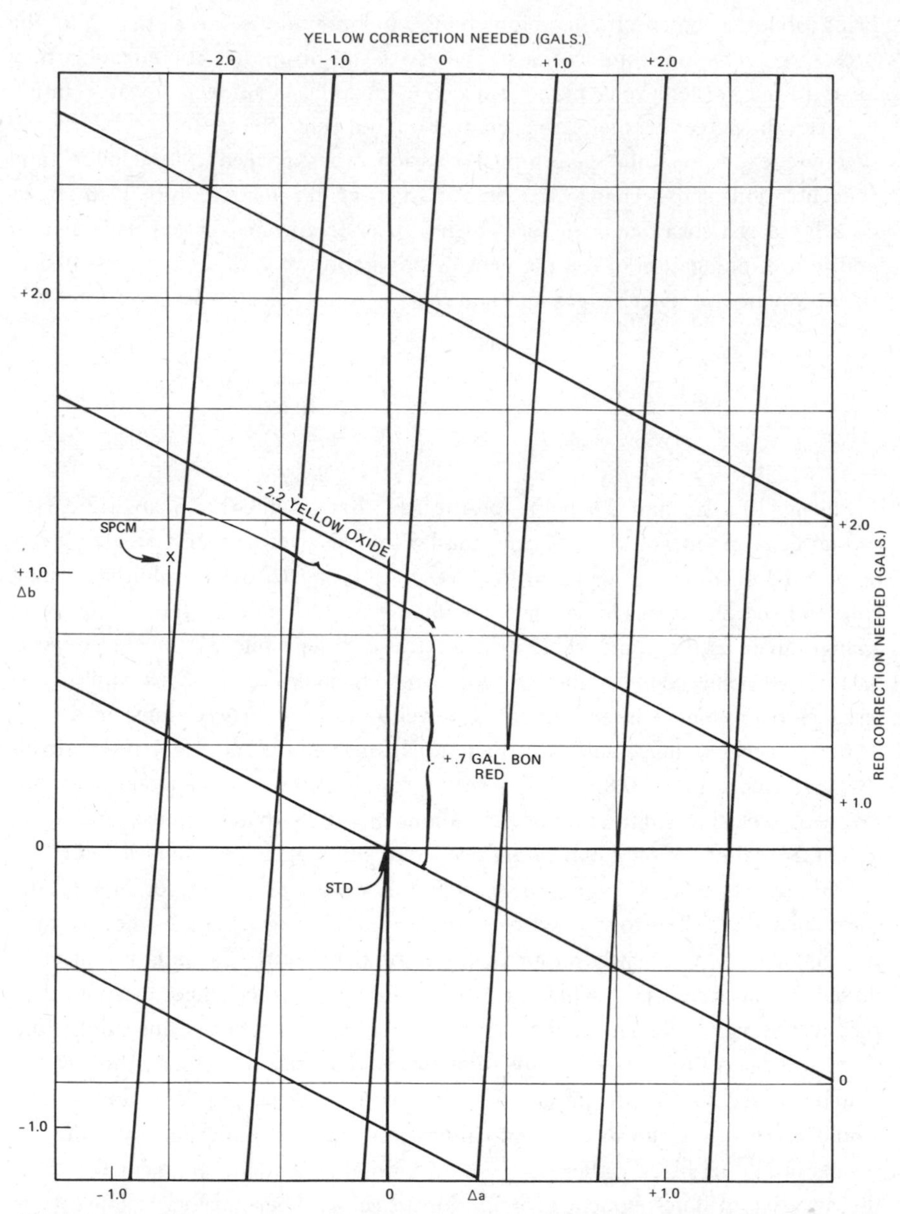


**Figure 17.6.** Adjustment of color of red spray enamel using *a* and *b* values. The grids for this chart relate to the size of the changes made in *a* and *b* by the addition of one gallon of red or one gallon of yellow pigment paste. Changes in *a* ($\Delta a$) and *b* ($\Delta b$) are indicated by the horizontal and vertical grids. The horizontal diagonal lines indicate the amount of red correction needed; the vertical diagonal lines indicate the amount of yellow correction needed to make a certain change in $\Delta a$ and $\Delta b$.

## Assessment of Product Performance in Service

Measurements of change in color or other appearance attributes that result from exposure are used to assess product performance under service conditions. Most of the measurements are concerned with lightfastness of textiles, plastics, and paints as shown by color change after exposure.

There is a series of colorfastness tests for textiles in which a textile product is rated for colorfastness after having been subjected to standard exposure procedures. (See Chapter 9.)

Gloss measurements also serve as a method of testing product performance in use. A change of gloss with exposure is an especially sensitive test for susceptibility to surface damage. Since gloss is directly related to the condition of the skin of the object surface, any damage to this skin immediately shows up as a change in surface gloss. Thus gloss is a useful indicator of surface damage of paints, plastics, and ceramics due to exposure of real or simulated weather conditions. This gloss change usually appears before a change of color. Figure 17.7 shows the change of gloss of a paint film from artificial weather exposure.

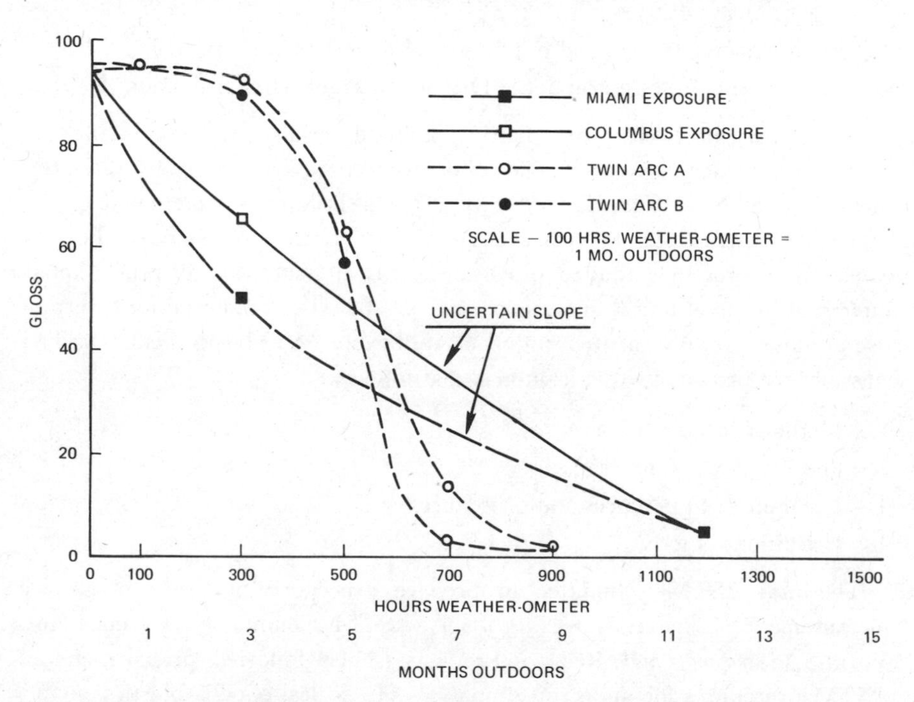


**Figure 17.7.** Change of 60° gloss of a paint film from artificial weather exposure. (Reprinted with permission, from the November 1965 Official Digest, a publication of the Federation of Societies for Paint Technology, Philadelphia, Pa.)

### Measurement of Nonappearance Variables

Frequently nonappearance attributes of materials are related to measurable appearance attributes. As an example, Figure 17.8 shows squares of clear plastic with circular abraded tracks applied by the Taber Abrader and used to test the resistance of these materials to abrasion. The actual abrasion damage caused by the device is measured by transmission haze. Another such example is the use made of a change of color in a reagent-impregnated paper to assess amount of blood sugar in the urine.

## TEST METHODS FOR APPEARANCE ANALYSIS AND MEASUREMENT

Instructions for making and applying analyses and measurement of appearance attributes have been prepared and published by:

1. Professional technical organizations
2. Government agencies
3. Private manufacturers and purchasers of materials in commerce
4. Committees of the International Organization for Standardization (ISO)

The American Society for Testing and Materials (ASTM) is the acknowledged leader in the worldwide effort to prepare instructions for these test methods (ASTM, 1972). The ASTM has a well-developed methodology for preparing instructions by what it terms its consensus method. This is essentially a procedure that contains safeguards to assure that all professionals interested in the technical analysis of any specific type of material for a certain use are given a voice in preparation of applicable ASTM documents. ASTM classifies the procedures that it publishes as follows:

1. Methods of test
2. Specifications of materials
3. Recommended practices and procedures
4. Definitions of terms

The first ASTM committee to prepare procedures for the appearance measurement of materials was probably ASTM Committee D-1 on Paint, Varnish, Lacquer, and Related Products. Then followed preparations of ASTM procedures for appearance analyses of textiles, papers, plastics, soaps, and waxes. Finally came procedures for appearance analyses of all materials generally. For this purpose ASTM established in 1948 a committee known as

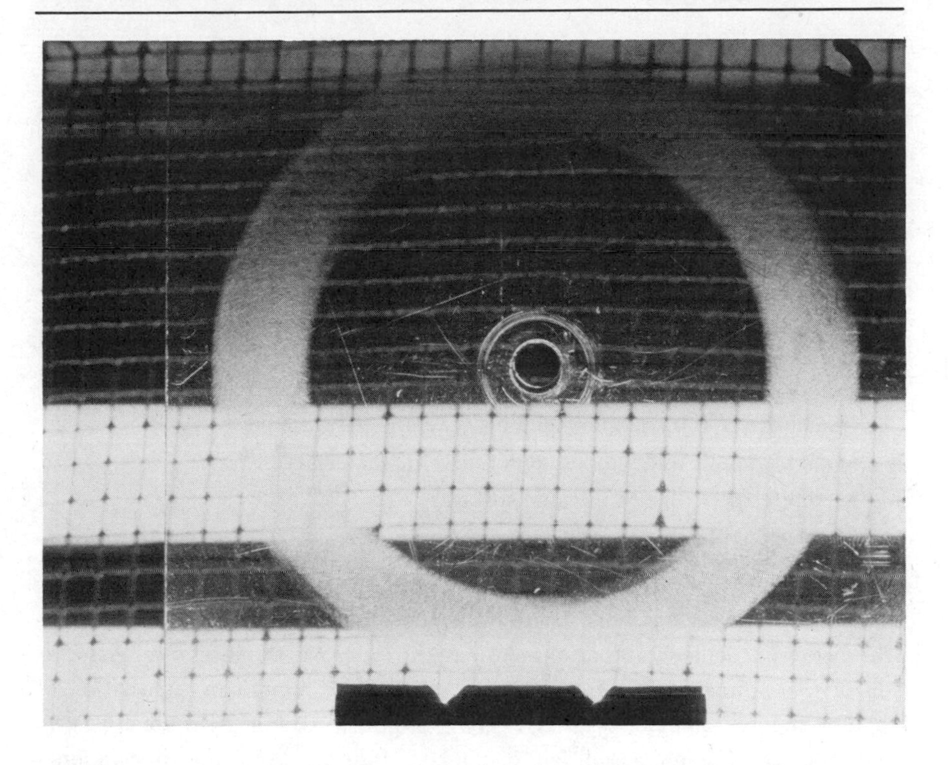

**Figure 17.8.** Tracks applied to a clear plastic disk by the Taber Abrader. Loss of transmission haze is used to measure surface damage as the result of abrasion.

Committee E-12 on the "Appearance of Materials," whose whole effort is devoted exclusively to evaluation of, and methods for, appearance analysis. Thus since 1948 ASTM has recognized appearance measurement as a separate specific methodology and has maintained a group working in the field and applying its efforts to all sorts of materials and problems in that area.

A published method for the measurement of appearance generally comprises the following sections:

1. An introduction giving scope and objectives
2. A summary of the method and references to pertinent documents
3. Definitions or descriptions of terms
4. Description of the apparatus required to conduct the test
5. Instructions for the collection and preparation of specimens for analysis
6. Instructions for making the measurements, including instructions for preparation of the apparatus, standardization, calibration, and so on

7. Procedures for reducing the data to the form required
8. Instructions for reporting the results

The published procedures for appearance analysis that are best known in the United States are identified in Table A.15. The dominance of the ASTM in this field is attested to by the fact that well over half of the tests shown in this table carry ASTM identifications. The methods in the table are arranged as follows:

1. General purpose methods applicable to a variety of materials.
   a. Those for color attributes.
   b. Those for geometric attributes.
   c. Those giving recommended practice and definitions.
2. Methods applicable to opaque nonmetallic materials.
3. Methods applicable to metallic surfaces.
4. Methods applicable to light-transmitting materials.

Federal and state government agencies have also prepared test methods for appearance analysis, as have trade organizations in industries for which appearance is particularly important. Among the latter are the Technical Association of the Pulp and Paper Industry (TAPPI), the American Association of Textile Chemists and Colorists (AATCC), and the Porcelain Institute (PEI).

On the international scene, within the last 10 or 15 years, the International Organization for Standardization (ISO) has developed rapidly the consensus procedures for preparing and publishing test procedures. The ISO is, at the moment, following the ASTM practices of more than 25 years ago by conducting for different materials completely separate efforts in the appearance measurement field. The committees that are at present operating within the ISO in developing international standards for specific types of materials are:

| | |
|---|---|
| ISO TC-6 | Paper, boards, and pulps |
| ISO TC-35 | Paints and varnishes |
| ISO TC-38 | Textiles |
| ISO TC-61 | Plastics |
| ISO TC-79 | Light metals and their alloys |
| ISO TC-80 | Safety colours |

Many procedures for the useful measurement of appearance attributes of specific products have been proposed but never developed. Many proposals are found in the technical literature for making certain appearance measurements of products. These suggested procedures have not, in many cases, been tested or proven by others, and thus cannot be said to be established. Other procedures for testing specific attributes of certain products have not been developed because there apparently exists no economic justification for doing the developmental work.

There are certain areas, however, where the undeniable needs for measurement procedures have not been met and should be. Hunter (1973) pointed out several of these needs to the ASTM in 1973, including

1. A simple reproducible method for surface-color measurement
2. An instrumental method of measurement of color difference that includes diagnostic methods for calibration of instruments
3. A simple instrumental procedure for the measurement of degree of metamerism
4. Reproducible methods for measurement of the colors of objects other than diffusely reflecting surfaces
5. A convenient readily reproducible method for high-gloss measurement
6. A standardized reproducible method of luster measurement
7. A method of analysis of surface and film uniformity—involving texture, sharpness in printing, and legibility

## SUMMARY

Appearance measurements are used throughout industry, government, and business, wherever appearance attributes are important enough to warrant detailed analyses. Specifically, the major uses for appearance measurements can be classified as follows:

1. To identify materials appearance for research, communication, and records
2. To test a product for conformity to specifications of appearance or for appearance match to standard
3. To provide scores for quality
4. To provide shade-sorting classifications so that samples of the same shade number can be used together in an assembled final product
5. To formulate product color to achieve desired final appearance
6. To assess product performance in service as measured by of change of appearance due to exposure
7. To make certain tests for properties of materials other than appearance that can be conveniently measured by appearance-measurement techniques

Because attributes of appearance have optically similar bases even when the materials are different, there is a growing trend toward methods for appearance analysis that are applicable to all optically similar materials. Formerly most of the published procedures followed in commerce were specific to the material tested. However, in the United States general-purpose appearance-measurement procedures are now coming into use. The benefit of these general-purpose methods is that expertise and high precision are achieved together with economy in the instrumentation cost through mass production.

The American Society for Testing and Materials is certainly the dominant factor in the preparation and publication of standardized procedures for evaluating appearance properties of materials. Other technical organizations, government agencies, and private organizations in commerce also publish and use these methods. Within the last 15 years there has been a rapid development of internationally recognized published procedures for appearance measurement through the International Organization for Standardization.

# APPENDIX

**TABLE A.1**

GLOSS MEASUREMENT PROCEDURES[a]

| Gloss, Specular Angle, and Specification | Applications | Field Angles, Degrees (First angle given is in plane of measurement) | | Specification of the Numerical Scale | Polished Black Glass $n = 1.54$ | Perfect White (diffuse correction factor) | Perfect Mirror |
|---|---|---|---|---|---|---|---|
| | | Source | Receptor | | | | |
| Sheen, 85° ASTM D523 FTM 141a/6103 | Sheen (shininess at grazing angles) of flat matte paints, camouflage coatings | 0.75 × 3.0 | 4.0 × 6.0 | $G_s$, black glass ($n$ = 1.567) = 100 | 99.7 | 0.03 | 161.5 |
| Specular gloss, 75° TAPPI T480 ASTM D1223 | Gloss of coated, waxed and glassine papers, printed cartons | 2.8 × 5.7 | 11.4 diam. | $G_s$, black glass ($n$ = 1.54) = 100 | 100.0 | 1.0 | 385.0 |
| Specular gloss, 60° ASTM D523 ASTM D2457 ASTM D1455 FTM 141a/6101 FTM 406/3051 NEMA LD1 2.13 ASTM C584 | Classify paints, plastics, as high, medium, low or intermediate in gloss. Measure gloss differences in all ranges but high and very low. | 0.75 × 3.0 | 4.4 × 11.7 | $G_s$, perfect mirror = 1000, or black glass ($n$ = 1.567) = 100 | 96.0 | 2.1 | 1000.0 |

| | | | | | | | |
|---|---|---|---|---|---|---|---|
| Specular gloss, 20°<br>ASTM D523<br>ASTM D2457<br>FTM 141a/6104 | Gloss of high gloss plastic film and appliance and automotive finishes | 0.75 × 3.0 | 1.8 × 3.6 | $G_s$, black glass ($n$ = 1.567) = 100 | 92.0 | 1.3 | 2032.0 |
| Specular gloss, 45°<br>ASTM D2457<br>ASTM C346<br>PEI-T2, T7, T18 | Gloss of polyethylene and other plastic films; change of porcelain finishes after use. | 1.4 × 3.0 | 8.0 × 10.0 | $G_s$, perfect mirror = 1000 | 55.9 | 5.4 | 1000.0 |
| Specular gloss, 20°<br>TAPPI T653<br>ASTM D1834 | Gloss of waxed, lacquered, and cast-coated papers; high-gloss ink films | 1 to 2.5 diam. | 5.0 diam. | $G_s$, black glass ($n$ = 1.54) = 100 | 100.0 | 3.8 | 2190.0 |
| Specular gloss, 30°<br>Distinctness-of-Image gloss, 30° ± 0.3<br>Narrow-angle haze, 32°<br>Wide-angle haze, 35°<br>ASTM E430 | Measures high-gloss, image-reflecting characteristics of nondiffuse metal finishes using abridged goniophotometer or goniophotometer | 0.44 × 7.0<br>0.44 × 7.0<br>0.44 × 7.0 | 0.38 × 4.0<br>0.14 × 4.0<br>0.4 × 4.0 | $G_s$, perfect mirror = 100 | 4.7 | 0.03 | 100.0 |
| 2-Parameter Specular gloss, 60°<br>ASTM D1471 | Tendency toward image formation by smooth paint surfaces | 0.75 × 3.0 | 4.4 × 11.7<br>2.0 × 4.5 | $G_s$, perfect mirror 1000 | 96.0<br>96.0 | 2.1<br>0.5 | 1000.0<br>1000.0 |

**TABLE A.1 (Continued)**

GLOSS MEASUREMENT PROCEDURES (CONT.)

| Gloss, Specular Angle, and Specification | Applications | Field Angles, Degrees (First angle given is in plane of measurement | | Specification of the Numerical Scale | Polished Black Glass $n = 1.54$ | Perfect White (diffuse correction factor) | Perfect Mirror |
|---|---|---|---|---|---|---|---|
| | | Source | Receptor | | | | |
| Contrast Luster $\frac{100\,(1 - R_{d,0°,45°})}{R_{d,45°,45°}}$ (no specification) | Luster of textiles, low-gloss papers and other materials | $2.8 \times 5.7$ | $8.0 \times 10.0$ | Perfect diffusor = 0.0 Any black glass = 100 | 100.0 | 0.0 | 100.0 |

[a] ASTM: American Society for Testing and Materials, Philadelphia, Pa.
TAPPI: Technical Association of the Pulp and Paper Industry, Atlanta, Ga.
PEI: Porcelain Enamel Institute, Washington, D.C.
NEMA: National Electrical Manufacturer's Association, New York, N.Y.
FTM: Federal Test Method, General Services Administration, Washington, D.C.

**TABLE A.2**

THE 1924 CIE LUMINOSITY FUNCTION

| Wavelength | Luminosity ($Y$) | Wavelength | Luminosity ($Y$) |
|---|---|---|---|
| 380 | 0.0000 | 600 | 0.6310 |
| 390 | 0.0001 | 610 | 0.5030 |
| | | 620 | 0.3810 |
| 400 | 0.0004 | 630 | 0.2650 |
| 410 | 0.0012 | 640 | 0.1750 |
| 420 | 0.0040 | | |
| 430 | 0.0116 | 650 | 0.1070 |
| 440 | 0.0230 | 660 | 0.0610 |
| | | 670 | 0.0320 |
| 450 | 0.0380 | 680 | 0.0170 |
| 460 | 0.0600 | 690 | 0.0082 |
| 470 | 0.0910 | | |
| 480 | 0.1390 | 700 | 0.0041 |
| 490 | 0.2080 | 710 | 0.0021 |
| | | 720 | 0.0010 |
| 500 | 0.3230 | 730 | 0.0005 |
| 510 | 0.5030 | 740 | 0.0003 |
| 520 | 0.7100 | | |
| 530 | 0.8620 | 750 | 0.0001 |
| 540 | 0.9540 | 760 | 0.0001 |
| | | 770 | 0.0000 |
| 550 | 0.9950 | 780 | 0.0000 |
| 560 | 0.9950 | | |
| 570 | 0.9520 | | |
| 580 | 0.8700 | | |
| 590 | 0.7570 | | |

**TABLE A.3**

THE 1931 CIE STANDARD OBSERVER FUNCTIONS

| Trichromatic Coefficients | | | Wave- | Tristimulus Specifications of the Equal-Energy Spectrum | | | Trichromatic Coefficients | | | Wave- | Tristimulus Specifications of the Equal-Energy Spectrum | | |
|---|---|---|---|---|---|---|---|---|---|---|---|---|---|
| $x$ | $y$ | $z$ | length, | $\bar{x}$ | $\bar{y}$ | $\bar{z}$ | $x$ | $y$ | $z$ | length, | $\bar{x}$ | $\bar{y}$ | $\bar{z}$ |
| 0.1741 | 0.0050 | 0.8209 | 380 | 0.0014 | 0.0000 | 0.0065 | 0.5125 | 0.4866 | 0.0009 | 580 | 0.9163 | 0.8700 | 0.0017 |
| 0.1740 | 0.0050 | 0.8210 | 385 | 0.0022 | 0.0001 | 0.0105 | 0.5448 | 0.4544 | 0.0008 | 585 | 0.9786 | 0.8163 | 0.0014 |
| 0.1738 | 0.0049 | 0.8213 | 390 | 0.0042 | 0.0001 | 0.0201 | 0.5752 | 0.4242 | 0.0006 | 590 | 1.0263 | 0.7570 | 0.0011 |
| 0.1736 | 0.0049 | 0.8215 | 395 | 0.0076 | 0.0002 | 0.0362 | 0.6029 | 0.3965 | 0.0006 | 595 | 1.0567 | 0.6949 | 0.0010 |
| 0.1733 | 0.0048 | 0.8219 | 400 | 0.0143 | 0.0004 | 0.0679 | 0.6270 | 0.3725 | 0.0005 | 600 | 1.0622 | 0.6310 | 0.0008 |
| 0.1730 | 0.0048 | 0.8222 | 405 | 0.0232 | 0.0006 | 0.1102 | 0.6482 | 0.3514 | 0.0004 | 605 | 1.0456 | 0.5668 | 0.0006 |
| 0.1726 | 0.0048 | 0.8226 | 410 | 0.0435 | 0.0012 | 0.2074 | 0.6658 | 0.3340 | 0.0002 | 610 | 1.0026 | 0.5030 | 0.0003 |
| 0.1721 | 0.0048 | 0.8231 | 415 | 0.0776 | 0.0022 | 0.3713 | 0.6801 | 0.3197 | 0.0002 | 615 | 0.9384 | 0.4412 | 0.0002 |
| 0.1714 | 0.0051 | 0.8235 | 420 | 0.1344 | 0.0040 | 0.6456 | 0.6915 | 0.3083 | 0.0002 | 620 | 0.8544 | 0.3810 | 0.0002 |
| 0.1703 | 0.0058 | 0.8239 | 425 | 0.2148 | 0.0073 | 1.0391 | 0.7006 | 0.2993 | 0.0001 | 625 | 0.7514 | 0.3210 | 0.0001 |
| 0.1689 | 0.0069 | 0.8242 | 430 | 0.2839 | 0.0116 | 1.3856 | 0.7079 | 0.2920 | 0.0001 | 630 | 0.6424 | 0.2650 | 0.0000 |
| 0.1669 | 0.0086 | 0.8245 | 435 | 0.3285 | 0.0168 | 1.6230 | 0.7140 | 0.2859 | 0.0001 | 635 | 0.5419 | 0.2170 | 0.0000 |
| 0.1644 | 0.0109 | 0.8247 | 440 | 0.3483 | 0.0230 | 1.7471 | 0.7190 | 0.2809 | 0.0001 | 640 | 0.4479 | 0.1750 | 0.0000 |
| 0.1611 | 0.0138 | 0.8251 | 445 | 0.3481 | 0.0298 | 1.7826 | 0.7230 | 0.2770 | 0.0000 | 645 | 0.3608 | 0.1382 | 0.0000 |
| 0.1566 | 0.0177 | 0.8257 | 450 | 0.3362 | 0.0380 | 1.7721 | 0.7260 | 0.2740 | 0.0000 | 650 | 0.2835 | 0.1070 | 0.0000 |
| 0.1510 | 0.0227 | 0.8263 | 455 | 0.3187 | 0.0480 | 1.7441 | 0.7283 | 0.2717 | 0.0000 | 655 | 0.2187 | 0.0816 | 0.0000 |
| 0.1440 | 0.0297 | 0.8263 | 460 | 0.2908 | 0.0600 | 1.6692 | 0.7300 | 0.2700 | 0.0000 | 660 | 0.1649 | 0.0610 | 0.0000 |
| 0.1355 | 0.0399 | 0.8246 | 465 | 0.2511 | 0.0739 | 1.5281 | 0.7311 | 0.2689 | 0.0000 | 665 | 0.1212 | 0.0446 | 0.0000 |

| | | | | | | | | | | | | | |
|---|---|---|---|---|---|---|---|---|---|---|---|---|---|
| 0.1241 | 0.0578 | 0.8181 | 470 | 0.1954 | 0.0910 | 1.2876 | 0.7320 | 0.2680 | 0.0000 | 670 | 0.0874 | 0.0320 | 0.0000 |
| 0.1096 | 0.0868 | 0.8036 | 475 | 0.1421 | 0.1126 | 1.0419 | 0.7327 | 0.2673 | 0.0000 | 675 | 0.0636 | 0.0232 | 0.0000 |
| 0.0913 | 0.1327 | 0.7760 | 480 | 0.0956 | 0.1390 | 0.8130 | 0.7334 | 0.2666 | 0.0000 | 680 | 0.0468 | 0.0170 | 0.0000 |
| 0.0687 | 0.2007 | 0.7306 | 485 | 0.0580 | 0.1693 | 0.6162 | 0.7340 | 0.2660 | 0.0000 | 685 | 0.0329 | 0.0119 | 0.0000 |
| 0.0454 | 0.2950 | 0.6596 | 490 | 0.0320 | 0.2080 | 0.4652 | 0.7344 | 0.2656 | 0.0000 | 690 | 0.0227 | 0.0082 | 0.0000 |
| 0.0235 | 0.4127 | 0.5638 | 495 | 0.0147 | 0.2586 | 0.3533 | 0.7346 | 0.2654 | 0.0000 | 695 | 0.0158 | 0.0057 | 0.0000 |
| 0.0082 | 0.5384 | 0.4534 | 500 | 0.0049 | 0.3230 | 0.2720 | 0.7347 | 0.2653 | 0.0000 | 700 | 0.0114 | 0.0041 | 0.0000 |
| 0.0039 | 0.6548 | 0.3413 | 505 | 0.0024 | 0.4073 | 0.2123 | 0.7347 | 0.2653 | 0.0000 | 705 | 0.0081 | 0.0029 | 0.0000 |
| 0.0139 | 0.7502 | 0.2359 | 510 | 0.0093 | 0.5030 | 0.1582 | 0.7347 | 0.2653 | 0.0000 | 710 | 0.0058 | 0.0021 | 0.0000 |
| 0.0389 | 0.8120 | 0.1491 | 515 | 0.0291 | 0.6082 | 0.1117 | 0.7347 | 0.2653 | 0.0000 | 715 | 0.0041 | 0.0015 | 0.0000 |
| 0.0743 | 0.8338 | 0.0919 | 520 | 0.0633 | 0.7100 | 0.0782 | 0.7347 | 0.2653 | 0.0000 | 720 | 0.0029 | 0.0010 | 0.0000 |
| 0.1142 | 0.8262 | 0.0596 | 525 | 0.1096 | 0.7932 | 0.0573 | 0.7347 | 0.2653 | 0.0000 | 725 | 0.0020 | 0.0007 | 0.0000 |
| 0.1547 | 0.8059 | 0.0394 | 530 | 0.1655 | 0.8620 | 0.0422 | 0.7347 | 0.2653 | 0.0000 | 730 | 0.0014 | 0.0005 | 0.0000 |
| 0.1929 | 0.7816 | 0.0255 | 535 | 0.2257 | 0.9149 | 0.0298 | 0.7347 | 0.2653 | 0.0000 | 735 | 0.0010 | 0.0004 | 0.0000 |
| 0.2296 | 0.7543 | 0.0161 | 540 | 0.2904 | 0.9540 | 0.0203 | 0.7347 | 0.2653 | 0.0000 | 740 | 0.0007 | 0.0003 | 0.0000 |
| 0.2658 | 0.7243 | 0.0099 | 545 | 0.3597 | 0.9803 | 0.0134 | 0.7347 | 0.2653 | 0.0000 | 745 | 0.0005 | 0.0002 | 0.0000 |
| 0.3016 | 0.6923 | 0.0061 | 550 | 0.4334 | 0.9950 | 0.0087 | 0.7347 | 0.2653 | 0.0000 | 750 | 0.0003 | 0.0001 | 0.0000 |
| 0.3373 | 0.6589 | 0.0038 | 555 | 0.5121 | 1.0002 | 0.0057 | 0.7347 | 0.2653 | 0.0000 | 755 | 0.0002 | 0.0001 | 0.0000 |
| 0.3731 | 0.6245 | 0.0024 | 560 | 0.5945 | 0.9950 | 0.0039 | 0.7347 | 0.2653 | 0.0000 | 760 | 0.0002 | 0.0001 | 0.0000 |
| 0.4087 | 0.5896 | 0.0017 | 565 | 0.6784 | 0.9786 | 0.0027 | 0.7347 | 0.2653 | 0.0000 | 765 | 0.0001 | 0.0000 | 0.0000 |
| 0.4441 | 0.5547 | 0.0012 | 570 | 0.7621 | 0.9520 | 0.0021 | 0.7347 | 0.2653 | 0.0000 | 770 | 0.0001 | 0.0000 | 0.0000 |
| 0.4788 | 0.5202 | 0.0010 | 575 | 0.8425 | 0.9154 | 0.0018 | 0.7347 | 0.2653 | 0.0000 | 775 | 0.0000 | 0.0000 | 0.0000 |
| 0.5125 | 0.4866 | 0.0009 | 580 | 0.9163 | 0.8700 | 0.0017 | 0.7347 | 0.2653 | 0.0000 | 780 | 0.0000 | 0.0000 | 0.0000 |
| Totals | | | | | | | | | | | 21.3713 | 21.3714 | 21.3715 |

**TABLE A.4**

1931 CIE $\bar{x}$, $\bar{y}$, AND $\bar{z}$ FUNCTIONS FOR ILLUMINANTS A, B, C, AND D6500

| Wave-length | CIE Standard Source A | | | CIE Standard Source B | | | CIE Standard Source C | | | CIE Standard Source D6500 | | |
|---|---|---|---|---|---|---|---|---|---|---|---|---|
| 380 | 1 | | 6 | 3 | | 14 | 4 | | 20 | 7 | | 31 |
| 390 | 5 | | 23 | 13 | | 60 | 19 | | 89 | 22 | 1 | 104 |
| 400 | 19 | 1 | 93 | 56 | 2 | 268 | 85 | 2 | 404 | 112 | 3 | 532 |
| 410 | 71 | 2 | 340 | 217 | 6 | 1,033 | 329 | 9 | 1,570 | 377 | 10 | 1,796 |
| 420 | 262 | 8 | 1,256 | 812 | 24 | 3,899 | 1,238 | 37 | 5,949 | 1,189 | 35 | 5,711 |
| 430 | 649 | 27 | 3,167 | 1,983 | 81 | 9,678 | 2,997 | 122 | 14,628 | 2,330 | 95 | 11,370 |
| 440 | 926 | 61 | 4,647 | 2,689 | 178 | 13,489 | 3,975 | 262 | 19,938 | 3,458 | 228 | 17,343 |
| 450 | 1,031 | 117 | 5,435 | 2,744 | 310 | 14,462 | 3,915 | 443 | 20,638 | 3,724 | 421 | 19,627 |
| 460 | 1,019 | 210 | 5,851 | 2,454 | 506 | 14,085 | 3,362 | 694 | 19,299 | 3,243 | 669 | 18,614 |
| 470 | 776 | 362 | 5,116 | 1,718 | 800 | 11,319 | 2,272 | 1,058 | 14,972 | 2,124 | 989 | 13,998 |
| 480 | 428 | 622 | 3,636 | 870 | 1,265 | 7,396 | 1,112 | 1,618 | 9,461 | 1,048 | 1,524 | 8,915 |
| 490 | 160 | 1,039 | 2,324 | 295 | 1,918 | 4,290 | 363 | 2,358 | 5,274 | 330 | 2,142 | 4,791 |
| 500 | 27 | 1,792 | 1,509 | 44 | 2,908 | 2,449 | 52 | 3,401 | 2,864 | 51 | 3,343 | 2,815 |
| 510 | 57 | 3,080 | 969 | 81 | 4,360 | 1,371 | 89 | 4,833 | 1,520 | 95 | 5,132 | 1,614 |
| 520 | 425 | 4,771 | 525 | 541 | 6,072 | 669 | 576 | 6,462 | 712 | 628 | 7,041 | 775 |
| 530 | 1,214 | 6,322 | 309 | 1,458 | 7,594 | 372 | 1,523 | 7,934 | 388 | 1,687 | 8,785 | 430 |
| 540 | 2,313 | 7,600 | 162 | 2,689 | 8,834 | 188 | 2,785 | 9,149 | 195 | 2,869 | 9,425 | 201 |
| 550 | 3,732 | 8,568 | 75 | 4,183 | 9,603 | 84 | 4,282 | 9,832 | 86 | 4,267 | 9,796 | 86 |
| 560 | 5,510 | 9,222 | 36 | 5,840 | 9,774 | 38 | 5,880 | 9,841 | 39 | 5,625 | 9,415 | 37 |
| 570 | 7,571 | 9,457 | 21 | 7,472 | 9,334 | 21 | 7,322 | 9,147 | 20 | 6,947 | 8,678 | 19 |
| 580 | 9,719 | 9,228 | 18 | 8,843 | 8,396 | 16 | 8,417 | 7,992 | 16 | 8,304 | 7,885 | 15 |
| 590 | 11,579 | 8,540 | 12 | 9,728 | 7,176 | 10 | 8,984 | 6,627 | 10 | 8,612 | 6,352 | 9 |

| | | | | | | | | | | | | |
|---|---|---|---|---|---|---|---|---|---|---|---|---|
| 600 | 12,704 | 7,547 | 10 | 9,948 | 5,909 | 7 | 8,949 | 5,316 | 7 | 9,046 | 5,374 | 7 |
| 610 | 12,669 | 6,356 | 4 | 9,436 | 4,734 | 3 | 8,325 | 4,176 | 2 | 8,499 | 4,264 | 3 |
| 620 | 11,373 | 5,071 | 3 | 8,140 | 3,630 | 2 | 7,070 | 3,153 | 2 | 7,089 | 3,161 | 2 |
| 630 | 8,980 | 3,704 | | 6,200 | 2,558 | | 5,309 | 2,190 | | 5,062 | 2,088 | |
| 640 | 6,558 | 2,562 | | 4,374 | 1,709 | | 3,693 | 1,443 | | 3,547 | 1,386 | |
| 650 | 4,336 | 1,637 | | 2,815 | 1,062 | | 2,349 | 889 | | 2,147 | 810 | |
| 660 | 2,628 | 972 | | 1,655 | 612 | | 1,361 | 504 | | 1,252 | 463 | |
| 670 | 1,448 | 530 | | 876 | 321 | | 708 | 259 | | 680 | 249 | |
| 680 | 804 | 292 | | 465 | 169 | | 369 | 134 | | 347 | 126 | |
| 690 | 404 | 146 | | 220 | 80 | | 171 | 62 | | 150 | 54 | |
| 700 | 209 | 75 | | 108 | 39 | | 82 | 29 | | 77 | 28 | |
| 710 | 110 | 40 | | 53 | 19 | | 39 | 14 | | 41 | 15 | |
| 720 | 57 | 19 | | 26 | 9 | | 19 | 6 | | 17 | 6 | |
| 730 | 28 | 10 | | 12 | 4 | | 8 | 3 | | 9 | 3 | |
| 740 | 14 | 6 | | 6 | 2 | | 4 | 2 | | 5 | 2 | |
| 750 | 6 | 2 | | 2 | 1 | | 2 | 1 | | 2 | 1 | |
| 760 | 4 | 2 | | 2 | 1 | | 1 | 1 | | 1 | | |
| 770 | 2 | | | 1 | | | 1 | | | 1 | | |
| Totals | 109,828 | 100,000 | 35,547 | 99,072 | 100,000 | 85,223 | 98,041 | 100,000 | 118,103 | 95,018 | 100,000 | 108,845 |
| | $x = 0.4476$ | $y = 0.4075$ | | $x = 0.3485$ | $y = 0.3518$ | | $x = 0.3101$ | $y = 0.3163$ | | $x = 0.3127$ | $y = 0.3291$ | |

**TABLE A.5**

WEIGHTED-ORDINATE METHOD FOR DERIVING $X$, $Y$, $Z$ VALUES FOR YELLOW SCHOOL BUS (ILLUMINANT C)

| | Standard Observer | | | Reflectance of Sample | Tristimulus Products | | |
|---|---|---|---|---|---|---|---|
| Wave-length | $(\bar{x}E)$ | $(\bar{y}E)$ | $(\bar{z}E)$ | $R$ (See Fig. 7.5) | $(\bar{x}ER)$ | $(\bar{y}ER)$ | $(\bar{z}ER)$ |
| 380 | 4 | | 20 | | | | |
| 390 | 19 | | 89 | | | | |
| 400 | 85 | 2 | 404 | 0.065 | 5 | | 26 |
| 10 | 329 | 9 | 1,570 | 0.066 | 21 | | 10 |
| 20 | 1,238 | 37 | 5,949 | 0.066 | 81 | 2 | 392 |
| 30 | 2,997 | 122 | 14,628 | 0.068 | 203 | 8 | 994 |
| 40 | 3,975 | 262 | 19,938 | 0.069 | 274 | 18 | 1,375 |
| 450 | 3,915 | 443 | 20,638 | 0.070 | 274 | 31 | 1,444 |
| 60 | 3,362 | 694 | 19,299 | 0.077 | 258 | 53 | 1,486 |
| 70 | 2,272 | 1,058 | 14,972 | 0.090 | 204 | 95 | 1,347 |
| 80 | 1,112 | 1,618 | 9,461 | 0.109 | 121 | 176 | 1,031 |
| 90 | 363 | 2,358 | 5,274 | 0.121 | 43 | 285 | 638 |
| 500 | 52 | 3,401 | 2,864 | 0.130 | 6 | 442 | 372 |
| 10 | 89 | 4,833 | 1,520 | 0.140 | 12 | 676 | 212 |
| 20 | 576 | 6,462 | 712 | 0.160 | 92 | 1,033 | 113 |
| 30 | 1,523 | 7,934 | 388 | 0.235 | 357 | 1,864 | 91 |
| 40 | 2,785 | 9,149 | 195 | 0.336 | 935 | 3,074 | 65 |
| 550 | 4,282 | 9,832 | 86 | 0.415 | 1,777 | 4,080 | 35 |
| 60 | 5,880 | 9,841 | 39 | 0.478 | 2,810 | 4,703 | 18 |
| 70 | 7,322 | 9,147 | 20 | 0.530 | 3,880 | 4,847 | 10 |
| 80 | 8,417 | 7,992 | 16 | 0.561 | 4,721 | 4,483 | 8 |
| 90 | 8,984 | 6,627 | 10 | 0.577 | 5,183 | 3,823 | 5 |
| 600 | 8,949 | 5,316 | 7 | 0.580 | 5,190 | 3,083 | 4 |
| 10 | 8,325 | 4,176 | 2 | 0.583 | 4,853 | 2,434 | 1 |
| 20 | 7,070 | 3,153 | 2 | 0.585 | 4,135 | 1,844 | 1 |
| 30 | 5,309 | 2,190 | | 0.583 | 3,095 | 1,276 | |
| 40 | 3,693 | 1,443 | | 0.584 | 2,156 | 842 | |

**TABLE A.5 (Continued)**

| | Standard Observer | | | Reflectance of Sample | Tristimulus Products | | |
|---|---|---|---|---|---|---|---|
| Wave-length | $(\bar{x}E)$ | $(\bar{y}E)$ | $(\bar{z}E)$ | $R$ (See Fig. 7.5) | $(\bar{x}ER)$ | $(\bar{y}ER)$ | $(\bar{z}ER)$ |
| 650 | 2,349 | 886 | | 0.589 | 1,383 | 521 | |
| 60 | 1,361 | 504 | | 0.595 | 809 | 299 | |
| 70 | 708 | 259 | | 0.611 | 432 | 158 | |
| 80 | 369 | 134 | | 0.632 | 233 | 84 | |
| 90 | 171 | 62 | | 0.665 | 113 | 41 | |
| 700 | 82 | 29 | | 0.694 | 56 | 20 | |
| 10 | 39 | 14 | | | | | |
| 20 | 19 | 6 | | | | | |
| 30 | 8 | 3 | | | | | |
| 40 | 4 | 2 | | | | | |
| 750 | 2 | 1 | | | | | |
| 60 | 1 | 1 | | | | | |
| 70 | 1 | | | | | | |
| Sums | 98,041 | 100,000 | 118,103 | Sums | 43,712 | 40,295 | 9,678 |

Tristimulus values $X = 43.7$ $Y = 40.3$ $Z = 9.8$

Chromaticity coordinates $x = \dfrac{X}{X+Y+Z}$ $y = \dfrac{Y}{X+Y+Z}$ $z = \dfrac{Z}{X+Y+Z}$

$= 0.466$ $= 0.430$ $= 0.104$

Tristimulus values relative to MgO as 100 $\dfrac{43{,}712}{98{,}041}$ $\dfrac{40{,}295}{100{,}000}$ $\dfrac{9{,}678}{118{,}103}$

$X_{\%} = 44.6$ $Y_{\%} = 40.3$ $Z_{\%} = 8.2$

**TABLE A.6**

THIRTY SELECTED WAVELENGTH ORDINATES FOR ILLUMINANTS A, B, AND C(CIE, 1931)[a]

| Ordinate | Source A | | | Source B | | | Source C | | |
|---|---|---|---|---|---|---|---|---|---|
| number | (*X*) | (*Y*) | (*Z*) | (*X*) | (*Y*) | (*Z*) | (*X*) | (*Y*) | (*Z*) |
| 1 | 444.0 | 487.8 | 416.4 | 428.1 | 472.3 | 414.8 | 424.4 | 465.9 | 414.1 |
| 2 | 516.9 | 507.7 | 424.9 | 442.1 | 494.5 | 422.9 | 435.5 | 489.4 | 422.2 |
| 3 | 544.0 | 517.3 | 429.4 | 454.1 | 505.7 | 427.1 | 443.9 | 500.4 | 426.3 |
| 4 | 554.2 | 524.1 | 432.9 | 468.1 | 513.5 | 430.3 | 452.1 | 508.7 | 429.4 |
| 5 | 561.4 | 529.8 | 436.0 | 527.8 | 519.6 | 433.0 | 461.2 | 515.1 | 432.0 |
| 6 | 567.1 | 534.8 | 438.7 | 543.3 | 524.8 | 435.4 | 474.0 | 520.6 | 434.3 |
| 7 | 572.0 | 539.4 | 441.3 | 551.9 | 529.4 | 437.7 | 531.2 | 525.4 | 436.5 |
| 8 | 576.3 | 543.7 | 443.7 | 558.5 | 533.7 | 439.9 | 544.3 | 529.8 | 438.6 |
| 9 | 580.2 | 547.8 | 446.0 | 564.0 | 537.7 | 442.0 | 552.4 | 533.9 | 440.6 |
| 10 | 583.9 | 551.7 | 448.3 | 568.8 | 541.5 | 444.0 | 558.7 | 537.7 | 442.5 |
| 11 | 587.2 | 555.4 | 450.5 | 573.1 | 545.1 | 446.0 | 564.1 | 541.4 | 444.4 |
| 12 | 590.5 | 559.1 | 452.6 | 577.1 | 548.7 | 448.0 | 568.9 | 544.9 | 446.3 |
| 13 | 593.5 | 562.7 | 454.7 | 580.9 | 552.1 | 450.0 | 573.2 | 548.4 | 448.2 |
| 14 | 596.5 | 566.3 | 456.8 | 584.5 | 555.5 | 451.9 | 577.3 | 551.8 | 450.1 |
| 15 | 599.4 | 569.8 | 458.8 | 588.0 | 559.0 | 453.9 | 581.3 | 555.1 | 452.1 |

| | | | | | | | | | |
|---|---|---|---|---|---|---|---|---|---|
| 16 | 602.3 | 573.3 | 460.8 | 591.4 | 562.4 | 455.8 | 585.0 | 558.5 | 454.0 |
| 17 | 605.2 | 576.9 | 462.9 | 594.7 | 565.8 | 457.8 | 588.7 | 561.9 | 455.9 |
| 18 | 608.0 | 580.5 | 464.9 | 598.1 | 569.3 | 459.8 | 592.4 | 565.3 | 457.9 |
| 19 | 610.9 | 584.1 | 467.0 | 601.4 | 572.9 | 461.8 | 596.0 | 568.9 | 459.9 |
| 20 | 613.8 | 587.9 | 469.2 | 604.7 | 576.7 | 463.9 | 599.6 | 572.5 | 462.0 |
| 21 | 616.9 | 591.8 | 471.6 | 608.1 | 580.6 | 466.1 | 603.3 | 576.4 | 464.1 |
| 22 | 620.0 | 595.9 | 474.1 | 611.6 | 584.7 | 468.4 | 607.0 | 580.5 | 466.3 |
| 23 | 623.3 | 600.1 | 476.8 | 615.3 | 589.1 | 470.8 | 610.9 | 584.8 | 468.7 |
| 24 | 626.9 | 604.7 | 479.9 | 619.1 | 593.9 | 473.6 | 615.0 | 589.6 | 471.4 |
| 25 | 630.8 | 609.7 | 483.4 | 623.3 | 599.1 | 476.6 | 619.4 | 594.8 | 474.3 |
| 26 | 635.3 | 615.2 | 487.5 | 628.0 | 605.0 | 480.2 | 624.2 | 600.8 | 477.7 |
| 27 | 640.5 | 621.5 | 492.7 | 633.4 | 611.8 | 484.5 | 629.8 | 607.7 | 481.8 |
| 28 | 646.9 | 629.2 | 499.3 | 640.1 | 619.9 | 490.2 | 636.6 | 616.1 | 487.2 |
| 29 | 655.9 | 639.7 | 508.4 | 649.2 | 630.9 | 498.6 | 645.9 | 627.3 | 495.2 |
| 30 | 673.5 | 659.0 | 526.7 | 666.3 | 650.7 | 515.2 | 663.0 | 647.4 | 511.2 |
| Factor: | 0.03661 | 0.03333 | 0.01185 | 0.03303 | 0.03333 | 0.02842 | 0.03268 | 0.03333 | 0.03938 |

[a] Reproduced from *The Science of Color*, 1963, courtesy of the Optical Society of America.

**TABLE A.7**

SELECTED-ORDINATE METHOD FOR DERIVING *X*, *Y*, *Z* VALUES FOR YELLOW SCHOOL BUS (ILLUMINANT C)

| | | | Reflectance of Sample (See Fig. 7.5) | | |
|---|---|---|---|---|---|
| $X$ | $Y$ | $Z$ | $x$ | $y$ | $z$ |
| 424.4 | 465.9 | 414.1 | 0.066 | 0.080 | 0.066 |
| 435.5 | 489.4 | 422.2 | 0.067 | 0.120 | 0.066 |
| 443.9 | 500.4 | 426.3 | 0.069 | 0.130 | 0.066 |
| 452.1 | 508.7 | 429.4 | 0.071 | 0.138 | 0.066 |
| 461.2 | 515.1 | 432.0 | 0.078 | 0.150 | 0.066 |
| 474.0 | 520.6 | 434.3 | 0.098 | 0.162 | 0.067 |
| 531.2 | 525.4 | 436.5 | 0.250 | 0.209 | 0.067 |
| 544.3 | 529.8 | 438.6 | 0.380 | 0.230 | 0.067 |
| 552.4 | 533.9 | 440.6 | 0.426 | 0.280 | 0.068 |
| 558.7 | 537.7 | 442.5 | 0.460 | 0.326 | 0.069 |
| 564.1 | 541.4 | 444.4 | 0.500 | 0.340 | 0.069 |
| 568.9 | 544.9 | 446.3 | 0.521 | 0.380 | 0.069 |
| 573.2 | 548.4 | 448.2 | 0.545 | 0.402 | 0.070 |
| 577.3 | 551.8 | 450.1 | 0.558 | 0.420 | 0.070 |
| 581.3 | 555.1 | 452.1 | 0.562 | 0.447 | 0.071 |
| 585.0 | 558.5 | 454.0 | 0.570 | 0.460 | 0.072 |
| 588.7 | 561.9 | 455.9 | 0.574 | 0.480 | 0.073 |
| 592.4 | 565.3 | 457.9 | 0.577 | 0.508 | 0.075 |
| 596.0 | 568.9 | 459.9 | 0.580 | 0.520 | 0.076 |
| 599.6 | 572.5 | 462.0 | 0.580 | 0.540 | 0.078 |
| 603.3 | 576.4 | 464.1 | 0.580 | 0.551 | 0.080 |
| 607.0 | 580.5 | 466.3 | 0.581 | 0.562 | 0.080 |
| 610.9 | 584.8 | 468.7 | 0.583 | 0.570 | 0.082 |
| 615.0 | 589.6 | 471.4 | 0.585 | 0.574 | 0.088 |
| 619.4 | 594.8 | 474.3 | 0.585 | 0.579 | 0.098 |
| 624.2 | 600.8 | 477.7 | 0.585 | 0.580 | 0.099 |
| 629.8 | 607.7 | 481.8 | 0.585 | 0.582 | 0.109 |

**TABLE A.7 (Continued)**

| $X$ | $Y$ | $Z$ | Reflectance of Sample (See Fig. 7.5) $x$ | $y$ | $z$ |
|---|---|---|---|---|---|
| 636.6 | 616.1 | 487.2 | 0.584 | 0.585 | 0.120 |
| 645.9 | 627.3 | 495.2 | 0.586 | 0.585 | 0.126 |
| 663.0 | 647.4 | 511.2 | 0.600 | 0.590 | 0.141 |
| Sums | | | 13.386 | 12.080 | 2.414 |
| Factor | | | .03268 | .03333 | .03938 |
| Tristimulus values | | | $X = 43.7$ | $Y = 40.3$ | $Z = 9.5$ |
| Chromaticity coordinates | | | $x = 0.467$ | $y = 0.431$ | $z = 0.102$ |
| Tristimulus values in percent specifications | | | $13.386 \div 30$ $X_{\%} = 44.6$ | $12.080 \div 30$ $Y_{\%} = 40.3$ | $2.414 \div 30$ $Z_{\%} = 8.0$ |

**TABLE A.8**

COMPARISON OF MAJOR COLOR SCALES

| Scale | Equations Relating to CIE or RGB Values | | | Applicability | Visual Data Basis of Scale |
|---|---|---|---|---|---|
| | Lightness Dimension | Red-Green Dimension | Yellow-Blue Dimension | | |
| CIE $Y, x, y$ | $Y$ | No opponent-colors scales | | All stimuli | See p. 83 |
| CIE $Y, \Lambda, p$ | $Y$ | No opponent-colors scales | | All objects | See p. 83 |
| Judd Maxwell Triangle $r, g, b$ (1935) | $Y$ | No opponent-colors scales | | All stimuli | Wavelength and purity interval discrimination |
| MacAdam $u, v$ (1937) | $Y$ | $u = \frac{2x}{6y-x+1.50}$ | $v = \frac{3y}{6y-x+1.50}$ | All stimuli | Revision of Judd Triangle |
| Breckenridge, Schaub RUCS (1939) | $Y$ | $y' = \frac{3.69700x - 5.07713y - 1.36896}{1.00000x - 7.05336y - 1.64023}$ | | All stimuli, but mainly light sources | Revision of Judd Triangle |
| | | $x' = \frac{0.82303x + 0.82303y - 0.82303}{1.00000x - 7.05336y - 1.64023}$ | | | |
| Hunter $L', \alpha', \beta'$ (1942) | $L' = K_l G^{1/2}$ ($K_l$ is often 100) | $\alpha' = \frac{221G^{1/4}(R-G)}{R+2G+B}$ | $\beta' = \frac{88.4G^{1/4}(G-B)}{R+2G+B}$ | Diffusing surfaces under Illuminant C | Revision of Judd Triangle, and Munsell spacing |

| | | | | | |
|---|---|---|---|---|---|
| Scofield $L_s$, $a_s$, $b_s$ (1943) | $L_s = 10G^{1/2}$ | $a_s = \frac{70G^{1/2}(R-G)}{R+2G+B}$ | $b_s = \frac{28G^{1/2}(G-B)}{R+2G+B}$ | Diffusing surfaces under Illuminant C | Revision of Hunter alpha-beta |
| Adams Chromatic Value (1942)[a] | $L_{VAL} = V_Y$ | $a_{VAL} = V_X - V_Y$ | $b_{VAL} = 0.4(V_Z - V_Y)$ | Diffusing surfaces under Illuminant C | Munsell color intervals |
| Adams Chromatic Valence (1943) | $L_{VNC} = V_Y$ | $a_{VNC} = \left(\frac{X_{\%}}{Y} - 1\right) V_Y$ | $b_{VNC} = 0.4\left(\frac{Z_{\%}}{Y} - 1\right) V_Y$ | Diffusing surfaces under Illuminant C | Munsell color intervals |
| Adams-Nickerson (Nickerson, 1950) | $L_A = 9.2V_Y$ | $a_A = 40\,(V_X - V_Y)$ | $b_A = 16\,(V_Y - V_Z)$ | Diffusing surface under Illuminant C | Revision of Adams chromatic value |
| Modified Adams-Nickerson (Glasser and Troy, 1952) | $L_M = 10V_Y$ | $a_M = 41.86(V_X - V_Y)$ | $b_A = 16.74(V_Y - V_Z)$ | Diffusing surfaces under Illuminant C | Revision of Adams chromatic value |
| Hunter $R_d$, $a_{Rd}$, $b_{Rd}$ (1958)[b] | $L_{Rd} = Yf(Y)$ | $a_{Rd} = 1.75fY(X_{\%} - Y)$ | $b_{Rd} = 0.70fY(Y - Z_{\%})$ | Diffusing surfaces under Illuminant C | Munsell color intervals |

Note: Tristimulus values above refer to perfect neutral white as 1.0 for $X_{\%}$, $Y$, and $Z_{\%}$ (also for $R$, $G$, $B$). Below, the values for the perfect neutral white are 100.0.

**TABLE A.8 (Continued)**

COMPARISON OF MAJOR COLOR SCALES (CONT.)

| Scale | Equations Relating to CIE or RGB Values | | | Applicability | Visual Data Basis of Scale |
|---|---|---|---|---|---|
| | Lightness Dimension | Red-Green Dimension | Yellow-Blue Dimension | | |
| Hunter $L_L, a_L, b_L$ (1958) | $L_L = 10\sqrt{Y}$ | $a_L = \frac{17.5(X_{\%} - Y)}{\sqrt{Y}}$ | $b_L = \frac{7.0(Y - Z_{\%})}{\sqrt{Y}}$ | Diffusing surfaces under Illuminant C | Munsell color intervals |
| Glasser Cube Root (Glasser et al., 1958) | $L_G = 25.29G^{1/3} - 18.38$ | $a_G = 106.0(R^{1/3} - Y^{1/3})$ (Note: $X_{\%}$ is widely used in place of $R$) | $b_G = 42.34(Y^{1/3} - Z_{\%}{}^{1/3})$ | Diffusing surfaces under Illuminant C | Munsell color intervals |
| Friele $R, G, B$ (1961) | $R + G$ | $R - G$ | $\frac{R + G}{2} - B$ | Diffusing surfaces | MacAdam (1942) threshold discriminations |
| CIE $U^*V^*W^*$ (Wyszecki, 1963) | $W^* = 25Y^{1/3} - 17$ | $U^* = 13W^* (u - u_o)$ ($u_o$ is $u$ value of illuminant) | $V^* = 13W^* (v - v_o)$ ($v_o$ is $v$ value of illuminant) | Diffusing surfaces | Revision of MacAdam $u, v$ |
| Friele-MacAdam, $P, Q, S$ (MacAdam, 1965) | $P + Q$ | $P - Q$ | $\frac{P - Q}{2} - S$ | Diffusing surfaces | Revision of Friele $R, G, B$ |

| | | | | | |
|---|---|---|---|---|---|
| Friele, MacAdam, Chickering (FMC) (Chickering, 1967) | $P + Q$ | $P - Q$ | $\frac{P + Q}{S} - S$ | Diffusing surfaces | Revision of Friele-MacAdam $P, Q, S$ |
| $L_Q$, $a_Q$, $b_Q$ scales for any illuminant $Q$ (Hunter, Christie, 1966)[c] | $L_Q = 10\sqrt{Y_Q}$ | $a_Q = \frac{175jx}{0.1L_Q} \times \left(\frac{1.02X_Q}{jx} - Y_Q\right)$ | $b_Q = \frac{70jz}{0.1L_Q} \times \left(Y_Q - \frac{0.847Z_Q}{jz}\right)$ | Diffusing surfaces | Revision of Hunter $L_L$, $a_L$, $b_L$ |
| $L'$, $a'$, $b'$ scales for clear liquids and metals | $L' = 10\sqrt{Y}$ | $a' = 175\left(\frac{1.02X}{Y} - 1\right)$ | $b' = 70\left(1 - \frac{0.847Z}{Y}\right)$ | Clear liquids and solids, metals (specular colors) | Hunter $L$, $a$, $b$ analogs of citrus colorimeter scales, (Hunter, 1967) |
| CIE 1976 ($L^*u^*v^*$) | $L^* = 25(100Y/Y_o)^{1/3} - 16$ | $u^* = 13L^*(u' - u_o')$<br>$u' = u$ (MacAdam 1937) | $v^* = 13L^*(v' - v'_o)$<br>$v' = 1.5v$ (MacAdam 1937) | Diffusing surfaces | Revision of CIE $U^*V^*W^*$ |
| CIE 1976 ($L^*a^*b^*$) | $L^* = 25(100Y/Y_o)^{1/3} - 16$ | $a^* = 500[(X/X_o)^{1/3} - (Y/Y_o)^{1/3}]$ | $b^* = 200[(Y/Y_o)^{1/3} - Z/Z_o)^{1/3}]$ | Diffusing surfaces | Revision of Adams-Nickerson |

[a] $V$ represents the Munsell value function for which $X_{\%}$, $Y_{\%}$, or $Z_{\%} = 1.2219V - 0.23111V^2 + 0.23951V^3 - 0.021009V^4 + 0.0008404V^5$.

[b] $f(Y)$ in Hunter $R_d$ equations $= .51\ (21 + 0.2Y)/(1 + 0.2Y)$.

[c] For any Illuminant $Q$, $X$, $Y$ and $Z$ computed in regular way; $jx = X_c/X_Q$ and $jz = Z_c/Z_Q$ for the ideal white standard.

**TABLE A.9**

MATHEMATICAL RELATIONSHIPS BETWEEN COLOR SCALES FOR ILLUMINANT C, 1931 2° STANDARD OBSERVER

| To Convert From To→ ↓ | $L, a, b$ | $X_{\%}$ $Y$ $Z_{\%}$ | CIE $X, Y, Z$ |
|---|---|---|---|
| $L, a, b$ | | $Y = 0.01L^2$<br>$X_{\%} = 0.01L^2 + \frac{aL}{175}$<br>$Z_{\%} = 0.01L^2 - \frac{bL}{70}$ | $Y = 0.01L^2$<br>$X = 0.9804\left(0.01L^2 + \frac{aL}{175}\right)$<br>$Z = 1.181\left(0.01L^2 - \frac{bL}{70}\right)$ |

| | | | |
|---|---|---|---|
| $X_{\%}, Y, Z_{\%}$ | $L = 10\sqrt{Y}$<br>$a = \frac{17.5(X_{\%} - Y)}{\sqrt{Y}}$<br>$b = \frac{7.0(Y - Z_{\%})}{\sqrt{Y}}$ | | $Y = Y_{\%}$<br>$X = 0.9804X_{\%} = \frac{X_{\%}}{1.02}$<br>$Z = 1.181Z_{\%} = \frac{Z_{\%}}{0.847}$ |
| CIE<br>$X, Y, Z$ | $L = 10\sqrt{Y}$<br>$a = \frac{17.5(1.02X - Y)}{\sqrt{Y}}$<br>$b = \frac{7.0(Y - 0.847Z)}{\sqrt{Y}}$ | $Y_{\%} = Y$<br>$X_{\%} = \frac{X}{0.9804} = 1.02X$<br>$Z_{\%} = \frac{Z}{1.181} = 0.847Z$ | Other Relationships:<br>$x = \frac{X}{X + Y + Z}$<br>$y = \frac{Y}{X + Y + Z}$<br>Hunter/Gardner<br>Multipurpose Reflectometer:<br>$G = Y_{\%}$<br>$A = R = 1.25X_{\%} - 0.25Z_{\%}$<br>$B = Z_{\%}$ |

**TABLE A.10**

COLORIMETRIC ATTRIBUTES OF WHITE SURFACES AND THEIR USES IN INDUSTRIES

| Industry | Lightness ($L$), or Luminous Reflectance $Y = R_d = G$ | Blue Reflectance $Z(\%) = B$, or Paper Brightness | Yellowness $YI^a$ | Whiteness $WI^b$ |
|---|---|---|---|---|
| Paint | Reflection efficiency, opacity | — | Yellowness, and yellowing in use | Product whiteness |
| Paper | Opacity | Pulp and paper brightness after bleaching | Supercalendar scorching | Product whiteness |
| Textiles | Raw fiber grading, soiling studies | Textiles after bleaching | Yellowness, yellowing in processing | Product whiteness |
| Soap, detergents, cleaners | Efficiency of cleaners and cleaning processes. Redeposition of soil on clean fabrics | Effectiveness of bleaches and optical brighteners in cleaners | Contribution of bleaches and optical brighteners to elimination of yellowness | Product whiteness after cleaning |
| Plastics | — | — | Yellowness, and yellowing in service | Product whiteness |
| Ceramics | Reflection efficiency, Opacity | — | — | Product whiteness |

[a] See text for alternate equations for Yellowness Index.

[b] See text for alternate equations for Whiteness Index.

**TABLE A.11**

VISUAL COLOR SCALES FOR THE YELLOWNESS SPECIFICATION OF TRANSPARENT LIQUIDS AND RESINS

| Scale Names & Reference | Attributes & Corresponding Quantities | Applications |
|---|---|---|
| 1. The Gardner Standards: ASTM D1544-68 | Yellowness by means of comparison with arbitrarily numbered glass standards between 1 and 18 | Color of drying oils, varnishes, resins, etc. |
| 2. Gardner-Holdt: Paint and Varnish Industry, Paint Manufacturing Assoc. of U. S., Circular 1939, 85 (1921) | Yellowness by means of comparison with arbitrarily numbered standard caramel solutions between 1 and 10 | Color of drying oils, varnishes, resins, etc. |
| 3. Union Lubricating Oil: ASTM D1500-65 | Yellowness by means of comparison with arbitrarily numbered glass standards between 0.5 and 8.0 | Color of lubricating oil |
| 4. AOCS, FAC Colors: Oil and Soap, Vol. 11, 46 (1934) | Yellowness by means of comparison with arbitrarily numbered glass standards | Color of oils and fats, both edible and nonedible |
| 5. Hazen Standards (APHA): *Am. Chem. J.* **14,** 300, 1892. American Public Health Asso. Inc. 10th ed. 1790 Broadway, New York, N.Y. | Yellowness by means of comparison with solutions of platinum cobalt arbitrarily numbered from 0 to 700 | Standard methods for examinating water, sewage and industrial wastes, nearly colorless solvents and other industrial chemicals |
| 6. Parlin Color Standards: ASTM D365-39 | Yellowness by means of comparison with 35 Standards, of which the first 10 are platinum cobalt solutions, and the rest are caramel solutions arbitrarily numbered from 4 to 500 | Color of nitrocellulose base solutions |
| 7. Barrett Color Standards: Physical and Chemical Examinations of Paints, 11th ed., Gardner and Sward, p. 96–97, 1950 | Yellowness by means of comparison with 14 standard solutions consisting of $CoCl_2 \cdot 6H_2O$, $FeCl_3 \cdot 6H_2O$, and $K_2Cr_2O_7$ | Benzene and coal tar solvents |

**TABLE A.11 (Continued)**

| Scale Names & Reference | Attributes & Corresponding Quantities | Applications |
|---|---|---|
| 8. Potassium Dichromate: *Ind. Eng. Chem.*, **16,** No. 1, 42–44, 1944 | Yellowness by means of comparison with 18 solutions of $K_2Cr_2O_7$ in $H_2SO_4$. The number of the standard corresponds to the grams of $K_2Cr_2O_7$(0.0039 – 3.0000) | Sulfuric acid colors |
| 9. Rosin Standards: USDA Miscellaneous, Circular No. 22, Supplement No. 7, December, 1935 JOSA, **30,** No. 4, 152–158 (1940) | Yellowness by means of comparison with 15 standard glass cubes designated by letters of the alphabet such as X, N, M, etc. | Color of rosin and resins |
| 10. British Paint Research Station Color Scales: Color Scales for Oils & Varnishes, D. L. Tilliard, *J. Oil Color Chem. Assoc.*, **20,** 124, 1937 | Yellowness by means of comparison with 20 glass standards ranging from ½ to 10 in half steps. (The half steps from curves drawn from Maxwell triangle charts) | Color of varnishes using Lovibond glasses |
| 11. Arny Color Standards: *J. Ind. Eng. Chem.*, 309, 1916 | Yellowness by means of comparison with 17 solutions of ferric chloride and cobaltous chloride in 2% HCl | For colors of all amber liquids |
| 12. Saybolt Colors: ASTM D156-64 | Yellowness by means of comparison with specified glass standards, ranging from +30 to −16 | Petroleum oils and lubricants |
| 13. Lovibond Colors: The Lovibond Color Systems, Scientific Papers of the NBS, No. 547. Colorimetry, a handbook of the Lovibond Tintometer, The Tintometer Ltd. Salisbury, England | Yellowness by means of comparison with 155 slides for each of the three colors, red, yellow and blue combined to make a match | Color of oils, other liquids, and some reflecting materials |

**TABLE A.11 (Continued)**

| Scale Names & Reference | Attributes & Corresponding Quantities | Applications |
|---|---|---|
| 14. Permanent Glass Color Standards for Maple Syrup: USDA Bulletin AIC-260, Feb., 1950 | | USDA ratings of maple syrup |
| 15. Permanent Glass Standards for Extracted Honey: USDA Bulletin AIC-307, May 1951 | | USDA ratings of honey |
| 16. Bryan Color Numbers: *J. Ind. Eng. Chem.*, **22,** 255, 1930 | | Color of maple syrup |
| 17. Hellige Standards: Hellige Tech. Bulletin 620C-80T1 | | Refined petroleum oils, other colored solutions |
| 18. Ferric Sulfate–Nickel Sulfate Standard: ASTM D29-65 | | Color of orange shellac |
| 19. Iodine Standard: Sec. 3.15 of "Official Methods of Analysis, Standards; Specifications and General Information on Shellac and Bleached Shellac of the American Bleached Shellac Mfrs. Assoc. Inc." | | Color of shellacs, varnish, and resins |

**TABLE A.12**

DIAGNOSTIC STANDARDS AND THEIR USES TO DETECT INSTRUMENT ERRORS

| | | Type of Measurement | | | |
|---|---|---|---|---|---|
| Type | Uses in Detecting Instrument Errors | Diffuse Reflection | Specular Reflection | Diffuse Transmission | Specular Transmission |
| Photometric scale accuracy set | Scale and photo-detector nonlinearity are detected by whether photometric intervals register properly | White, gray, black series of reflectance standard tiles | Series of metal mirrors | — | Series of neutral filters |
| Zero scale accuracy set | Permits reading on zero value scale in order to detect zero scale errors such as stray light responses | Polished black glass | Black cavity | Clear air | Opaque object |
| Spectral accuracy set | Identifies correct responses to wave-lengths | | | | |

| | | | | | |
|---|---|---|---|---|---|
| a. Center of gravity pairs | Pairs show whether average wavelength response is too high or too low | A gray, and a brown, tan, or yellow tile of same $X$, $Y$, or $Z$ value | — | — | Neutral and brown filter pairs |
| b. Metameric pairs | Pairs detect difference in shapes of wavelength response curves | Metameric pairs of tiles having same $X$, $Y$, $Z$, $a$ or $b$ value | — | — | Metameric spectral filter pairs |
| Geometric accuracy set | Identifies correct combinations of directions of incidence and view | | | | |
| a. Geometric metamers | Pairs detect differences in geometric response to light | A ceramic gray and diffuse, aluminum panel of about same diffuse reflectance factor | High and low distinctness-of-image pair of same specular reflectance factor | Wide-and narrow-angle films of same diffuse transmittance factor (haze) | High and low distinctness-of-image pair of same specular transmittance factor |
| b. Translucency error | Pairs detect differences in response to penetration of light into specimen | Porcelain enamel and translucent glass panels of same diffuse reflectance factor | | Thick and thin volumes of same diffuse transmittance factor | |

**TABLE A.13**

STANDARDS USED FOR THE MEASUREMENT OF DIFFUSE REFLECTION

| Class | Materials Used | Chief Applications | Notes |
|---|---|---|---|
| Ideal reference | Does not exist | Provides 100% reference for reflectance factor measurements. Serves as reference for calibration of primary standards. | Reflectance factor is the ratio of flux reflected from a specimen for specified geometric and spectral conditions of radiation and view to that reflected from a perfect diffuse reflector under the same conditions. |
| Primary standards | Pressed powder tablets | Used to calibrate master standards | Fresh, reflectance-standard grade MgO and $BaSO_4$ powders are measured by using auxiliary sphere or Taylor (1920) method to arrive at absolute reflectance value |
| Secondary standards | Porcelain enamel panels, opaque glass panels, ceramic wall tile, acrylic painted chips, Munsell or other painted chips | | |
| Master | | Used to calibrate other secondary standards | These standards are kept carefully in boxes and used only for calibration of secondary standards |
| Working | | Used as reference for calibration of a group of laboratory standards | Although ceramic materials most often used are quite stable, standards must nevertheless be treated with care and kept clean |
| Instrument | | Used to maintain calibration of one specific instrument | |
| Hitching post | | Used instead of the white standard to standardize the instrument | This standard is chosen to be close in color to samples to be measured |

**TABLE A.13 (Continued)**

| Class | Materials Used | Chief Applications | Notes |
|---|---|---|---|
| Specific calibration | | Used as reference to measurements of an impermanent specific product | Because of unintended or intentional differences between instruments, it is often necessary to assign numbers to a standard so that one can obtain on a specific product numbers from a second instrument comparable to those given by a first or master instrument |
| Material comparison standards (uncalibrated) | Usually a specimen of product to be evaluated | Used to represent target color, or limit color, for product being evaluated | Standard samples of commerce. Freezer or bone-dry storage is used to keep comparison standards of paint, paper, etc. for a long time |

## TABLE A.14

STANDARDS USED FOR THE MEASUREMENT OF SPECULAR REFLECTION

| Class | Materials Used | Chief Application | Notes |
|---|---|---|---|
| Ideal reference (perfect mirror) | Does not exist | Provides 100% reference for specular reflectance measurement | A goniophotometer in which incident beam is aimed at receptor gives same signal as perfect mirror |
| Primary standards | Polished black glass, or liquids on ground glass | Used to calibrate master standards | Surface gloss is computed by using the refractive index of the black glass and the Fresnel equation |
| Master standards | Polished black glass for high gloss, diffuse porcelain enamel or tile for intermediate or low gloss, evaporated metal on glass for metallic reflectance | Used to calibrate secondary standards | These standards are kept carefully in boxes and used only for calibration of secondary standards |
| Secondary standards | | Used for day-to-day operation of glossmeters both in setting scales for every series of observations, and for periodic calibrations | Nonmetal standards are used for gloss of nonmetal paints, plastics, papers, etc., metal-film standards are used for gloss of base metal surfaces |
| Material comparison standards (uncalibrated) | Specimens of products to be evaluated | Used to represent target or limit gloss of product under examination | Careful handling and storage needed to avoid change of surface and therefore of gloss. Tolerances for gloss are generally greater than those for color and diffuse reflectance. |

**TABLE A.15**

MAJOR U.S. PUBLISHED PROCEDURES FOR THE ANALYSIS OF APPEARANCE

PART 1. GENERAL PURPOSE METHODS APPLICABLE TO APPEARANCE

a. General Purpose Methods for Color Attributes

| Type of Measurement | ASTM | TAPPI | Federal Test | Other |
|---|---|---|---|---|
| Practice for spectrophotometry and description of color in CIE 1931 system | E308 | T442 | | |
| Indexes of whiteness and yellowness of near-white opaque materials | E313 | | | |
| Yellowness index of plastics | D1925 | | 141a/6131 | |

b. General Purpose Methods for Geometric Attributes

| Type of Measurement | ASTM | TAPPI | Federal Test | Other |
|---|---|---|---|---|
| Goniophotometry of transmitting objects and materials | E166 | | 406/3031 | |
| Goniophotometry of reflecting objects and materials | E167 | | 406/3031 | |
| 45°, 0° Directional reflectance of opaque specimens by filter photometry | E97 | | 141a/4252, 6121, 6122, 6131, 6242 | |

c. General Purpose Recommended Practices and Definitions for Appearance

| Type of Measurement | ASTM | TAPPI | Federal Test | Other |
|---|---|---|---|---|
| Definition of terms relating to appearance of materials | E284 | 017.04 | | |
| Selection of geometric conditions for measurement of reflectance and transmittance | E179 | | | |
| Preparation of reference white reflectance standards | E259 | | | |
| Absolute calibration of reflectance standards | E306 | | | |
| Illuminants for visual evaluation of color differences of opaque materials | D1729 | T508 | | |
| Lighting cotton classing rooms for color grading (recommended practice) | D1684 | | | |

**TABLE A.15 (Continued)**

PART 2. METHODS FOR APPEARANCE ANALYSIS OF OPAQUE NONMETALLIC MATERIALS, COLOR OF WHICH IS SEEN PRIMARILY BY DIFFUSELY REFLECTED LIGHT

a. Methods for Color Attributes of Opaque NonMetallic Materials

| Type of Measurement | ASTM | TAPPI | Federal Test | Other |
|---|---|---|---|---|
| Specifying color by the Munsell system | D1535 | | | |
| Color of paper and paperboard by Hunter *L*, *a*, *b* colorimetry | | T524sm | | |
| Color of paper and paperboard in CIE *Y*, *x*, *y* or *Y*, dominant wavelength and excitation purity | | T527sm | | |
| Instrumental evaluation of color differences of opaque materials | D2244 | RC348 | 141a/6123 | |
| Evaluating change in color with a gray scale | D2616 | | | AATCC Eval. Proc. 1 |
| Color change of white architectural enamel | D1543 | | | |
| Small color differences between ceramic wall or floor tile | C609 | | | |
| Color of strength of color pigments by use of a miniature sand mill | D3022 | | | |
| Reflectance, blue, and whiteness of bleached fabric | | | | AATCC 110 |
| Yellowness index (by reflection) | D1925 | | 141a/6131 | |
| Indexes of whiteness and yellowness of near-white opaque materials | E313 | | | |

**TABLE A.15 (Continued)**

PART 2b. Methods for Geometric Attributes of Opaque NonMetallic Materials

| Type of Measurement | ASTM | TAPPI | Federal Test | Other |
|---|---|---|---|---|
| GLOSS | | | | |
| Specular gloss of nonmetallic surfaces | | | | |
| 60° (medium gloss) | D523<br>D1455 | | 141a/6101<br>406/3051 | NEMA<br>LD1-2.13 |
| 20° (high gloss) | D523 | | 141a/6104 | |
| 85° (sheen) | D523 | | 141a/6103 | |
| Specular gloss of plastic films (60°, 20°, 45°) | D2457 | | | |
| 75° Specular gloss of paper and paperboard | D1223 | T480os | | |
| 60° Specular gloss of emulsion floor polish | D1455 | | | |
| Two-parameter, 60° specular gloss | D1471 | | | |
| 20° Specular gloss of waxed paper | D1834 | T653os | | |
| 45° Specular gloss of ceramic materials | C346 | | | PEI T2, T7, T18 |
| Abrasion resistance of porcelain enamels | C448 | | | PEI T2, T18 |
| Image gloss of porcelain enamel surfaces | C540 | | | |
| 60° Specular gloss of glazed ceramic whitewares and related products | C584 | | | |
| DIFFUSE REFLECTANCE | | | | |
| Reflectivity and coefficient of scatter of white porcelain enamels | C347 | | | PEI T13 |
| Light reflectance of acoustical materials by the integrating sphere reflectometer | C523 | | | |

**TABLE A.15 (Continued)**

PART 2b. Diffuse Reflectance of Nonmetallic Materials (Continued)

| Type of Measurement | ASTM | TAPPI | Federal Test | Other |
|---|---|---|---|---|
| Opacity of paper (TAPPI) | D589 | T425m | 141a/4121, 4122 | |
| Diffuse opacity of paper (printing opacity) | | T519su | | |
| 45° 0° directional reflectance for blue light (brightness) of paper (and pulp) | D985 | T452, T646, T217 | | |
| Instrumental tinting strength of white pigments | D2745 | | | |
| Hiding power of paints | D2805 | | 141a/4121, 4122 | |
| Measuring soil removal and reflectance retention of fabrics | D3050 | | | |
| Soil resistance of floor finishes | D3206 | | | |
| Night visibility of traffic paints | D1011 | | | |
| Determination of dry brightness of pulverized limestone | | | | PLA 5 |
| 45°, 0° Directional reflectance of opaque specimens by filter photometry | E97 | | | |

**TABLE A.15 (Continued)**

PART 3. METHODS FOR APPEARANCE ANALYSIS OF METALLIC AND LIGHT-TRANSMITTING MATERIALS

a. Methods for Color Attributes of Metallic and Light-Transmitting Materials

| Type of Measurement | ASTM | TAPPI | Federal Test | Other |
|---|---|---|---|---|
| YELLOWNESS, PRIMARILY | | | | |
| Saybolt color of petroleum products (Saybolt Chromometer Method) | D156 | | | |
| Color of clear liquids (platinum-cobalt scale) | D1209 | | | |
| ASTM color of petroleum products (ASTM color scale) | D1500 | | | |
| Color of transparent liquids (Gardner Color Scale) | D1544 | | 141a/4248 | |
| Color of solid aromatic hydrocarbons and related materials in the molten state (platinum-cobalt scale) | D1686 | | | |
| Color of halogenated organic solvents and their admixtures (platinum-cobalt scale) | D2108 | | 141a/4243.1 | |
| Color of films from water emulsion floor polishes | D3210 | | | |
| Measurement of color of low-colored clear liquids using the Hunterlab Color Difference Meter | E450 | | | |
| Yellowness index of plastic | D1925 | | | |

**TABLE A.15 (Continued)**

PART 3b. Methods for Geometric Attributes of Metallic and Light-Transmitting Materials

| Type of Measurement | ASTM | TAPPI | Federal Test | Other |
|---|---|---|---|---|
| Measurement and calculation of reflecting characteristics of metallic surfaces using integrating sphere instruments | E429 | | | |
| Gloss of high-gloss metal surfaces using abridged goniophotometer or goniophotometer | E430 | | | |
| Haze and luminous transmittance of transparent plastics | D1003 | . | 406/3022 | |
| Diffuse light transmission factor of reinforced plastics panels | D1494 | | | |
| Transparency of plastic sheeting | D1746 | | | |
| Solar energy transmittance and reflectance (terrestrial) of sheet materials | E424 | | | |
| Specular gloss of plastic films (60°, 20°, 45°) | D2457 | | | |
| Surface irregularities of flat transparent plastic sheets | D637 | | | |
| Deviation of line of sight through transparent plastics | D881 | | | |
| Transparency of paper | | T522 | | |

# BIBLIOGRAPHY AND INDEX OF AUTHORS


Adams, E. Q., A Theory of Color Vision, *Psychol. Rev.,* **30,** 56, 1923, 117.

Adams, E. Q., *X-Z* Planes in the 1931 I.C.I. System of Colorimetry, *J. Opt. Soc. Am.,* **32,** 168–173, 1942, 106, 117, 119–121.

Adams, E. Q., Chromatic Valence as a Correlate of Munsell Chroma, *J. Opt. Soc. Am.,* **33,** 683 A, 1943, 117, 119.

Adams, J. M., *Optical Measurements in the Printing Industry,* Pergamon Press, Oxford, 1965, 114, 187.

Allen, E., Digital Computer Color Matching, Am. Dyest. Rep., **54,** 57–63, 1965, 275.

Allen, E., Basic Equations Used in Computer Color Matching, *J. Opt. Soc. Am.,* **56,** 1256–1259, 1966, 275.

ASTM, Recommendations on Form of ASTM Standards, American Society for Testing and Materials, 13-000001-00, Philadelphia, Pa., 1972, 280.

Ayers, J. W., Inert Pigments and their Effects in Flat Paints, Scien. Sect. Circ. No. 568, Nat. Paint, Varnish and Lacquer Assn., 109, 1938, 73.

Balinkin, I. H., Measurement and Designation of Small Color Differences, *Bull. Am. Ceram. Soc.,* **20,** 392–402, 1941, 137.

Barkman, E. F., Specular and Diffuse Reflectance Measurements of Aluminum Surfaces, Metallurgical Research Report, 571-13A, Reynolds Metal Co., April 1959, 75.

Barkman, E. F., Appearance of Metallic Surfaces, ASTM STP 478, American Society for Testing and Materials, 46–58, 1970, 75.

Barnes, B. T., A Four-filter Photoelectric Colorimeter, *J. Opt. Soc. Am.,* **29,** 448, 1939, 241.

Baumgartner, G. R., A Light-sensitive Cell Reflectometer, *General Electric Review,* 525–527, November 1937, 224, 248.

Bellamy, B. R., and S. M. Newhall, Attributive Limens in Selected Regions of the Munsell Color Solid, *J. Opt. Soc. Am.,* **32,** 465–473, 1942, 137.

Billmeyer, F. W., Jr., The Significance of Recent CIE Recommendations for Color Measurement, *Color Eng.,* **6,** No. 1, 34–38, 1968, 127.

Billmeyer, F. W., Jr., The MacAdam Color-Differences Metrics, *Opt. Spectra,* 64–70, November/December 1969, 142.

Billmeyer, F. W., Jr., Comparative Performance of Color-Measuring Instruments, *Appl. Opt.,* **8,** 775–783, 1969, 132.

Billmeyer, F. W., Jr., and M. Saltzman, *Principles of Color Technology,* John Wiley & Sons, New York, 1966, 87.

Billmeyer, F. W., Jr., and R. Smith, Optimized Equations for MacAdam Color-Difference Calculation, *Color Eng.,* **5,** No. 6, 28–29, 1967, 141.

Breckenridge, F. C., and W. R. Schaub, Rectangular-uniform-chromaticity-scale Coordinates, *J. Opt. Soc. Am.*, **29**, 370–380, 1939, 111.

Brown, W. R. J., The Influence of Luminance Level on Visual Sensitivity to Color Differences, *J. Opt. Soc. Am.*, **41**, 686–680, 1951, 141.

Brown, W. R. J., Color Discrimination of Twelve Observers, *J. Opt. Soc. Am.*, **47**, No. 2, 137–143, 1957, 149.

Brown, W. R. J., and D. L. MacAdam, Visual Sensitivities to Combined Chromaticity and Luminance Differences, *J. Opt. Soc. Am.*, **39**, 808–834, 1949, 141.

Chickering, K. D., Optimization of the MacAdam-Modified 1965 Friele Color Difference Formula, *J. Opt. Soc. Am.*, **57**, 537–541, 1967, 126, 141.

Chickering, K. D., Perceptual Significance of the Differences Between CIE Tristimulus Values, *J. Opt. Soc. Am.*, **59**, 986–990, 1969, 141.

Christie, J. S., *Instruments for the Measurement of Metallic Appearance,* ASTM, Philadelphia, Pa., 1970, 75, 230.

CIE, International Commission on Illumination, Proceedings of the Sixth Session, Geneva, 1924, 81.

CIE, International Commission on Illumination, Proceedings of the Eighth Session, Cambridge, England, 1931, 50, 85, 296.

CIE, International Commission on Illumination, *International Lighting Vocabulary,* 3rd ed., CIE No. 17 (E-1.1.), 1970, 247.

CIE, International Commission on Illumination, Colorimetry, CIE No. 15 (E-1.3.1.), 1971, 50.

Coblentz, W. W., and W. B. Emerson, *Relative Sensibility of the Average Eye to Light of Different Colors and Some Practical Applications of Radiation Problems,* U.S. Bureau of Standards Bull. 14, 1918, 81.

Considine, D. M., and S. D. Ross, Eds., *Handbook of Applied Instrumentation,* McGraw-Hill Book Co., New York, 1964, 23.

Coppock, W. A., The Chemstrand Whiteness Scale, *Am. Dyest. Rep.*, 343–346, 1965, 164.

Dana, R., H. S. Bayer, and G. W. McElrath, Color and Shade Control of Ceramic Tile, *Ind. Qual. Control,* 608–614, June, 1965, 277.

Davidson, H. R., and E. Friede, The Size of Acceptable Color Differences, *J. Opt. Soc. Am.*, **43**, 581–589, 1953, 141.

Davidson, H. R., and J. J. Hanlon, Use of Charts for Rapid Calculation of Color Difference, *J. Opt. Soc. Am.*, **45**, 617–620, 1955, 141.

Davidson, H. R., and H. Hemmendinger, Color Prediction Using the Two-Constant Turbid-Media Theory, *J. Opt. Soc. Am.*, **56**, 1102–1109, 1966, 275.

Davies, W. E. R., and G. Wyszecki, Physical Approximation of Color Mixture Functions, *J. Opt. Soc. Am.*, **52**, 679–685, 1962, 210.

Davis, M. N., The Brightness Tester, *Tech. Assoc. Pap.*, **17**, 131–134, 1934, 161.

Dresler, A., and H. G. Frühling, Über ein Photoelektrisches Dreifarbenmessgerät, *Das Licht,* **8**, 238, 1938, 241.

Evans, R. M., *An Introduction to Color,* John Wiley & Sons, New York, 1948, 5.

Foss, C. E., D. Nickerson, and W. C. Granville, Analysis of the Ostwald Color System, *J. Opt. Soc. Am.*, **34**, 361–381, 1944, 183.

Foss, C. E., and G. G. Field, The Foss Color Order System, GATF Research Progress Report No. 96, 1973, 173.

Foster, R. S., Color Speed Computing Charts, A New Simplified System of Charts for Rapid Color Difference Calculations, *Color Eng.*, **4,** January-February 1966, 141.

Friele, L. F. C., Analysis of the Brown and Brown-MacAdam Colour Discrimination Data, *Die Farbe,* **10,** 193–224, 1961, 106, 125.

Friele, L. F. C., Further Analysis of Color Discrimination Data, *J. Opt. Soc. Am.,* **55,** 1314–1319, 1965, 125.

Friele, L. F. C., Preliminary Analysis of the Munsell Colour System in Terms of the Mueller Theory, *Die Farbe,* **20,** 215–229, 1971, 125, 147.

Gage, H. P., Color Filters for Altering Color Temperature, Pyrometer Absorption and Daylite Glasses, *J. Opt. Soc. Am.,* **23,** 46–54, 1933, 52.

Gage, H. P., Glass Color Filters for Special Applications, *J. Opt. Soc. Am.,* **27,** 159, 1937, 210.

Gibson, K. S., Photoelectric Photometers and Colorimeters, *Instruments,* **9,** 309, 335, 1936. See also NBS Letter Circular LC545, Photoelectric Colorimeters, March 8, 1939, 241.

Gibson, K. S., F. K. Harris, and I. G. Priest, The Lovibond Color System, I. A spectrophotometric analysis of the Lovibond glasses, NBS Sci. Paper 547, **22,** 1927, 172.

Gibson, G. L., and D. A. Neubrech, Automatic Photoelectric Colorimeter for Direct Reading of Munsell Coordinates, *J. Opt. Soc. Am.,* **44,** 703–712, 1954, 127.

Gibson, K. S., and E. P. Tyndall, Visibility of Radiant Energy, NBS Sci. Paper 475, 1923, 81.

Glasser, L. G., A. H. McKinney, C. D. Reilly, and P. D. Schnelle, Cube-Root Color Coordinate System, *J. Opt. Soc. Am.,* **48,** 736–740, 1958, 123, 124, 139.

Glasser, L. G., and D. J. Troy, A New High-Sensitivity Differential Colorimeter, *J. Opt. Soc. Am.* **42,** 652–660, 1952, 120, 139.

Godlove, I. H., Improved Color-Difference Formula, with Applications to the Perceptibility and Acceptability of Fadings, *J. Opt. Soc. Am.,* **41,** 760–772, 1951, 136, 138.

Granville, W. C., and E. Jacobson, Colorimetric Specification of the Color Harmony Manual from Spectrophotometric Measurements, *J. Opt. Soc. Am.,* **34,** 382–395, 1944, 185.

Grum, F., S. Saunders, and T. Wightman, Artificial Light Sources to Simulate Illuminant D, *TAPPI,* **53,** No. 7, July 1970, 207.

Guild, J., The Instrumental Side of Colorimetry, *J. Sci. Instr.,* **11,** 69, 1934, 241.

Hale, N., Color Tolerances, ISCC Newsletter No. 204, January-February, 1970, 151.

Hardy, A. C., A New Recording Spectrophotometer, *J. Opt. Soc. Am.,* **25,** 305–311, 1935, 234.

Hardy, A. C., *Handbook of Colorimetry,* The Technology Press, Massachusetts Institute of Technology, Cambridge, Mass., 1936, 94, 98.

Harrison, V. G. W., *Definition and Measurement of Gloss,* W. Heffner & Sons, Ltd., Cambridge, England, 1945, 226.

Haupt, G. W., and F. L. Douglas, Chromaticities of Lovibond Glasses, J. Res. NBS, **39,** 11, 1974; also *J. Opt. Soc. Am.,* **37,** 698, 1947, 172.

Haupt, G. W., J. C. Schleter and K. L. Eckerle, The Ideal Lovibond Color System for CIE Standard Illuminants A and C Shown in Three Colorimetric Systems, NBS Technical Note 716, 1972, 172.

Helmholtz, H. *Treatise on Physiological Optics,* 3rd ed., 1909, 205.

Horning, S. C., and M. P. Morse, The Measurement of the Gloss of Paint Panels, *Off. Dig., Fed. Soc. Paint Technol.,* **266,** 153, 1947, 73.

Huey, S. J., Low Temperature Storage of Color Standards Panels, *Color Eng.,* **3,** 24–27, September-October 1965, 250.

Hunter, R. S., The Glossmeter, Circular No. 456, Scientific Section, National Paint, Varnish and Lacquer Assn. Inc., 1934, 66, 225.

Hunter, R. S., Methods of Determining Gloss, NBS Research Paper RP 958, *J. Res. NBS,* **18,** January 1937, 67.

Hunter, R. S., A Multipurpose Photoelectric Reflectometer, NBS Research Paper RP 1345, November 1940, 221, 240.

Hunter, R. S., Photoelectric Tristimulus Colorimetry with Three Filters, NBS Circ. C429, 1942, 106, 113, 134, 163, 164, 240, 241.

Hunter, R. S., Accuracy, Precision and Stability of New Photoelectric Color-Difference Meter., *J. Opt. Soc. Am.,* **38,** 1094A, 1948, 122, 241.

Hunter, R. S., Gloss Evaluation of Materials, ASTM Bulletin No. 186, 48, 1952, 73.

Hunter, R. S., Photoelectric Color Difference Meter, *J. Opt. Soc. Am.,* **48,** 985–995, 1958, 163, 164, 206, 216, 241.

Hunter, R. S., Standardization of Test for Specular Gloss of Paper at 75°, *Tappi,* **41,** No. 8, 385–396, 1958, 73, 122.

Hunter, R. S., New Reflectometer and Its Use for Whiteness Measurement, *J. Opt. Soc. Am.,* **50,** No. 1, 44–48, 1960, 222.

Hunter, R. S., Development of the Citrus Colorimeter, *Food Technol.,* **21,** No. 6, 100–105, 1967, 169.

Hunter, R. S., High Resolution Goniophotometer, *Modern Aspects of Reflectance Spectroscopy,* Plenum Press, New York, 1968, 220.

Hunter, R. S., *Appearance Attributes of Metallic Surfaces,* ASTM, Philadelphia, Pa., 1970, 75, 230.

Hunter, R. S., Visually Perceived Attributes of the Appearance of Materials and ASTM Progress Toward Their Measurement, Sensory Evaluation of Appearance of Materials, ASTM STP 545, 18–34, 1973, 283.

Hunter, R. S., and J. S. Christie, Improved Natick Laboratories Colorimeter for Textile Fabric Inspection, U.S. Army Natick Lab, Technical Report 66-19-CM, February 1966, 128, 154, 210.

Hunter, R. S., and G. L. Gibson, Improvement of the Accuracy of Photoelectric Tristimulus Colorimeters, *Color Eng.,* March-April 1969, 244.

Hunter, R. S., and D. B. Judd, Development of a Method of Classifying Paints According to Gloss, ASTM Bull. No. 97, 11, March 1939, 73, 226.

Hunter, R. S., and C. Lofland, A Gloss Test for Waxed Paper, *Tappi,* **39,** 833–841, 1956, 74.

Hunter, R. S., and J. N. Yeatman, Direct-Reading Tomato Colorimeter, *J. Opt. Soc. Am.,* **51,** 552–554, 1961, 168.

Ingersoll, L. R., A Means to Measure the Gloss of Paper, *Electr. World,* **63,** No. 12, 645, 1914, 71, 225.

Ingersoll L. R., The Glarimeter: An Instrument for Measuring the Gloss of Paper, *J. Opt. Soc. Am.,* **5,** 213, 1921, 225.

Institute of Paper Chemistry, Instrumentation Studies V, Report on Gloss, Paper Trade J. 104: TS6, January 7, 1937, 73, 226.

ISO/R105/I Parts 2 and 3, International Organization for Standardization, Central Secretariat, Geneva, Switzerland, 1964, 150.

ISO/TC38/SCI, Doc. N418 (USA Proposal—Amendment to Gray Scale for Assessing Change in Colour and Gray Scale for Assessing Staining), International Organization for Standardization, Manchester, 1974, 150.

Ives, H. E., A Precision Artificial Eye, *Phys. Rev.,* **6,** 334, 1915, 85, 240.

Jacobsen, A. E., Non-Adaptability of the ICI System to Some Near-Whites Which Show Absorption in the Far Blue Region of the Spectrum, *J. Opt. Soc. Am.,* **38,** 442–444, 1948, 99, 161.

Jacobson, E., *The Color Harmony Manual,* Container Corporation of America, Chicago, 1942, 185.

Jaeckel, S. M., Utility of Color-Difference Formulas for Match-Acceptability Decisions, *Appl. Opt.,* **12,** 1299–1316, 1973, 145.

Johnston, R. M., Colorimetry of Transparent Materials, *J. Paint Technol.,* **43,** No. 553, 42–50, 1971, 167.

Jones, L. A., The Gloss Characteristics of Photographic Paper, *J. Opt. Soc. Am.,* **6,** 140, 1922, 218.

Judd, D. B., A Maxwell Triangle Yielding Uniform Chromaticity Scales, *J. Opt. Soc. Am.,* **25,** 24–38, 1935, 106, 109, 139.

Judd, D. B., A Method for Determining Whiteness of Paper, *Paper Trade J.,* **103,** TS 154, 38–44, 1936, 164.

Judd, D. B., Specification of Color Tolerances at the National Bureau of Standards, *Am. J. Psychol.,* **52,** 418, 1939, 134, 139.

Judd, D. B., A Comparison of Direct Colorimetry of Titanium Pigments with their Indirect Colorimetry based on Spectrophotometry and a Standard Observer, *J. Opt. Soc. Am.,* **39,** 945–950, 1949, 100, 162, 163.

Judd, D. B., Response Functions for Types of Vision According to the Mueller Theory, *J. Res. NBS,* **42,** 1–16, 1949, 45.

Judd, D. B., Fundamental Studies of Color Vision from 1860 to 1960, *Proc. Natl. Acad. Sci.* **55,** No. 6, 1313–1330, 1966, 45.

Judd, D. B., D. L. MacAdam, and G. Wyszecki, Spectral Distribution of Typical Daylight as a Function of Correlated Color Temperature, *J. Opt. Soc. Am.,* **54,** 1031–1040, 1964, 51.

Judd, D. B., and G. Wyszecki, *Color in Business, Science, and Industry,* 2nd ed., John Wiley & Sons, New York, 1963, 32, 50, 78, 83, 85, 95, 100, 125, 151, 167, 178, 180, 186.

Kaufman, J. E., Ed., *IES Lighting Handbook,* 4th ed., Illuminating Engineering Society, 1966, 20, 22, 196, 198.

Kelly, K. L., A Universal Color Language, *Color Eng.,* **3,** 2–7, March-April, 1965, 150.

Kelly, K. L., K. S. Gibson, and D. Nickerson, Tristimulus Specification of the Munsell Book of Color from Spectrophotometric Measurements, *J. Opt. Soc. Am.,* **33,** 355–376, 1943, 181.

Kelly, K. L., and D. Judd, The ISCC-NBS Method of Designating Colors and a Dictionary of Color Names, NBS Circ. 553, November 1955, 156, 181.

Kramer, A., and B. A. Twigg, *Fundamentals of Quality Control for the Food Industry,* Avi Publishing Co., Westport Conn., 1970, 171.

Kubelka, P., New Contributions to the Optics of Intensely Light Scattering Materials, Part I, *J. Opt. Soc. Am.,* **38,** 448–457, 1948, 239.

Kubelka, P., and F. Munk, Ein Beitrag zur Optik der Farbanstricke, *Z. Tech. Phys.,* **12,** 593, 1931, 32.

Kuehni, R., Acceptability Contours and Small Color Difference Formulas, *J. Color Appear.,* **1,** 1971, 145.

MacAdam, D. L., The Specification of Whiteness, *J. Opt. Soc. Am.,* **24,** No. 7, 188–191, 1934, 163, 164.

MacAdam, D. L., Maximum Visual Efficiency of Colored Materials, *J. Opt. Soc. Am.,* **25,** No. 11, 361–367, 1935, 98.

MacAdam, D. L., Projective Transformations of ICI Color Specifications, *J. Opt. Soc. Am.*, **27,** 294–299, 1937, 110.

MacAdam, D. L., Visual Sensitivities to Color Differences in Daylight, *J. Opt. Soc. Am.*, **32,** 247–274, 1942, 106, 134, 141.

MacAdam, D. L., Specification of Small Chromaticity Differences, *J. Opt. Soc. Am.*, **33,** 18–26, 1943, 134, 141, 142.

MacAdam, D. L., Chromatic Adaptation, *J. Opt. Soc. Am.*, **46,** 500–513, 1956, 95.

McLaren, K., Scaling Factors in Color Difference Formulas, *Color Eng.*, 38–45, December 1969, 145.

McNicholas, H. J., Equipment for Measuring the Reflective and Transmissive Properties of Diffusing Media, NBS Research Paper RP704, August 1934, 220.

Maerz, A., and M. R. Paul, *A Dictionary of Color*, McGraw-Hill Book Co., New York, 1st ed. 1930; 2nd ed. 1950, 188.

Maxwell, J. C., On the Theory of Compound Colours and the Relations of the Colours to the Spectrum. *Proc. R. Soc. Lond.*, **10,** 404, 484, 1860, 85.

Middleton, W. E. Knowles, and A. G. Mungall, An Instrument for the Measurement of Distinctness-of-Image Gloss, *Can. J. Technol.*, **31,** 160–167, 1953, 230.

Moon, P., and D. E. Spencer, A Metric Based on the Composite Color Stimulus, *J. Opt. Soc. Am.*, **33,** 270–277, 1943, 127.

Morton, T. H., $\Delta E$ by Differential Colorimetry in Production Dyeing, *Die Farbe*, **18,** 164–170, 1969, 145.

Mueller, G. E., Über die Farbenempfindungen. *Z. Psychol. Ergaenz.*, 17, 18, 1930, 45, 125.

Munsell, A. H., *A Color Notation*, Boston, 1905, 117.

*Munsell Book of Color*, standard ed., Munsell Color Company, Baltimore, Md., 1929, 117, 179.

Munsell, A. E. O., L. L. Sloan, and I. H. Godlove, Neutral Value Scales. I. Munsell Natural Value Scale, *J. Opt. Soc. Am.*, **23,** 394–411, 1933, 104.

Newhall, S. M., P. W. Burnham, and J. R. Clark, Comparison of Successive with Simultaneous Color Matching, *J. Opt. Soc. Am.*, **47,** 43–45, 1957, 150.

Newhall, Sidney M., D. Nickerson, D. B. Judd, Final Report of the O.S.A. Subcommittee on the Spacing of the Munsell Colors, *J. Opt. Soc. Am.*, **33,** 385–418, 1943, 117, 118, 181.

Nickerson, D., The Specification of Color Tolerance, *Tex. Res.*, **6,** 505–514, 1936, 134, 137.

Nickerson, D., History of the Munsell Color System and Its Scientific Application, *J. Opt. Soc. Am.*, **30,** 575–586, 1940, 181.

Nickerson, D., Summary of Available Information on Small Color Differences, *Am. Dyest. Rep.*, **33,** May-July, 1944, 138.

Nickerson, D., *A Handbook on the Method of Disk Colorimetry*, USDA Publ. 580, March 1946, 178.

Nickerson, D., Interrelation of Color Specifications, *Pap. Trade J.*, **125,** TS 219, 1947, 188.

Nickerson, D., Tables for Use in Computing Small Color Differences, *Am. Dyest. Rep.*, **39,** August 1950, 120, 138.

Nickerson, D., History of the Munsell Color System, *Color Eng.*, **7,** 42–51, September-October 1969, 181.

Nickerson, D., and F. K. Stultz, Color Tolerance Specification, *J. Opt. Soc. Am.*, **34,** 550–570, 1944, 137.

Nimeroff, I., A Two-Parameter Gloss Method, *J. Res. NBS*, **48,** No. 3, 127, 1957, 73.

Nimeroff, I., and J. A. Yurow, Degree of Metamerism, *J. Opt. Soc. Am.,* **55,** 185, 1965, 152.

Nimeroff, I., NBS Monograph 104, *Colorimetry,* 1968, 110, 172.

Optical Society of America Committee on Colorimetry, *The Science of Color,* Washington, D.C., 1963, 297.

Park, R. H., and E. I. Stearns, Spectrophotometric Evaluation, *J. Opt. Soc. Am.,* **34,** 112–113, 1944, 32, 275.

Pearson, M. L., I. Pobboravsky, and J. A. C. Yule, Computation of Halftone Color Gamuts of Process Inks, Graphic Arts Research Center, Report No. 29, 1968, 176.

Perry, J. W., The Objective Measurement of Colour, *J. Sci. Instr.,* **15,** 270, 1938, 240.

Pfund, A. H., The Measurement of Gloss, *J. Opt. Soc. Am.,* **20,** 23–26, 1930, 73, 225.

Plochere, Gladys, and G. Plochere, *Plochere Color System,* Fox Printing Co., Los Angeles, 1948, 183.

Priest, I. G., and F. G. Brickwedde, The Minimum Perceptible Colorimetric Purity as a Function of Dominant Wavelength with Sunlight as Neutral Standard, *J. Opt. Soc. Am.,* **13,** 306(A), 1926, 108.

Ridgway, R., *Color Standards and Color Nomenclature,* A. Hoen & Co., Baltimore, 1912, 188.

Sanders, C. L., and G. Wyszecki, Correlate for Lightness in Terms of CIE Tristimulus Values, *J. Opt. Soc. Am.,* **47,** 398–404, 1957, 105.

Saunderson, J. L., and B. J. Milner, Modified Chromatic Value Color Space, *J. Opt. Soc. Am.,* **36,** 36–42, 1946, 127.

Schofield, R. K., The Lovibond Tintometer Adapted by Means of the Rothamsted Device to Measure Colours on the CIE System, *J. Sci. Instr.,* **16,** 74, 1939, 172.

Schultze, W., and L. Gall, Experimentelle Überprüfung mehrerer Farbabstandsformeln bezüglich der Helligkeits- und Sättigungs-differenzen bei gesättigten Farben, *Die Farbe,* **18,** 131, 1969, 145.

Scofield, F., A Method for Determination of Color Differences, Circ. 664, Natl. Paint, Varnish and Lacquer Assn., Inc., July 1943, 106, 115, 140.

Simon, F. T., and W. J. Goodwin, Rapid Graphical Computation of Small Color Differences, Bakelite Company, New York, 1957, 141, 143.

Stearns, E. I., *The Practice of Absorption Spectrophotometry,* John Wiley & Sons, New York, 1969, 239.

Stensby, P. S., Optical Brighteners and Their Evaluation, *Soap Chem. Spec.,* 1967, 164.

Stieg, F. B., The Geometry of White Hiding Power, *Official Digest J. Paint Technol. Eng.,* **34,** No. 453, October 1962, 32.

Strocka, D., Color Difference Formulas and Visual Acceptability, *Appl. Opt.,* **10,** No. 6, 1308–1313, 1971, 145.

Taguti, R., and M. Sato, Exponential Color Coordinate System, *Acta Chromatica,* **1,** No. 1, 19, 1962, 127.

Taylor, A. H., A Simple Portable Instrument for the Absolute Measurement of Reflection and Transmission Factors, Sci. Pap. NBS, No. 405, November 1920, 221, 224, 248, 312.

Tingle, W. H., and D. J. George, Measuring Appearance Characteristics of Anodized Aluminum Automotive Trim, Report 650513, Society of Automotive Engineers, May 1965, 230.

Tingle, W. H., and F. R. Potter, New Instrument Grades for Polished Metal Surfaces, *Prod. Eng.,* March 1961, 230.

Van den Akker, J. A., Chromaticity Limitations of the Best Physically Realizable Three-filter Photoelectric Colorimeter, *J. Opt. Soc. Am.,* **27,** 401, 1937, 241.

White, L. S., and A. E. Jacobsen, New Blue Filter for the Colormaster Differential Colorimeter Based on Judd's Modified Standard Observer, *J. Opt. Soc. Am.,* **55,** 177–184, 1965, 161.

Wickstrom, W. A., and M. Horner, Closed-Loop Color Control for Printing Papers, *Tappi,* **53,** 784–791, 1970, 269.

Winch, G. T., and E. H. Palmer, A Direct Reading Photoelectric Trichomatic Colorimeter, *Trans. Illum. Eng. Soc.,* Lond., **2,** 137, 1937, 241.

Wright, W. D., *The Measurement of Colour,* 4th ed., Alger Hilger Ltd., London; Van Nostrand Reinhold Co. United States sales rights, 1969, 84, 85, 152.

Wright, W. D., and F. H. G. Pitt, Hue Discrimination in Normal Color-vision, *Proc. Phys. Soc.* (Lond.), **46,** 459, 1934, 108.

Wright, H., and G. Wyszecki, Field Trials of 10° Color-Mixture Functions, *J. Opt. Soc. Am.,* **50,** 647–650, 1960, 177.

Wyszecki, G., Proposal for a New Color-Difference Formula, *J. Opt. Soc. Am.,* **53,** 1318A, 1963, 115, 140.

Wyszecki, G., Recent Agreements Reached by the Colorimetry Committee of the Commission Internationale de l'Eclairage, *J. Opt. Soc. Am.,* **58,** 290–297, 1968, 127.

Wyszecki, G., Development of New CIE Standard Sources for Colorimetry, *Die Farbe,* **19,** No. 1/6, 1970, 207.

Wyszecki, G., Technical Note, CIE Colorimetry Committee—Working Program on Color Difference, *J. Opt. Soc. Am.,* **64,** 896–897, 1974, 116, 122, 139.

Wyszecki, G., and W. S. Stiles, *Color Sci.,* John Wiley & Sons, New York, 1967, 108, 116.

Yeatman, J. N., A. P. Sidwell, and K. H. Norris, Derivation of a New Formula for Computing Raw Tomato Juice Color from Objective Color Measurement, *Food Technol.,* **14,** 16–20, 1960, 168.

Yule, J. A. C., Principles of Color Reproduction, John Wiley & Sons, New York, 1967, 173.

# GLOSSARY-INDEX

AATCC: American Association of Textile Chemists and Colorists, 136, 138, 150–151, 161, 282, 316

Absolute reflectance value: reflectance value relative to the perfectly reflecting and perfectly diffusing surface as 1.0, 224, 248, 312, 315

Absorptance: the ratio of absorbed to incident radiant energy (equal to unity minus the reflectance and transmittance)

Absorption: process by which light or other electromagnetic radiation is converted into heat or other radiation when incident on or passing through material, 26–35, 43, 48, 173, 234, 236, 238, 239, 254

Absorption coefficient: the capacity of unit thickness of a material to absorb spectral light energy, 32–33, 78, 155, 272

Accuracy: conformity of a measured result to an accepted reference value or scale, 210, 235, 238, 240, 244–247, 249, 310–311

Achromatic color: a neutral color, such as white, gray or black, that has no hue (also termed nonchromatic), 7–12

Adams Chromatic Valence: a color scale very similar to Adams Chromatic Value System, but utilizing X/Y and Z/Y ratios instead of X-Y and Z-Y differences, 106–107, 117–122, 130, 136, 138, 301

Adams Chromatic Value: an opponent-colors scale based on E. Q. Adams' theory of color vision in which the Munsell Value function is applied to the three visual receptor responses, and differences between the X and Y functions give redness-greenness, and differences between the Y and Z functions give yellowness-blueness, 106–107, 117–122, 127, 129, 130, 136, 138, 144, 301

Adams-Nickerson L,a,b (ANLab) Scale: an opponent-colors scale for color and color difference proposed by Nickerson and based on Adams Chromatic Value parameters, 107, 120, 122, 135–139, 150, 301

Adams-Nickerson 1950 Scale: a scale based on Adams Chromatic Value scale, with constants suggested by Dorothy Nickerson, 107, 119–120, 136, 138–139, 150, 301

Adams-Nickerson Modified 1952 Scale: a scale based on Adams Chromatic Value scale, with constants suggested by Glasser and Troy, 107, 119–120, 136, 138–139, 301

Adams Theory of Color Vision: a theory devised by E. Q. Adams combining the tristimulus Young-Helmholtz theory and the Hering opponent-colors theory, on the basis of a nonlinear assumed photometric response from each of the three retinal photoreceptors, 106, 117–119, 129, 138, 163

Adaptation: process by which the eye becomes accustomed to magnitude or spectral character of illumination, 131, 148

Additive colorimeter, *see* Visual additive colorimeter

Additive color mixture: superposition or other nondestructive combination of lights of different chromaticities, the visual effects of which depend only on their chromaticities and not on their spectral distributions, 82–83, 93, 96, 166, 177–179

Angle of incidence: the angle between the axis of an impinging light beam and perpendicular to the specimen surface, 26, 33, 35–43, 48–49, 53–57, 68–76, 199–206, 218–233, 238–239, 286–288

Angle of view: the angle between axis of observation and perpendicular to specimen surface, 26, 33, 35–43, 48–49, 53–57, 68–76, 199–206, 218–233, 238–239, 286–288

ANLab, *see* Adams-Nickerson L,a,b Scale

AOCS, FAC Color Standards: a series of 26 solutions of inorganic salts developed by the Fats Analysis Committee (FAC) of the American Oil Chemists' Society (AOCS) and the American Chemical Society for the specification of animal fats, 167, 307

Aperture color: color perceived through aperture so that position in space and form of object cannot be recognized, 5–6, 83, 102, 177–178
Aperture mode of viewing: perceiving a light stimulus through an aperture so that object form and position in space cannot be recognized, 5–6, 83, 94, 102–104, 147, 177–178, 192
Aperture stop, *see* Stop
APHA Hazen Standards: American Public Health Association platinum-cobalt solutions developed to measure the color of natural waters, 167, 307
Apparent reflectance, *see* Reflectance factor
Appearance: the aspect of visual experience by which things are recognized, 3–17, 27, 28, 43, 47, 58, 264–265, 280–281, 286–288, 315
Army Color Standards for Liquid Colors: a series of solutions used for specification of the color of liquids, 308
Artificial illuminants: a synthetic light source of spectral distribution as close as possible to that of the natural illuminant (usually daylight) to be duplicated, 49–53, 56, 207
ASTM: American Society for Testing and Materials, 73–75, 77, 134, 136–138, 150, 163, 164, 167, 226, 228, 231, 248, 253, 280–283, 307–309, 286–288, 315–320
Attribute: distinguishing characteristic of a sensation, perception or mode of appearance; distinction is made between chromatic and geometric appearance attributes, 3–17, 27–28, 58–60, 66–72, 286–288

Baffle: a form of shield used in an optical instrument to intercept unwanted light, 201, 224
Balinkin Index of Fading: a measure of color difference devised in 1939 and 1944, based on distance between points in Munsell color space, using a geometry which is Euclidean for small hue differences, 136–137
Barrett Color Standards: a series of solutions used for specification of the color of benzine and coal tar solvents, 167, 307
Bausch and Lomb instruments, 226, 235
Beer's Law: transmittance of stable solution is exponential function of product of concentration and length of path in solution, 239
Blackbody: thermal radiator of uniform temperature whose radiant exitance in all parts of spectrum is maximum obtainable from any thermal radiator at same temperature. (The designation blackbody is appropriate because such a body will absorb all incident flux.), 21–22
Blackbody radiator; a body which absorbs all the radiation falling on it and reflects none, 21–22, 195–196
Bleaching: the treatment, usually chemical, of a product to remove blue-light-absorbing constituents and thereby to improve its whiteness, 158–160, 306
Bloom: the scattering of light in directions near the specular direction by a deposit or excretion on a specimen. Bloom can be removed by rubbing or polishing, 68–70
Bouguer's Law: equal layers will absorb equal fractions of radiant flux entering them
Breckenridge and Schaub RUCS (Rectangular-Uniform-Chromaticity Scales): a uniform-chromaticity scale devised by Breckenridge and Schaub in 1939 as a modification of Judd's 1935 UCS diagram placing the chromaticity of an equal-energy illuminant at the origin of the coordinate system, 107, 111–113, 300
Brightener, optical, *see* Fluorescent whitening agent
Brightness:
optics and appearance measurement; the attribute of visual sensation by which an observer aware of differences in luminance; lightness
pigments, dyes and colored products; the attribute of color which corresponds to its

perceived difference from the color of dirt; vividness, 7, 9–11, 13
metals; freedom of metallic surfaces from reflection haze and texture, 75
lighting; the luminous intensity of any surface in a given direction per unit of projected area of the surface as viewed from that direction; luminance
dyeing; the color quality, a decrease in which corresponds to the effect of the addition of a small quantity of neutral gray dye to the dyestuff; strength
*see also* TAPPI brightness

British Paint Research Station Color Scales: combinations of Lovibond glasses used for grading of color of oils and varnishcs, 308

Broad-band wavelength selection: allowing the transmission of light of a wide range of, but not all, wavelengths, usually through the use of an absorption filter, 206–210

Bryan Color Numbers: a scale used to specify the color of maple syrup, 309

Bundesanstalt fur Materialprufung, (BAM) Germany: Federal Institute for the Testing of Materials, 248

Candela (formerly candle): the luminous intensity in the perpendicular direction of a surface of 1/600,000 $m^2$; one candela produces 1 lm of luminous flux per steradian of solid angle measured from the source

Cascade photometry: a method of photometry in which successive comparisons of similar chromaticities are made, and the relative luminances of very different chromaticities are calculated as products of ratios of luminances of intervening pairs, 82

Ceramics: a product made from earth, and baked or fired in a furnace, often used as a standard of color or gloss, 31, 67, 225, 243, 250, 255, 266, 272, 279, 286–288, 306, 312, 316, 317

Chroma, *see* Munsell chroma

Chromatic: perceived as having a hue; not white, gray, or black, 7–17

Chromatic attributes: those attributes associated with the spectral distribution of light; hue and saturation, 7–12, 15–17, 33–35, 58–61, 192, 234–245

Chromaticity: that part of a color specification which does not involve luminance. Chromaticity is two dimensional and is specified by pairs of numbers such as dominant wavelength and purity, 33–35, 58–61, 82, 91–101, 108–128, 241, 254

Chromaticity coordinates, *see* CIE chromaticity coordinates

Chromaticity diagram, *see* CIE (x,y) chromaticity diagram

Chromaticness: the visually perceived qualities of hue and saturation taken together, 7–12, 15–17

CIE, Commission Internationale de l'Eclairage; in English, the International Commission on Illumination: the main international organization concerned with problems of color and color measurement, 50, 81–104, 116, 122, 127, 139, 141, 161, 207, 247, 289–306

CIE chromaticity coordinates (trichromatic coefficients or trilinear coordinates): the ratios of each of the tristimulus values of a color to the sum of the tristimulus values. In the CIE systems they are designated by x,y, and z, 91–94, 96, 102–103, 290–291, 293, 295, 299–305

CIE (x,y) chromaticity diagram: the plane diagram found by plotting the CIE chromaticity coordinates x and y against each other, 92–98, 110, 113–114, 117–118, 123–125, 128, 141–142, 167, 175, 185–187

CIE Color Solid or Color Space: a 3-dimensional coordinate system in which colors are located by their Y,x,y values, 98–99, 102–103, 134, 137

CIE luminosity function ($\bar{y}$): a plot of the relative magnitude of the visual response as a function of wavelength from about 380 to 770nm, adopted by CIE in 1924, 19, 21, 41, 44, 81–82, 86, 102–106, 213, 220, 225, 289, 300–303

CIE 1976 L*a*b* Color Space: a uniform-color space utilizing an Adams-Nickerson cube root formula, suggested in 1974 for adoption by the CIE in 1976 for use in the measurement of small color differences, 107, 122, 136, 139, 303

CIE 1976 L*u*v* Color Space: a modification of the CIE 1964 U*V*W* color space, suggested in 1974 for adoption by the CIE in 1976 for use in measurement of large color differences, 107, 116, 136, 141, 303

CIE Standard Observer: a hypothetical observer having the tristimulus color-mixture data recommended in 1931 by the CIE for a $2^\circ$ field of vision ($\bar{x},\bar{y},\bar{z}$ or $\bar{x}_2,\bar{y}_2,\bar{z}_2$). A supplementary observer for a larger $10^\circ$ field ($\bar{x}_{10},\bar{y}_{10},\bar{z}_{10}$) was adopted in 1964, 58, 81–103, 109, 130, 145, 161, 163, 177, 206, 208, 210, 216, 239–244, 290–306, 315–316

CIE Standard Sources: standard sources for which the CIE in 1931 specified the spectral energy distribution as follows:

Illuminant A: a tungsten filament lamp operated at a color temperature os 2854K, approximating a blackbody operating at that temperature, 50, 88–89, 93–95, 154, 172, 210, 292–293, 296–297

Illuminant B: an approximation of noon sunlight having a correlated color temperature of approximately 4870K and obtained by a combination of Source A and a special filter, 50, 88–89, 93–94, 172, 292–293, 296–297

Illuminant C; an approximation of overcast daylight having a correlated color temperature of approximately 6770K and obtained by a combination Source A and a special filter, 50, 88–89, 93–95, 98–99, 128, 154, 172–173, 206, 210, 244, 292–305

The CIE recommended in 1965 new illuminants to supplement Sources A, B and C designated by D and a number corresponding to its color temperature. The most important of this series is D6500, with a correlated color temperature of 6500K, 50–51, 88–89, 128, 154, 206–210

CIE tristimulus values: the amounts of the three reference or matching stimuli required to give a match with the color stimulus considered, in a given trichromatic system. The symbols recommended for the tristimulus values are X,Y,Z in the CIE (1931) Standard Colorimetric System and $X_{10},Y_{10},Z_{10}$ in the CIE (1964) Supplementary Colorimetric System, 86–103, 123, 154–157, 164, 166–169, 173–177, 185–187, 216, 221, 238–244, 290–306

X,Y,Z, 86–103, 111, 117–123, 295, 299

$X_{\%},Y_{\%},Z_{\%}$, 95–96, 101, 119, 122, 304–305

Y,x,y, 93–94, 98–103, 106, 131, 159, 172, 316

CIE-UCS diagram (1960): MacAdams' 1937 u,v uniform-chromaticity-scale diagram adopted by the CIE in 1959 as the provisional standard UCS diagram, 115–116

CIE U*V*W* Color Space: a uniform-color space devised by Wyszecki in 1963 and provisionally recommended by the CIE in 1964, combining the 1960 CIE-UCS chromaticity diagram and a cube root lightness function, 107, 115–116, 127, 129, 136, 140–141, 144 302

Circumferential illumination: light incident as though along the surface of a cone, directed toward the apex, rather than in discrete beams. In colorimeters or reflectometers, such a system is of great value in minimizing variations in readings caused by change of sample orientation, 200, 242–243, 255, 257, 261

Citrus redness (CR): a special chromatic scale used for citrus juices, 168–170

Citrus yellowness (CY): a special chromatic scale used for citrus juices, 168–170

Clarity: the characteristic of a transparent material whereby distinct images may be observed through it, 15, 16, 39–41, 60, 77, 231–232, 266

Cleaning materials: soaps, detergents and polishes used for cleaning; also, washing machines

Light: electromagnetic radiation in the spectral range detectable by the normal human eye (approximately 380 to 780 nm), 4–6, 18–21, 26–43, 48–57
Light beam: a bundle of parallel, converging, or diverging light rays, 200–201
Lightfastness: resistance to change of appearance caused by exposure to light, 150–151, 279
Lightness: perception by which white objects are distinguished from gray objects and light from dark colored objects, 8–17, 46, 60, 104–106, 113, 115, 122–128, 140–143, 148, 155, 160, 216, 306
Light pipe: transparent material, usually drawn into a cylindrical or conical form, through which light is channeled from one end to the other by total internal reflection, as in light fibers, 204–205, 244
Light source: that element in an instrument or in the visual observing situation that furnishes radiant energy in the form of light, 18–25, 41, 48–57, 193–199, 226, 242, 300
Limestone, test for appearance
Liquids: substances in a fluid state, 127, 166–169, 231, 243, 253, 255, 261–262, 307–309, 319
Lovibond tintometer: visual instrument for evaluating color of transmitting material by comparison with colors of glasses of Lovibond system, 166, 169–173, 175–176, 308
Lumen: unit of luminous flux, defined as flux emitted through unit solid angle (one steradian) from directionally uniform point source of one candela
Luminance: the luminous intensity of any surface in a given direction per unit of projected area of the surface as viewed from that direction
Luminosity function, *see* CIE luminosity function
Luminous: adjective indicating that radiant flux is evaluated by weighting according to the luminous efficiency function of the CIE 1931 standard observer
Luminous attribute: the color attribute associated with amount of light. It can be considered to be a geometric as well as a chromatic attribute, 8–17, 60
Luminous efficiency: a measure of efficiency in the production of light; lumens per watt measures conversion of electric power to light; in the conservation of light, reflectance factor of reflector surfaces, 160, 225, 306
Luminous efficiency function, *see* CIE luminosity function
Luminous flux, *see* Lumen
Luminous reflectance: ratio of luminous flux reflected by object to that incident on it
Luminous transmittance: ratio of luminous flux transmitted by object to that incident on it
Luster (contrast gloss): gloss associated with contrasts of bright and less bright adjacent areas of the surface of an object. Luster increases with increased ratio between light reflected in the specular direction and that reflected in diffuse directions which are adjacent to the specular direction, 15, 35, 54, 65, 68–70, 218, 221, 225, 229, 261, 283, 288

MacAdam ellipses: ellipses drawn in the CIE x,y chromaticity diagram as a representation of the distribution of equiluminous colors which are barely perceptibly different in color from a fixed color, 106–107, 123–125, 128, 136, 137, 141–143, 302
MacAdam limits: the loci in a color solid of colors of maximum achievable luminous efficiency and chromaticity, 98–99
MacAdam Unit of Color Difference: just perceptible difference in surface color or chromaticity according to MacAdam experiment, 134, 136, 138, 141–143
MacAdam u,v Chromaticity Diagram: a transformation of Judd's 1935 uniform-chromaticity-scale diagram, using rectangular instead of triangular coordinates, 107, 110–111, 115–116, 136, 140–141, 300
Macbeth-Taylor reflectometer: an early visual reflectance meter, 221

Refraction: the bending of light rays as they pass from one medium into another having a different index of refraction, 27
Refractive index (n): ratio of sines of angles of incidence and refraction in transparent material, 26, 248, 286–288
Reilly Cube Root, *see* Glasser Cube Root
Repeatability: the degree to which a single instrument gives the same reading on the same specimen at different times, 246–247
Reproducibility: the agreement attainable between measurements performed in different laboratories, 246–247
Resins, 166–167, 262, 307–309
Ridgway Color System, an early system of 1115 color chips often used in horticulture and chosen to represent by relatively equal steps a wide range of variation in hue, saturation and lightness, 188
Rosin Standards, USDA: a series of 15 glass cubes used to identify color of rosins and resins, 167, 308

Sample: a lot of a material from which specimens are selected for measurement, 251
Saturation: the attribute of color perception that expresses the degree of departure from the gray of the same lightness, 7–17, 60, 96, 105, 146, 167, 254
Saunderson-Milner Zeta Color Space: a uniform-color space devised by Saunderson and Milner in 1946 to modify the Adams Chromatic Value system so as to bring it more into line with Munsell spacing, 127
Saybolt color: color identification of a column of liquid such as refined petroleum oil obtained by comparison with specified glass standards, 308, 319
Scattering: the process by which light passing through granular, fibrous or rough surface matter is redirected throughout a range of angles, 26–33, 239, 254
Scattering coefficient: rate of increase of reflectance with thickness, 32–33, 78, 155, 272
Scattering power: the product of multiplying the scattering coefficient by the thickness, 78
Scofield L,a,b Color Space: a modification of the Hunter $L', \alpha', \beta'$ color coordinate system devised by Scofield in 1943 as a simplification, 106, 115, 129, 136, 138, 140, 144, 301
Secondary standards: all standards other than primary standards, 248–249, 312–313
Selected ordinate method of integrating X,Y,Z values: a method of computing tristimulus color values in which the usual numerous multiplications are avoided by summation of the spectral values at specially selected, non-uniformly spaced wavelengths, 89, 91, 96, 296–299
Selective absorption: absorption which varies with wavelength, 28–35, 48
Sensation: primitive awareness or uninterpreted conscious response to stimulation of a sense receptor
Shade, *see* Hue; Lightness; and Shade sorting
Shade sorting: the grouping together of similarly colored materials so that the materials within each group may be used together in a finished product such as an article of clothing, 268, 272, 274
Sheen: specular gloss at a large angle of incidence for an otherwise matt specimen; usual angle of measurement is 85°, 15, 68–69, 71, 286
Shellac, 256, 309
Simon-Goodwin charts: a graphical procedure developed by Simon and Goodwin in 1957 for the rapid calculation of small color differences using the 1942 MacAdam chromaticity difference ellipses, 136, 141–143
Single-number measurement scale: a scale which provides a partial but useful measure of color using one number only instead of the usual three, 159–169

Soil removal: in cleaning of textiles, floors, etc., 160–161, 225, 266, 306, 318
Society of Dyers and Colourists: British society of dye chemists, 138
Source, *see* Light source
Spatial: considered with relation to space, *see* Geometric attributes
Specific calibration standard: permanent specimen with values that relate to measurements made of an impermanent specific product on a standard instrument, 249, 313
Specimen: the pieces or part of a sample actually measured, 203–204, 251–263
Specimen holder: any apparatus or means by which a specimen is held in the best position for repeatable measurement when presented to the viewing window of an instrument, 256–262
Specimen preparation and handling: techniques by which objects are made as flat, uniform and opaque as possible, and are presented for instrumental measurement so that the most valid and repeatable results are obtained, 147, 203–204, 245, 252–263
Specimen viewing window of instrument: that part of an instrument where the specimen is placed for measurement, 192, 203–204, 226–227, 238, 242–243, 252–263
Spectral: pertaining to the visible spectrum, hence, having to do with color, 19, 22–23, 48, 193, 206
Spectral conditions for visual evaluation: the spectral characteristics of the light source and the surround, 48–57
Spectral components of an instrument: the components of an appearance-measuring instrument which affect its color-measuring characteristics, such as the color of the light source, the wavelength sensitivity of the detectors, and the wavelength selectivity of prisms, filters and monochromators, 41, 193, 206–215, 235–238, 246, 310–311
Spectrophotometer: an instrument used for measuring the transmittance and reflectance of specimens as a function of wavelength, 33–35, 41, 60, 90, 192, 206–208, 218, 234–239, 242, 315
Spectrophotometry, abridged: measurement of spectral transmittance or reflectance at a limited number of wavelengths, usually by using filters for spectrum isolation rather than a monochromator, 208, 230, 234–235
Spectrum: spatial arrangement of electromagnetic energy in order of wavelength. For visible radiation, the spectrum is a band of color produced by breaking white light into its component colors, 18–20
Specular: having the qualities of a speculum or mirror: having a smooth reflecting surface, 12–17, 65, 286–288
Specular angle: the angle between the perpendicular to the surface and the reflected ray that is numerically equal to the angle of incidence and that lies in the same plane as the incident ray and the perpendicular but on the opposite side of the perpendicular to the surface, 36–43, 48, 53–57, 65–77, 202–203, 219–233, 286–288
Specular reflection: process by which incident light is redirected at the specular angle, as from a mirror, without diffusion, 12–17, 26–32, 36–43, 48, 53–57, 65–77, 192–194, 199–203, 219–233, 247, 286–288, 310–311, 314
Specular transmission; process by which incident light is transmitted through an object in a rectilinear, straight-through manner, without diffusion, 12–17, 39–41, 77, 192, 199, 231, 232, 247, 310–311
Specular transmittance: ratio of flux transmitted without change in image-forming state, to incident flux, 39–41, 59–61, 77, 247, 311
Standard observer, *see* CIE Standard Observer
Standard: a reference against which instrumental measurements are made, 226–227, 245–251, 310–315
Stop: any window or diaphragm which can restrict passage of light rays in an optical

218–219, 231–232, 255, 258, 287, 311, 320

Transmittance: ratio of transmitted to incident flux (adjectives are used to identify spectral and geometric weighting of flux), 39–41, 59–61

Transmittance factor: the ratio of the flux transmitted through a specimen for specified geometric and spectral conditions of radiation and view to that transmitted by a perfect diffuse transmitter under the same conditions. The term directional transmittance has been used in the USA in this sense, 59–61, 247, 311

Transparency: the property of a material by which a negligible portion of the transmitted light undergoes scattering, 12–17, 28, 39–41, 65, 77, 221, 231–232, 319–320

Transparent materials, colorimetry of, *see* ISCC subcommittee for Problem 14

Trichromatic coefficients, *see* CIE chromaticity coordinates

Tristimulus colorimeter: instrument which senses tristimulus values and converts them to chromaticity components of color, 132, 206–210, 234, 239–245

Tristimulus reflectometer: instrument which measures tristimulus reflectance values, 234, 239–241

Tristimulus values, *see* CIE tristimulus values

Tungsten lamp, *see* Incandescent filament lamp

Turbidity: loss of transparency due to diffusion caused by presence of particulate matter, 15, 77, 231–232, 266

Two-beam instrument: instrument in which light from a single source is directed simultaneously to both specimen and standard, 205–206, 246

Twyman and Perry colorimeter: an early photoelectric tristimulus colorimeter, 240

Uniform color scale: a color space or color solid in which the differences between points correspond to the perceptual (visual) differences between the colors represented by these points, 102–132, 147, 159

Union Lubricating Oil Standards: a series of glass standards used for identifying colors of lubricating oils, 307

USDA Standards for Food Colors: plastic, photographic and other standards used by the U. S. Department of Agriculture for visually grading the quality of food products, 170–171, 309

U. S. Rosin Standards, *see* Rosin Standards, USDA

U*V*W*, *see* CIE U*V*W* Color Space

Value, *see* Munsell Value Function

Varnish, 256, 307–309

Viewing conditions, *see* Observing conditions

Vignetting: in an optical instrument, edge losses of light rays by interception at stops other than the aperture stop and field stops which properly control the light rays passing through, 201, 203, 205, 222, 226

Visual additive colorimeter: a visual colorimeter in which a color match is obtained between an unknown light illuminating one part of a divided photometric field and the sum of three fixed lights illuminating the other half, 177–179

Visual evaluation: analyses (usually comparative) of appearance attributes of materials by eye, 47–57, 252, 264–265

Visual subtractive colorimeter: a visual colorimeter in which a match to an unknown color is produced by controlled subtraction of spectral light from the standard source by the use of filters, 169–173

Vividness, *see* Brightness

Wavelength: the distance, measured along the line of propagation, between two points that are in phase on adjacent waves. Wavelength distribution determines the color of light. Wavelengths of visible light range from about 400 to about 700 nanometers (nm), 18–21, 32

Wavelength, dominant, *see* Dominant wavelength

Wavelength selection device: the element in a color measuring instrument that controls the energy measured. Narrow-band and broad-band wavelength selection devices are used, 195, 206–210, 234–238

Wax, 228, 256, 287

Weber's Law: at normal stimulus levels, a just noticeable difference in stimulus is a constant fraction of the stimulus intensity, 45–46

Weber-Fechner relationship: the intensity of the sensory response is proportional to the logarithm of the stimulus intensity, 45–46, 104

Weighted ordinate method of deriving X,Y,Z values: a method of computing X,Y,Z tristimulus values by multiplication at equal wavelength intervals of values of spectral reflectance (or transmittance) by weighting factors that are products of spectral energy of illuminant and spectral tristimulus color mixture data of the CIE Observer, followed by addition of these products, 88–89, 95–96, 294–295

Whiteness: preference rating for the color of white materials. Ideal white is probably bluish, and departures from this ideal white toward yellow reduce whiteness rating about four times as much as departures toward gray, 155–165, 222–223, 264, 306, 315–318

Window, *see* Stop

Working standards: secondary standards which are calibrated against master standards and used as references for the calibration of laboratory instruments, 249, 312

Xenon arc source: a lamp utilizing the ionization of Xenon, a heavy, colorless, inert gaseous element, 53, 198, 207–209

XYZ, X%Y%Z%, *see* CIE tristimulus values

Yarns, 243, 260–261

Yellowness: the attribute by which an object color is judged to depart from a preferred white toward yellow, 155–167, 253, 257, 264, 306–309, 315–316, 319

Young-Helmholtz theory of color vision: a theory which suggests that there are three kinds of retinal receptors that when separately stimulated give rise to the color sensations of red, green and blue, respectively, 45, 110